ELECTRONIC PROTOTYPE CONSTRUCTION

By
Stephen D. Kasten

Howar
4300 WEST 6

FIRST EDITION
FIRST PRINTING — 1983

International Standard Book Number: 0-672-21895-X
Library of Congress Catalog Card Number: 82-62201

Edited by: *C. Herbert Feltner*
Illustrated by: *Jill E. Martin*

Printed in the United States of America.

Preface

Many books are being written in today's wide-open field of electronics for the hobbyist and experimenter concerning the design of IC-based circuits, particularly circuits related to microcomputers and computer interfacing. However, very little practical information is available concerning construction techniques for converting schematics and ideas into functional electronic prototype units. Therefore, this book was written as a guide and a reference for persons interested in learning modern electronic construction methods, particularly for the experimenter or design laboratory worker who would like to assemble prototype equipment for evaluation and testing.

The text material can be roughly divided into four areas: wire-wrap assembly, printed circuit board design/fabrication, graphic techniques, and hardware packaging. A certain amount of overlap exists among all of these areas, but they are organized as follows:

Wire-Wrap — Chapter 1 deals with wire-wrap and related techniques, such as solder pad and perforated board assembly. In general, wire-wrap techniques are well-suited for one-of-a-kind construction projects using integrated circuit components, because they give the designer a quick route to a finished circuit board that is easily modified. Tools and equipment are fairly simple and hardware accessories for wire-wrap are readily available.

Printed Circuit Boards — The PCB is the modern basis for electronic construction using ICs and other miniaturized components; therefore, the greatest emphasis in this book is on printed circuit techniques. Chapter 2 starts with a general discussion of modern printed circuit technology, and concludes with a proposed simplified approach developed in the author's laboratory for fabrication of double-sided PCB prototypes. Chapter 3 is concerned with PCB design and preparation of camera-ready artwork. In Chapter 5, simple photo resist techniques are presented for transferring PCB patterns to blank circuit boards. The actual etching process is explained in Chapter 6. Chapter 7 is concerned with the art of electroplating, particularly as it relates to plating gold and nickel coatings on PCB edge connector fingers. The chapter also includes a procedure for coating protective solder on etched PCB traces. Chapter 9 deals with the final machining, soldering, assembly, and cleaning of PCBs; techniques for modification and repair are also presented. Finally, Chapter 10 discusses a novel approach for designing and constructing high-density PCBs as an extension of double-sided PCB fabrication techniques.

Graphic Techniques — Modern electronic construction techniques are heavily based on graphic arts and photography, part of a general approach known as photofabrication. Chapter 4 develops general graphic arts photographic principles, and includes procedures for generating photo masks using 35 mm photography and simple darkroom equipment. These masks are needed in Chapter 5 for the photo resist image transfer process. In Chapter 8, general principles and techniques for screen printing are presented; screen printing is useful for both PCB fabrication and enclosure labels/designs. A general discussion for graphic techniques for panel designs is given at the end of Chapter 11.

Hardware Packaging — The total electronic system is of interest in packaging, rather than individual electronic assemblies at the circuit board level. Chapter 11 includes basic information on hardware packaging: design considerations, enclosures, parts layout, wiring, front panels, fabrication tools, and machining holes.

Throughout this book, the reader will discover some novel approaches aimed at simplifying electronic construction and minimizing the need for expensive equipment. This objective is particularly evident in the various chapters concerning PCB fabrication. The reader should find the following PCB techniques especially interesting and useful:

- The Print-Etch-Drill-Print-Etch approach
- 35 mm photography for masks using Kodalith film
- Flow-coating of KPR 3 photo resist
- Manual solder coating and solder reflow
- Two-layer high density PCBs with clearance holes
- Simple plating baths for gold and nickel

A single book cannot cover all areas of electronic construction in detail, so frequent references to other books, magazines, and commercial sources are given throughout the text. The reader will find that much up-to-date and useful information is available in the form of trade journals and product literature from manufacturers, rather than formal publications.

I would like to thank Dr. Stephen N. Falling for his help with photographic illustrations, which are a great help in displaying the various tools, techniques, and equipment used in electronic construction. Many other companies and individuals supplied materials, figures, photographs, and information; these are acknowledged throughout the text. The assistance from Dr. Jonathan A. Titus of the Blacksburg Group in planning, organizing, and editing the book is greatly appreciated.

STEPHEN D. KASTEN

DEDICATION

This book is dedicated to my wife Janice.

Contents

CHAPTER 6

CHAPTER 7

CHAPTER 8

CHAPTER 9

CHAPTER 10

CHAPTER 11

APPENDIX A

APPENDIX B

APPENDIX C

APPENDIX D

Chapter 1
Wire-Wrapping

INTRODUCTION • In this chapter we will explore methods used to assemble electronic circuits using techniques *other* than traditional printed circuit board (PCB) design and construction. In addition to wire-wrap, these techniques will include related methods such as pencil wiring, pad soldering, and perforated board assembly, which may be used in various combinations with wire-wrap assembly.

Throughout the chapter, frequent references will be made to specific sources for supplies and equipment. These are not necessarily the only sources, but rather ones with which the author is familiar. The reader will benefit greatly by obtaining catalogs from distributors and manufacturers, since they supply a wealth of information, drawings, and photographs to promote the sale and use of their products. Addresses are found in Appendix B. Catalogs from the following companies are particularly useful for wire-wrap construction reference:

Augat (Interconnection Systems Div.)
Digi-Key Corporation
Douglas Electronics
GC Electronics
Jameco Electronics
Newark Electronics
OK Machine & Tool Corp.
Radio Shack (Tandy Corp.)
Vector Electronic Co., Inc.

OBJECTIVES • The objectives of this chapter are:

- Familiarize the reader with the various methods, tools, and equipment used in wire-wrap construction, and closely related techniques.
- Provide references to commercial suppliers of wire-wrap hardware.
- Demonstrate the basic considerations that go into a good wire-wrap design and assembly.

TYPICAL APPLICATIONS •

Commercial • An important advantage of wire-wrap relative to PCB construction is the high density of interconnections achievable with no conductor crossover limitations. For this reason, commercial electronic equipment is on the market today with wire-wrapped connections, avoiding the design and costly manufacture of multilayer PCBs. An example is shown in Fig. 1-1, a computer backplane. In addition to providing high-density connections, wire-wrapping allows this backplane to be configured in unlimited ways based on which particular circuit boards are plugged into the system. A second example is illustrated in Fig. 1-2, which shows a microprocessor-based instrument controller. The manufacturer probably chose wire-wrap construction due to low-volume production, and with a view towards easily making design changes in the future. The wiring side of this 9 by 17 inch (229 by 432 mm) perforated board is shown in Fig. 1-3. Although wire-wrap is not always considered as a commercial construction method, the connections are sturdy and suitable for industrial installations; automated machinery for high-volume wire-wrap assembly is also available.

Hobbyist • To the electronic hobbyist using modern integrated circuit (IC) components, wire-wrapping is unquestionably the easiest approach to building one-of-a-kind devices. Less preliminary design is required relative to PCBs, and no photography or chemical etching is necessary. Drilling is minimized; circuits can be easily modified; the final product is fairly compact and neat; and the assembly can fit into standard electronic enclosures. For the computer hobbyist, microcomputer "plugboards" are available for wire-wrapping to all major bus systems, such as S-100, TRS-80®, and Apple II®, greatly simplifying the task of computer interface construction. Modern methods for wire-

Fig. 1-1. Wire-wrapped computer backplane from Univac.

wrap have become standardized and are supported by major manufacturers, making the necessary hardware widely available through electronic distributors, mail-order supply houses, and hobby outlets such as Radio Shack. (This standardization of hardware can be directly attributed to the development of the modern IC Dual-In-Line Package (DIP), with all pins spaced on multiples of 0.1 inch (2.5 mm) centers.)

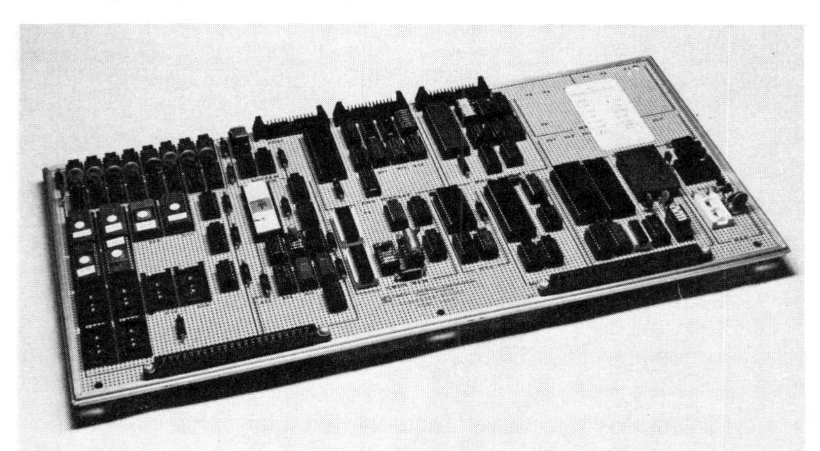

Fig. 1-2. Microprocessor-based instrument controller from Omega Controls Corporation.

Fig. 1-3. Typical wire-wrap construction, with pins projecting underneath the board.

Prototype Design • A situation where wire-wrap is widely used by professional designers is in the assembly of the first test version, or *prototype,* of a new electronic device. Once a circuit design has been worked out and is considered basically sound, the next step is to assemble a prototype and check out its operation. Wire-wrap is an excellent method for construction, since some modifications are always expected, and assembly is rapid; the circuit is also easily accessible for testing, with all signals available on wire-wrap pins.

Frequently, a principal concern in prototype testing is to determine if any noise problems exist in the external connection system. This requires that the test board be constructed in approximately the same *modular form* as the final envisioned commercial product, usually a PCB, and the same connection scheme must, therefore, be followed. Modular equivalence in construction allows a prototype to be placed in the proper *environment* (electrical and physical) for testing. As will be seen later in this chapter, a good selection of wire-wrap connectors is available, making the translation of prototype board to PCB straightforward. It should be noted that most finished electronic products are actually marketed as PCB assemblies for the following important reasons:

1. PCBs are least expensive for automated, high-volume production.
2. PCBs are compact in the vertical dimension (no wire-wrap pins hanging down underneath the board, for example).
3. Noise problems in high-frequency circuits can be reduced or eliminated in PCB wiring with ground planes and proper choice of conductor sizes and spacings. However, wire-wrapped circuits are more susceptible to variable noise difficulties due to the random positioning of signal connections.
4. Wiring errors and bad connections during assembly are minimized with PCB construction.

METHODS AND TOOLS FOR WIRE-WRAP •

Standard and Modified Wraps • The standard wire-wrap connection is shown at the left of Fig. 1-4A. Wire-wrap terminals are 0.025 inch (0.64 mm) square posts with sharp corners that are press-fitted in the pre-drilled 0.042 inch (1 mm) holes, or mechanically secured by gluing, staking, or soldering to a pad. Solid wire is wrapped under tension around the post, creating a good electrical and mechanical connection as the sharp post corners cut into the wire under pressure. The electrical connection is extremely sound, with a contact resistance less than 0.01 ohm when properly made with six to eight turns of wire; no soldering is required to ensure a good connection. The wire can also be unwrapped, allowing easy circuit alterations. At the right of Fig. 1-4A is seen the "modified wrap," which includes at least one initial turn of insulated wire at the bottom of the wrap. An insulation loop provides more flexibility for the wire as it meets a terminal, preventing mechanical damage from vibration or stress. The modified wrap is generally recommended, since it also reduces the chance for electrical shorting of closely spaced connections, and it provides a way of taking up slack by wrapping extra insulated turns as required.

Some examples of incorrectly formed wraps are found in Fig. 1-4B. The leftmost wrap is too loose, with adjacent spiral turns not touching. Both center connections are properly made modified wraps, but the rightmost wrap has overlapping turns. Another condition to avoid is having too many connections on the same pin. To maintain ease of circuit alterations, no more than *three* separate wraps should be made on one post; otherwise, it will be invariably discovered that the connection which requires removal is at the bottom of a post.

(A) Standard and modified wraps.

(B) Incorrectly formed wraps on left and right posts.

Fig. 1-4. Wire-wrap methods.

Conventional Wrap Tools and Wire • A simple, versatile wire-wrap tool is shown in Fig. 1-5, the WSU-30M from OK Machine & Tool Corporation. This tool may be used for wrapping and unwrapping in either clockwise or counterclockwise directions on 0.025 inch (0.64 mm) square posts, and includes a wire-stripping feature for removing insulation from 30 AWG (American Wire Gauge) solid wire. The most popular type of wire in present use is 30 AWG Kynar[1]. The insulation is available in a variety of colors (red, yellow, white, orange, blue, black, green) and can be obtained in pre-cut, pre-stripped kits with many different lengths, or in bulk rolls from mail-order supply houses, such as Digi-Key Corporation. Illustrated in Fig. 1-6 are:

- Pre-cut, pre-stripped wire packages.
- 50-foot roll of AWG 30 (0.25 mm) Kynar wire.
- A refillable wire dispenser which cuts wire to desired length and strips a measured 1 inch of insulation from AWG 30 wire.

Getting back to the WSU-30M tool, its tip is illustrated in detail in Fig. 1-5B. Two holes are seen in the tip: one central hole that fits over the 0.025 inch (0.64 mm) wrap post, and one off-center hole, the "wire tunnel," that feeds the wrapping wire. A short cross-sectional gap is present in the wire tunnel approximately one-fourth inch away from the tip, and this gap allows the wire to be correctly positioned for

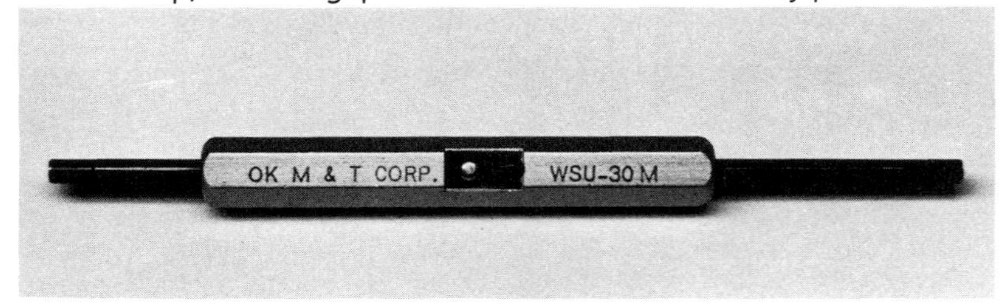

(A) Photograph.

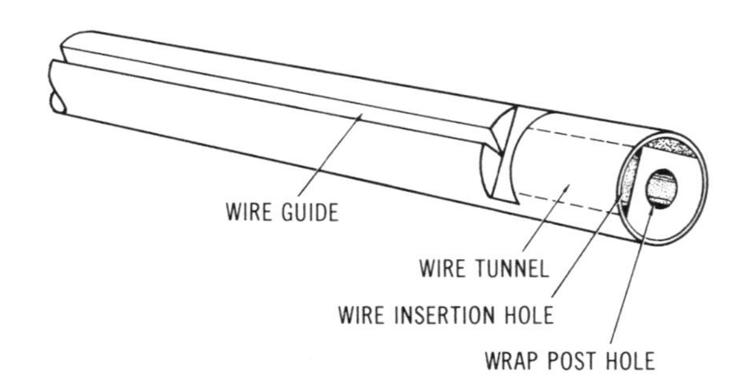

(B) Rough sketch.

Fig. 1-5. Conventional manual wire-wrap tool, WSU-30M from OK Machine & Tool Corporation.

the modified wrap. A series of steps is shown in Fig. 1-7 for making wire-wrap connections with the WSU-30M tool:

Fig. 1-7A — Wire is cut to length.
Fig. 1-7B — About 1 inch of insulation is stripped from the end to be wrapped.
Fig. 1-7C — Wire is inserted into the wire tunnel and pushed up into the gap until insulation comes into view.
Fig. 1-7D — Exiting wire is bent 90°, tool is positioned over the post and pushed down to the desired level for wrapping. Wire is then held at 90° relative to the tip as the tool is turned firmly with slight downward pressure to make 6–8 wrap turns on the post.

During wire-wrapping, as the wire exits from the tunnel at a 90° angle and is rotated under tension around the square post, it presses tightly against the sharp corners and makes good electrical contact. By unwrapping a connection (the other end of the WSU-30M tool is designed to do this operation) close examination will show that the wire and the post corners are actually deformed slightly from the great contact pressure; wrapped joints are often referred to as "gas-tight." Unwrapping is accomplished by pushing the tool down firmly on the connection, and turning it in the opposite direction of the original wrap.

If it is desired to use the modified wrap to take up slack in a wire, the wire should *not* be bent or firmly held as wrapping begins; the tool will generally pull in several turns of insulation if the exiting wire is held loosely, or until the wire begins to tighten up under increasing tension.

Fig. 1-6. Kynar-insulated 30 AWG wire for wire-wrapping.

Nonstandard Wraps • Conventional wire-wrap tools similar to the one illustrated in Fig. 1-5 are marketed by companies such as Vector Electronic and Radio Shack. These tools can also be used to make *nonstandard* wrapped connections on any odd-sized posts, round or square, that will fit up into the central hole of the tool. The author has drilled out his own WSU-30M tool slightly larger to allow for wrapping around most resistor, capacitor, diode, and transistor leads. However, these leads *do not* have the sharp corners of standard wrap posts; therefore, the wrap may appear tight, but a thin oxide layer on wire or terminal leads can cause a poor connection. It is, therefore, advisable to solder any wrap connections made on nonstandard terminals; this type of joint will be referred to as *pencil wired* in a later section.

Daisy Chaining • Several variations of wire-wrap tools have been developed to allow for *daisy chaining,* or continuous wrapping (see Fig. 1-8). The main advantage of daisy chaining over conventional wrapping is speed; there is no need to measure, cut, and strip each wire. Wire ends do not have to be individually inserted in the tool, and wires are cut only at the end of a run. Using regular tools, daisy chaining is only possible with a long length of bare wire fed down through the handle

(A) Wire is cut to length.

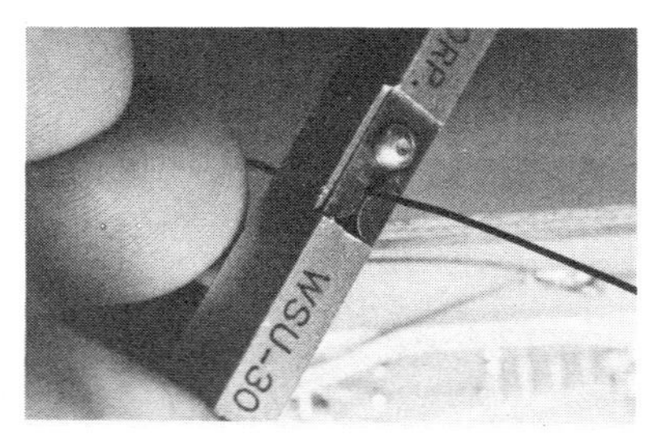

(B) Insulation is stripped.

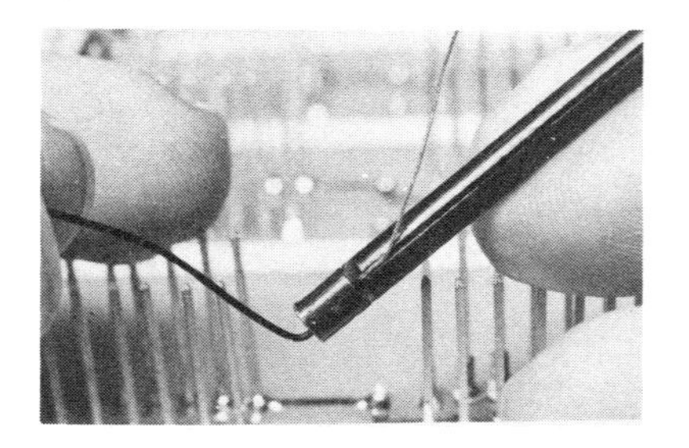

(C) Wire end is inserted in tool.

(D) Wire is bent 90°, tool is pushed over the post, and 6-8 turns are made.

Fig. 1-7. Steps in making a conventional wire-wrap connection.

of the tool; in fact, Vector sells a spool assembly (P160-5B) designed to mount on their P160-series conventional tools to allow for bare wire "strapping," as they refer to daisy chain wiring.

Unfortunately, bare wire is really only suitable for wiring ground or power supply lines that do not physically cross on the board. Another possibility for daisy chaining is to wrap with wire having solder-through insulation, such as the polyurethane-nylon type that is available in small gauges, However, the need to solder every wire-wrap connection eliminates most of the advantages of this technique with regard to making circuit changes.

The basic approach that manufacturers have taken to allow for continuous daisy chained wrapping with *insulated* wire is to design the tip of the tool so that wire insulation is opened up as it wraps onto a post. Vector refers to this method as *Slit-N-Wrap* with their P180- and P184-series tools, which contain a replaceable cutting edge in the tip to slit insulation prior to wrapping. *Just Wrap*, a tool by OK Machine & Tool, compresses the wire against the post as it is wrapped, depending more on the sharp post corners to open up insulation. Both of these tools are illustrated in Fig. 1-9. In either case, the wire is mounted on a spool at the top of the tool, and feeds continuously through the handle into the wrapping tip. Insulation is not opened when wire is pulled straight out of the tool, or at a slight angle when wire is routed from post to post; the insulation-opening action occurs *only* when wire passes at right angles through the wrapping tip as it is turned clockwise.

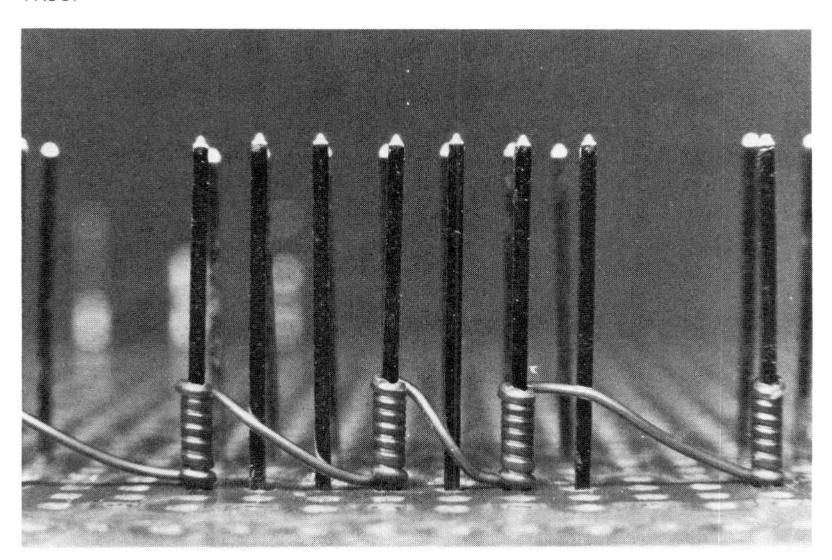

Fig. 1-8. Daisy chain wire-wrap connections, also known as strapping.

The Just-Wrap tool (Fig. 1-9A) uses standard 30 AWG Kynar-insulated wire, and includes a built-in feature to cut wire at the end of a run; it performs satisfactorily when sharp-cornered, high quality wrap posts are used. The daisy chain wraps illustrated in Fig. 1-8 were made with this tool.

The similar Vector Slit-N-Wrap tools exist in two versions, and are advertised as meeting Mil-Std-1130A for reliable gas-tight wrap joints. The P180-series tools use a thin (0.0008 inch or 0.02 mm) coating of polyurethane-nylon insulation on solid AWG 28 wire. This insulation is easily slit and makes for very flexible, compact wrap connections. The idea behind polyurethane-nylon wire in this case is that it permits transition from slit daisy chained wire-wrap posts to soldered connections on component leads and socket tails, all without any wire-stripping required, since the insulation can be soldered-through at 750–850°F where it vaporizes. Unfortunately, there are two serious disadvantages: the insulation is fairly easily rubbed off when it is routed around a board and contacts various sharp-cornered wrap posts (causing shorts), and it is difficult to safely solder other connections close by. For these reasons, this type of wire is not recommended for general wire-wrap use; it is discovered too frequently that a new wrap post needs to be soldered in place next to 15 wires already tightly routed on the board.

The P184-series Slit-N-Wrap tools from Vector (see Fig. 1-9B) are generally recommended, since they use the tougher Tefzel-insulated AWG 28 wire, with a high temperature rating. Tefzel is a durable wire-wrap insulation which provides some protection for routing small gauge wire around sharp corners. Kynar wire can also be used with the P184 tools; Vector states in their literature that slitting-tip life is reduced to about one-half using Kynar versus Tefzel insulation, indicating that

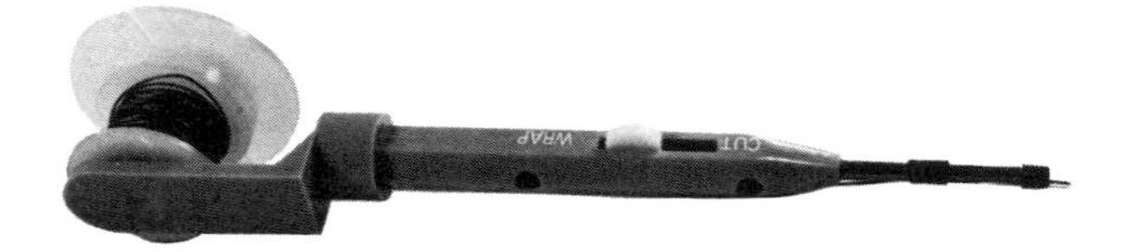

(A) Just-Wrap daisy chain tool from OK Machine & Tool Corporation.

(B) Slit-N-Wrap daisy chain tool from Vector Electronic Company.
Fig. 1-9. Just Wrap and Slit-N-Wrap tools.

Kynar is still somewhat tougher. The cutter bit is ordinarily rated for 7000 7-turn wraps before it requires replacement, and slitting action should be verified periodically by unwrapping and closely examining the wire from a wrap.

The development of tools for insulated daisy chain wiring has led to descriptions ranging from "unique" to "revolutionary." Although it may appear on the surface to be the wire-wrap method of general choice, several disadvantages of daisy chaining should be noted:

1. Connections are more difficult to remove than conventional wraps, because the wire exits from the top of the post. Daisy chains must, therefore, be cut before unwrapping, a difficult operation in close quarters.
2. After cutting, wire cannot be re-wrapped on the same post. However, with conventional wraps using stripped ends, it is often possible to carefully unravel the end and re-use it following removal from a post. There is an advantage in being able to do this, because the other end of the wire may be at the bottom of 3 wraps on another post, compounding the problem.
3. Wire from conventional wraps tends to lie close to the board because it enters at the bottom of each wrap. However, half of all daisy chain connections must exit from the top of a wrap, and this leads to an up-and-down see-saw effect which raises the general wire level higher on the board; wire routing becomes clumsier. For this reason, daisy chain wire is sometimes spiraled loosely back down each post before continuing on to the next connection, another clumsy operation.
4. Insulation-opening daisy chained wraps can only be made in one direction, generally clockwise, since the tools are not designed to open insulation in both directions. However, it is often neater to be able to wrap in *either* direction depending upon which way the wire approaches a post.
5. Wire used with a daisy chain tool tends to be all of the same color, unless you take time to change spools frequently, or unless you can afford to have six loaded tools on the bench. The use of random colors in wire-wrap is a great help in labeling signals and tracing connections.

Power Tools • Conventional wrapping tools and insulation-opening daisy chain tools can all be powered by electrical hand-held driver units, such as the Radio Shack battery-powered model illustrated in Fig. 1-10. In the Vector series, all manual tools can slip up into the spindle of a power unit for automated operation. Motorized wrapping provides fast assembly for large, tedious jobs, and also ensures a high-quality, uniform connection with proper tension.

An example of the state-of-the-art in hand-held wire-wrap tools is Vector's new P184-7 automated Slit-N-Wrap device, shown in Fig. 1-11. The P184-7 has internal logic to automatically wrap a preset number of wire turns on each post as the daisy chain proceeds from point to point, all with adjustable, regulated tension; the operator need only select the number of turns (3–9), place the tool over a post, and press the trigger, which completes a Slit-N-Wrap connection in about 1 second. The tool is powered by 117-V ac, and holds a 300-foot spool of AWG-28 Tefzel wire. The set-screw mounted bit is replaceable and is

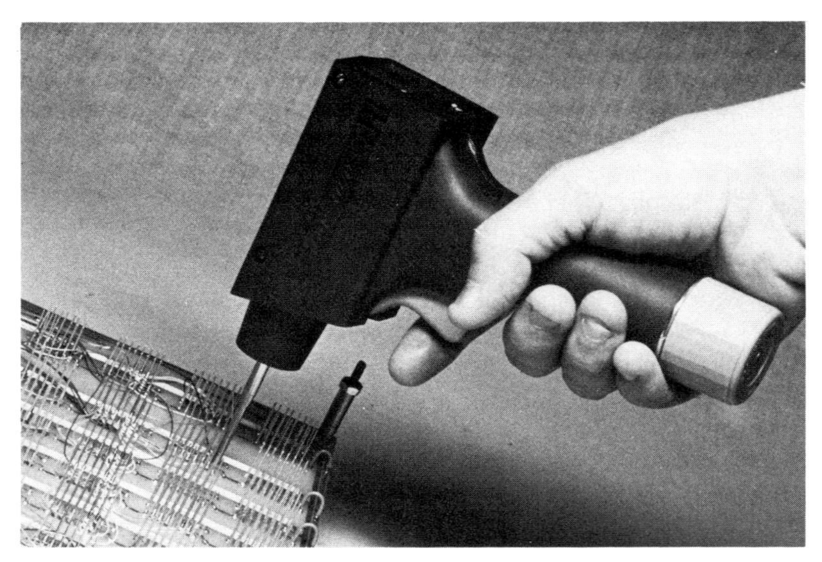

Fig. 1-10. Battery-powered conventional wrapping tool from Radio Shack.

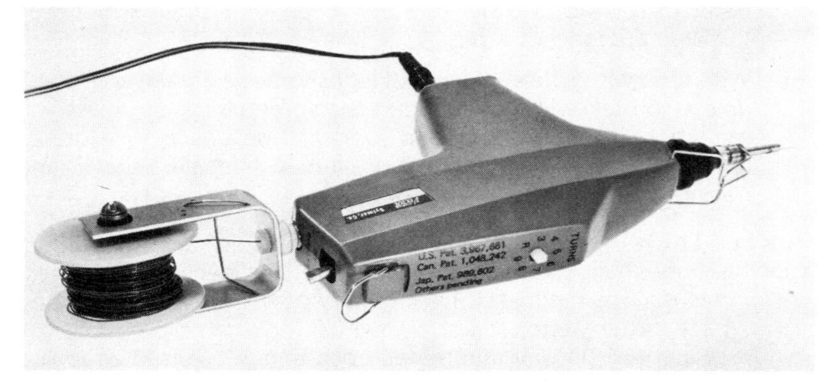

Fig. 1-11. Automated P184-7 Slit-N-Wrap power tool. *(Courtesy Vector Electronic Co.)*

guaranteed for the same life as in the manual slitting tool.

Unless a large job is being attempted, the actual wrapping time saved by a power tool may be insignificant compared to the total time spent on wiring. There are many tedious operations involved: locating pins, routing wires neatly, double-checking schematic diagrams, and correcting errors. A bulkier power tool may not be preferred over a light, pencil-thin manual wrap tool which can be deftly inserted into a maze of pins and wires to make a connection. However, it should be noted that a power tool is always welcome for *unwrapping* when a large number of connections must be quickly removed for circuit modifications, or for complete rewiring of a board. Using the Radio Shack power tool in Fig. 1-10, batteries are simply reversed for unwrapping; the bit is pushed firmly down around each post before squeezing the trigger for wire removal.

PENCIL WIRING ● *Pencil wiring* will be defined as any type of wrapped connection requiring *solder* to ensure proper electrical continuity, and it is the modern equivalent of point-to-point wiring methods. As was previously mentioned, standard wire-wrap tools can be used to make such connections on most component leads, or on any other type of terminal that will fit inside the central hole of the tool. However, the term "pencil wiring" actually gets its name from a special tool known as the wiring pencil, such as the Vector P178-1 shown in Fig. 1-12. The wiring pencil feeds wire from a spool, through the handle, and out through the point of the needle which is used to make manual wraps around any type of terminal. The P178-1 is designed to use 36 AWG wire with solder-through insulation, allowing continuous daisy chain operation; at the end of a run, the wire is cut by twisting the sharp end of the needle, and all connections are soldered at their terminals. As previously noted, the polyurethane-nylon insulation is not very durable, and this disadvantage must be balanced against the ability to make long, unbroken chained runs without stripping, cutting, or measuring wire. Two other advantages also make pencil wiring useful:

1. Wrap posts and sockets are not required, and less attention need be paid to the manner in which components are terminated. Any component lead or lug can be pencil wired, and IC pins can be

Fig. 1-12. Wiring pencil P178-1. *(Courtesy Vector Electronic Co.)*

connected directly. The price paid is in loss of flexibility for circuit alterations later, since all connections are soldered, and desoldering can be tedious.

2. The assembly is more compact and resembles a PCB; no long wrap posts protrude through the bottom of the board.

It is still necessary to mechanically secure pencil wired components to the board in some manner, either by bending leads, soldering to etched pads, cementing with adhesive, or attaching to press-fitted terminals. Mounting parts is less of a worry with conventional wire-wrapping, where almost all components are secured by attachment to square-post terminals and sockets. Terminals, sockets, and connectors will be discussed in more detail in the Hardware section of this chapter.

Wire-wrap assemblies often include a certain amount of pencil wired connections, particularly if many non-IC components are involved. The author prefers to make these connections with standard 30 AWG Kynar wire, using wire-wrap tools or small needle nose pliers; each wire end must be stripped. A manual wrap tool for standard 0.025 inch (0.635 mm) square posts will accommodate many leads, and a larger tool designed for wrapping 0.045 inch (1.14 mm) posts is also available (GC Electronics 41-063.) The needle nose pliers are used to pencil wire connectors that won't fit up into the wrap tools, such as relay lugs with large solder eyelets. Large components, such as transformers, fuses, capacitors, and relays are mechanically secured with machine screws, and/or with their leads soldered directly to "flea clips" pushed into the board's 0.042 inch (1.07 mm) diameter holes. More information on flea clip terminals is found in the Hardware section of this chapter. The flea clips can be pencil wired back to wire-wrapped IC sockets with 30 AWG wire; in some cases where heavier wire is required (such as in the power supply section) 24 AWG insulated wire can be soldered point-to-point between the flea clip tails.

SOLDER PAD WIRING • A common problem with wire-wrapping is the extra space consumed by wrap pins hanging underneath the board in the vertical dimension. This may not be troublesome in a roomy cabinet, but the wire-wrap assembly may be mounted in a tight space, such as a computer card cage filled with other circuit boards. Space problems occur frequently when the wrapped board is a prototype of what will eventually become a compact PCB; as was previously noted, proper testing requires that a prototype be mounted in the same physical environment as the final PCB.

To solve space problems, an important variation of PCB prototyping is the *solder pad* wiring method. Basically, solder pad construction is a form of point-to-point wiring on a printed circuit board, with a special etched pattern of 2–4 connected pads in rows drilled on 0.1 inch (2.54 mm) centers to accommodate modern IC packaging.

Components and sockets are mounted on the board by soldering them to pads on the reverse side of the board, just as PCB components are normally soldered into place. The wiring is accomplished by routing insulated conductors from point to point on the component side, and soldering them to pads on the reverse side. An example of a finished solder pad board is seen in Fig. 1-13A, a dense microcomputer PCB prototype. A portion of the reverse side of the prototype is seen in Fig. 1-13B illustrating the 3-pad pattern of soldered leads. Wiring is made simplest with 3 pads per conductor, since one pad is soldered to the component (IC, resistor, capacitor, etc.) and two are available for soldering to wire leads. The two uncommitted pads allow a daisy chain to continue: one wire in, another continuing out to the next point. Solid 24 AWG hookup wire with colored insulation is commonly used

(A) Solder pad construction on a dense microcomputer PCB prototype. (Courtesy Larry Johnson of Microproducts and Systems, Kingsport, TN)

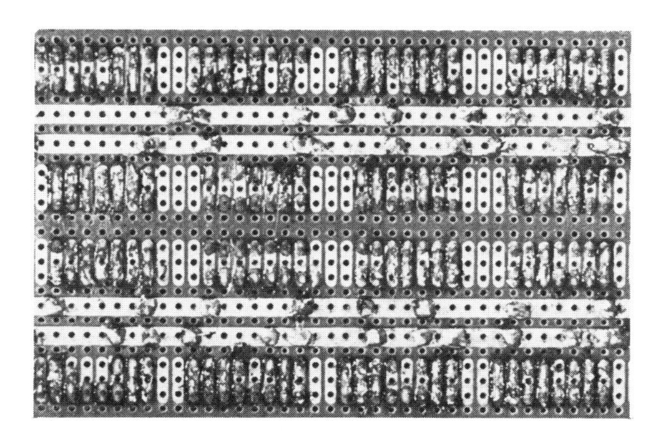

(B) Reverse side of solder pad board.
Fig. 1-13. Both sides of a solder pad board.

for wiring; for greater density, 30 AWG Kynar can be used with only 2 pads per conductor pattern, and several wires may be inserted in each hole before soldering. Wires are generally routed between IC socket rows as seen in Fig. 1-13A. The board occupies the same vertical space as a standard PCB, with no wires or pins hanging below, and the component side is only as high as the ICs mounted on sockets.

Advantages of the solder pad system over wire-wrapping are obvious: the vertical space problem is solved, while still preserving a high degree of flexibility as far as circuit alterations. If sockets are used to mount ICs and multiple-pin components, circuit changes are straightforward. Thus, ICs can be moved around and replaced, and conductor changes are as simple as heating up a solder pad and pulling the wire out, or pushing it back in. All solder pads are readily accessible from the back of the board, although the component side may look like a bird's nest. Another less obvious advantage of solder pad wiring over wire-wrap assembly is the fact that wiring is faster and more accurate from the *component* side of a board, where each device can be seen as the wires are routed. Disorientation and confusion are problems when viewing a mass of wire-wrap pins from the back of a board, with no components in sight! (Stick-on wire-wrap labels are now available to help with this problem, but they tend to become obscured with wire.)

The main disadvantage of solder pad prototyping is the amount of soldering required. No great skill is involved, but printed circuit pads will eventually lift off their laminated base material after repeated soldering and desoldering of conductor leads. A small, low-wattage soldering iron (such as the 25-watt model shown in Fig. 1-31) is recommended for all printed circuit work, and particularly for solder pad wiring alterations.

REVERSE WRAP PINS • Another elegant solution to the vertical space problem uses *reverse wrap* pins, that is, wire-wrap posts extending *above* the board rather than *below* it. This prototyping method requires a PCB with an etched pattern similar to the one just discussed for solder pad wiring, with typically 3–4 pads per conductor spaced on 0.1 inch (2.54 mm) centered rows. Standard IC sockets and components are soldered into place as before, but each lead is then brought back out on the component side of the board with a wire-wrap post soldered to the next connecting pad. An example of construction with reverse wrap pins is shown in Fig. 1-14. Wiring is accomplished by standard mechanical wire-wrapping on the *component* side, with no space lost below the board. To achieve greatest vertical density, short wire-wrap pins may be used that allow only two levels of wrapping, and these pins extend very little distance past the height of components.

Wire-wrap boards with reverse pins give all of the advantages of the solder pad method, but without having to solder interconnecting wires. A great deal of work is required initially to install necessary pins and sockets; for that reason, commercial reverse pin boards are available with this soldering already accomplished. Such a board from Douglas Electronics is shown in Fig. 1-20; Augat is another well-known manufacturer of similar wiring boards. Reverse pin wire-wrap prototypes are generally regarded as the neatest and best approximations to conventional PCBs, while still retaining the ability to make circuit alterations easily. Removal and installation of components is simple because the soldered reverse side is completely free of obstructions. Testing is particularly convenient because every signal is available as a pin on top of the board. The assembly of reverse pin prototypes with dense IC packages can be accomplished rapidly using commercial, off-the-shelf, fully socketed boards like the ones supplied by Douglas Electronics. If the circuit will contain many discrete components, assembly can become more tedious; a blank board with the appropriate etched patterns may be purchased, and all sockets and pins installed by the prototyper. An alternative is to mount discrete components on *headers* that plug into IC sockets, as will be shown later.

Because it is such an outstanding method for producing PCB prototypes, the step-by-step design and construction of a reverse pin wire-wrap board is presented as an example at the end of this chapter.

BOARDS FOR WIRE-WRAP •

General Purpose • The boards used for wire-wrap assembly consist of pre-drilled epoxy-glass or phenolic-base material available in a

Fig. 1-14. Reverse wrap pins for wire-wrap construction.

wide variety of sizes, hole patterns, thicknesses, and etched patterns to make wiring more convenient. The most widely used hole pattern is known as pattern "P," illustrated in Fig. 1-15A with one-sixteenth inch bare perforated board from Vector. The round holes are 0.042 inch (1.07 mm) in diameter spaced on 0.1 inch (2.54 mm) centers, accommodating push-in terminals and all integrated circuits and devices packaged with leads on 0.1 inch (2.54 mm) multiples. When bare perforated board is used for wire-wrapping, everything must be anchored by methods other than soldering — cement, screws, or staking. For that reason, the same basic 0.1 inch (2.54 mm) pattern board is available with an *etched pad* surrounding each 0.042 inch (1.07 mm) hole, as illustated in Fig. 1-15B with a general-purpose wire-wrap board from Radio Shack. This *pad-per-hole* board allows a great deal of flexibility

(A) Bare perforated P-pattern board from Vector.

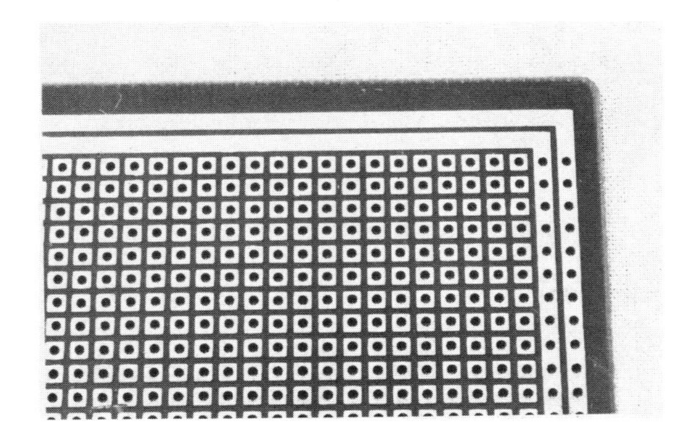

(B) Pad-per-hole board from Radio Shack.
Fig. 1-15. Wire-wrap boards.

with no space or holes committed for any particular purpose; all devices can be secured by soldering leads, terminals, and posts to pads.

Making connections to power supply and ground can be a tedious job using general-purpose wire-wrap boards with no committed etched patterns. These connections are needed often for every IC, and they must be made in such a way as to avoid noise problems caused by high-impedance common supply lines, or *ground loops*. (A ground loop is created when common ground-return pathways have enough impedance so that connecting lines are at varying potentials as current demands change.) Illustrated in Fig. 1-16 is the general-purpose 4066 *plugboard* from Vector which has some useful features obtained with the use of etched patterns in its 0.1 inch (2.54 mm) array of punched holes as follows:

- Edge-card connector (36/72 contacts), a widely used system for bringing signals into and out of circuit boards. In addition to providing a large number of contact points for external signals, an edge-connector finger gives physical support to the board after it is plugged into the mating connector. Fingers are nickel/gold-plated to ensure proper connection, and are brought out to holes on the board to install wrap pins as needed.
- Zig-zag power and ground traces distribute these lines throughout the board, at most 1 or 2 holes away from any IC pin. These patterns are etched on the reverse side of the board. A wide border

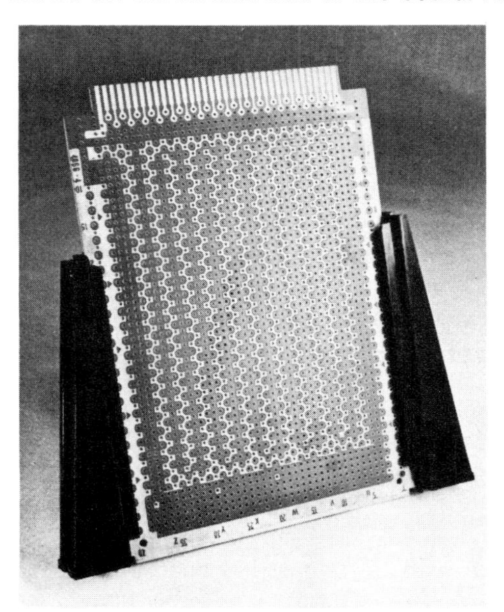

Fig. 1-16. General-purpose 4066 plugboard from Vector.

strip on both sides provides low-impedance power and ground signals to the zig-zag patterns as needed.

- A sheet ground plane on the component side surrounds all holes but remains isolated by a short distance from inserted pins. Ground planes are desirable in radio-frequency circuits to eliminate instability and to improve high-frequency response, and in digital circuits to reduce noise levels and ringing.

In addition to these features, the 4066 board has a blank space at one end for mounting large discrete components and connectors. The etched patterns allow installation of various combinations of 8, 14, 16, 18, 24, 36, and 40 pin DIP packages; up to thirty-four 16-pin ICs may be installed, for example. Many tips and suggestions are provided by Vector Electronic Co. in their literature accompanying prototyping boards, which include over 50 models in their catalog. Any reader new to wire-wrap construction is urged to obtain a copy of this catalog as an excellent reference to tools, hardware, accessories, and construction techniques.

Microcomputer Boards • The majority of wire-wrap applications today involve constructing microcomputer interface circuits and other computer-related assemblies. To simplify this task, a wide selection of wire-wrap boards are available that are designed to plug into *specific* microcomputer bus systems. The following photographs show some typical general-purpose computer interface boards for wire-wrapping:

- Fig. 1-17 shows the California Computer Systems CCS 7510A solder-tail board for the Apple II computer, with a three-hole pattern for solder pad or reverse wrap pin construction. Many other prototyping boards for the Apple II computer are advertised in microcomputer catalogs, ranging from simple wire-wrap boards with "pad-per-hole" patterns, to more specialized interfacing boards with I/O edge connector hardware. The CCS boards have plated-through-holes which simplify signal connections.
- Fig. 1-18 shows an S-100 board from Vector, the 8800V. This board has etched patterns for power lines, as well as a special area at both ends to install heat sinks and TO-220 regulators for on-board power regulation. The board is a standard S-100 size of 10 inches (254 mm) wide by 5 inches (127 mm) high, and can hold many combinations of DIP sockets, some of which are shown installed. The maximum capacity is fifty-two 16-pin sockets.
- Fig. 1-19 shows an Intel SBC-80 compatible board from Douglas Electronics, with 10 rows of 109 quadruple pads in each row, and interspersed power supply lines. This board is particularly well-suited for solder pad construction or for use with reverse wrap

pins on each IC socket and component; all holes are plated-through.

- Fig. 1-20 shows a board from Douglas Electronics for the DEC LSI-11 dual-width bus. This company supplies their boards in many

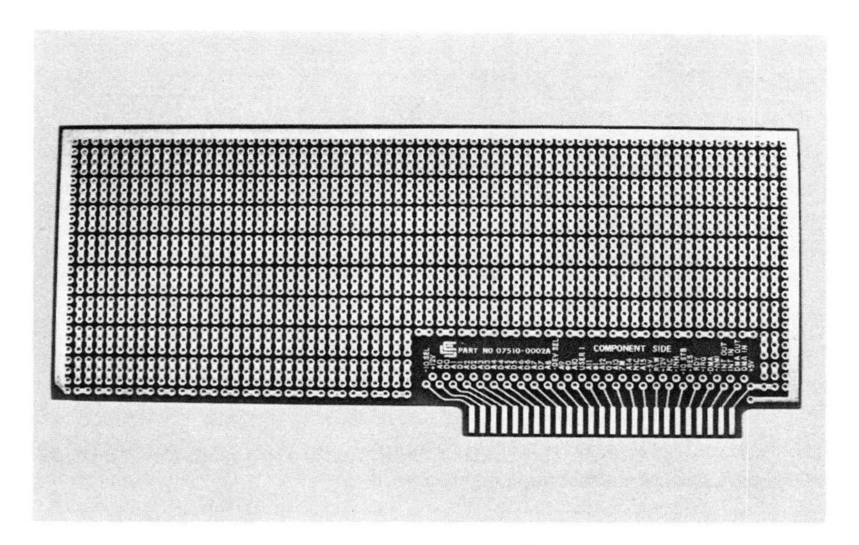

Fig. 1-17. Prototyping board for the Apple II computer, the model 7510A solder-tail board from CCS.

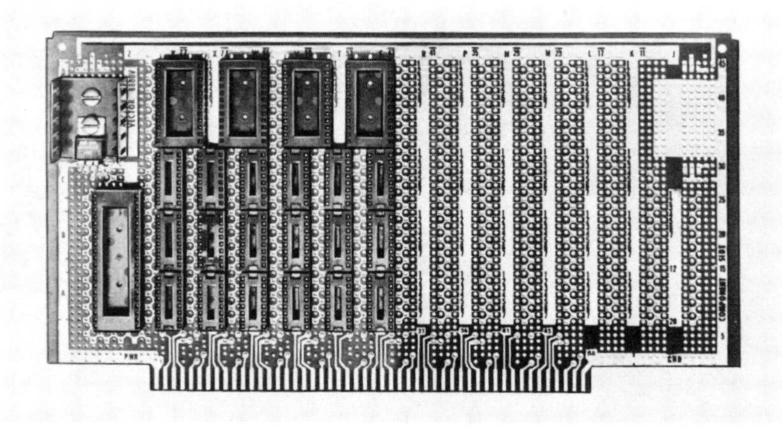

Fig. 1-18. Wire-wrap board 8800V for the S-100 bus. ***(Courtesy Vector Electronic Co.)***

"loaded" versions, with sockets, pins, and bypass capacitors already installed; the 11-DE-11-H shown here is fully loaded for reverse wrap construction.

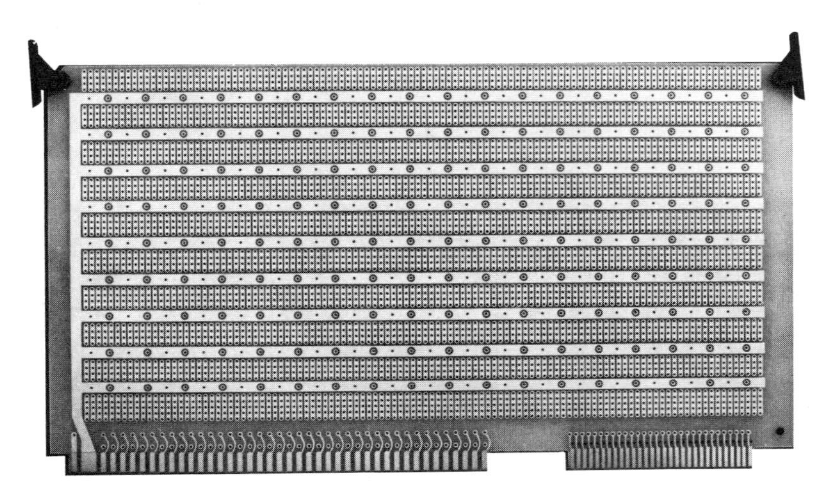

Fig. 1-19. Intel SBC-80 compatible Multibus breadboard, the 12-DE-16. ***(Courtesy Douglas Electronics)***

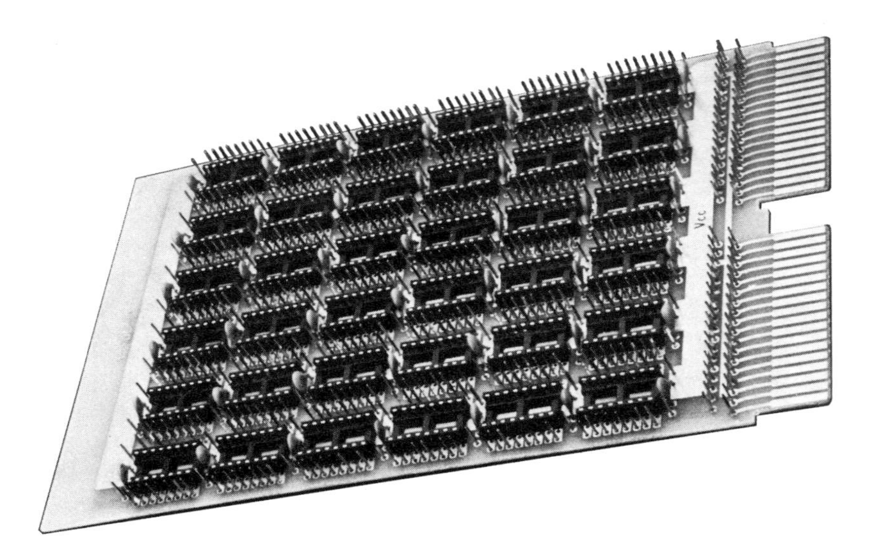

Fig. 1-20. DEC LSI-11 wire-wrap board, the 11-DE-11-H. ***(Courtesy Douglas Electronics)***

Combining Wire-Wrap and Printed Circuit Techniques • As will be seen in later chapters, etching single-sided printed circuit boards, or etching double-sided PCBs with the same pattern on both sides, is a relatively easy task with few registration or photographic problems. Previous examples of commercial boards have shown how a fairly simple etched pattern can greatly improve the design and assembly of wire-wrap boards, as compared to using blank perforated base material; why not etch your own wire-wrap patterns for specific prototyping applications, starting with blank copper-clad boards? Blank boards with edge connector patterns already etched and plated for various bus systems are available, thus avoiding the problems of having to etch, plate, and machine connector fingers to exacting specifications yourself. If a wire-wrap assembly does not need edge connectors, the task is simplified further by using blank copper-clad PCB stock material.

Preparation of artwork for printed circuit wire-wrap boards is quick and simple, because the difficult task of laying out signal connections is avoided completely; these connections will be made by wire-wrap techniques. The combined printed circuit/wire-wrap patterns shown in Fig. 1-21 were prepared by the author and illustrate how a dual approach (PCB and wire-wrap) can be valuable. This board interfaces a 61-key electric piano to a parallel computer I/O port using complementary metal-oxide semiconductor (CMOS) logic. The assembly is fairly densely populated with ICs, and requires a large number of connections to multiplex the piano key switches, since the circuit was designed to both read (detect if notes are being played) and write (play notes). To achieve the required density and fit this circuit into the electric piano's housing, a multilayer PCB would have been required to carry all of the logic signals.

Instead, components were laid out in a sketch, and artwork was prepared on a 1:1 scale to provide a pad for each component, IC socket, and connector to be soldered into the single-sided board. Power supply and ground buses were then taped in between the socket rows, and a few input/output signals were taped in from the four ribbon cable connectors. Artwork was prepared in one evening. No camera photography was necessary to make a negative from this 1:1 artwork on clear plastic; a contact negative was made in 30 minutes using Kodalith film, and the single-sided circuit board was etched from blank stock using simple negative techniques with Kodak photo resist type 3 (KPR-3). (Kodalith and KPR-3 products are discussed later in Chapters 4 and 5.) Note that if positive photoresist were used, the board could have been exposed directly from 1:1 artwork prepared on clear Mylar film, without the need to go through a contact negative.

Although the electric piano circuit could have been assembled with the same high density on some general-purpose wire-wrap board, several advantages of etching your own wrap patterns are evident:

- Components are easily attached to the board, with a solder pad intentionally provided for each lead. Odd-sized connectors and components are thus mounted with no problems, and extra pads for attaching wire-wrap posts can be included.
- Power and ground buses are available exactly as you desire.
- Board is shaped in any geometry you require.
- Board is inexpensive. Hobbyists may find commercially available etched wire-wrap boards to be fairly costly.

Two disadvantages of etching your own wrap boards are the need for drilling holes, and the decreased ability to make unforeseen circuit changes that require extra components and ICs. This second problem can be overcome to some extent by leaving extra space, and/or by etching a few extra IC pad patterns. By learning the printed circuit board techniques described in later chapters, the reader will find that his approach to generating wire-wrap designs is greatly expanded; the combination of wire-wrap assembly and etched single-sided PCBs is versatile and opens up new possibilities for creating unique electronic prototypes.

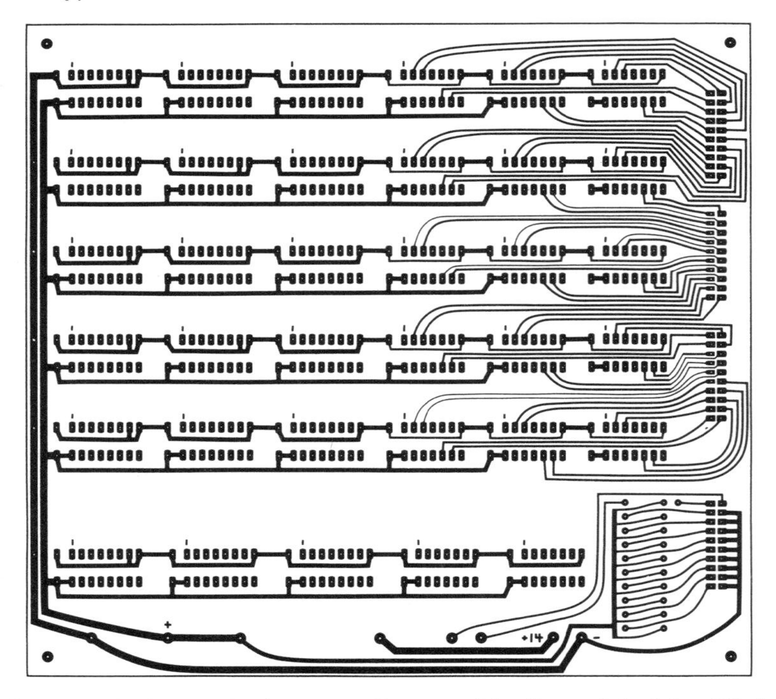

Fig. 1-21. Custom etched patterns for a combination printed circuit/wire-wrap board.

HARDWARE •

Sockets • Integrated circuit DIP sockets for wire-wrap are available from many sources, such as Digi-Key Corp., Jameco Co., and Radio Shack. These sockets should have gold-plated contacts to ensure reliable electrical connection of ICs and wrapping wire. Wrap pins extend 0.6 inch (15.24 mm) below a one-sixteenth inch board to provide 3 levels of wire-wrap connections. Four principal methods are used to attach sockets to boards as follows:

- Adhesive — Quick-set epoxy may be used for a permanent bond, or a nonhardening pliable cement, such as Permatex adhesive, allows sockets to be easily removed.
- Screws — Some sockets have recessed center holes to allow for screw mounting. This requires that appropriate holes be drilled in the board.
- Soldering — Solder pads provide the most convenient method of securing sockets. If no electrical connection to the pads is needed, it is only necessary to solder two diagonal corner pins to mount a socket; socket removal then only requires minimal desoldering.
- Wire-wraps themselves will hold a socket down fairly well. The socket is pushed firmly against the board while several blank wraps are made flush against the opposite side.

Construction techniques such as solder pad wiring, pencil wiring, and reverse wrapping do not require IC sockets with wire-wrap pins. Standard DIP sockets used for general PCB installation of ICs have short pins and are usually mounted by soldering to pads.

Terminals and Pins • Terminals and pins used for breadboarding and wire-wrapping are shown in Fig. 1-22A with items from Vector and OK Machine & Tool. Most of these pins are installed by press-fitting into standard 0.042 inch (1.07 mm) holes on 0.1 inch (2.54 mm) grid-pattern boards, although pins for larger holes are available. The same group of pins pictured in Fig. 1-22A is shown installed in Fig. 1-22B. Needle nose pliers provide the simplest installation method, although special terminal-installing tools are often used for rapid insertion of a large number of pins. In some cases, pins are designed to fit tightly into holes after press-fitting, and do not require any additional mechanical attachment. For example, some have a knurled center body flange to "bite" hole edges. However, it is advisable to solder all terminals to the underside of a board if solder pads are present. An *alignment block* makes it easier to install long wire-wrap pins exactly perpendicular to the surface of the board. Such a block can be made by simply gluing together 10 pieces of blank 0.1 inch (2.54 mm) grid perforated board in a stack.

Terminals and pins have two basic applications:

1. Stand-alone wrap posts. These pins may mount on a power bus, ground bus, or a solder pad. They may also feed signals from one side of the board to the other, as in reverse wrapping. Fig. 1-23 illustrates some wrap posts soldered into edge connector finger pads, carrying signals out of the connector and into the board. Similar pins are soldered into commercial PCBs to act as convenient test and troubleshooting points for important signals.
2. Support and attachment of discrete components, such as resistors, capacitors, transistors, diodes, and hookup wire. These terminals may or may not be wrapped on the back side, but their primary function is to attach and support a component lead. The following figures illustrate typical methods of component mounting:
   (a) In Fig. 1-24A components are soldered into forked terminals, and also pushed into three-pronged "clip" terminals, which are all wire-wrapped on the opposite side. (Connections are shown prior to soldering.)
   (b) Fig. 1-24B shows individual socket wire-wrap pins arranged to form a 3-lead transistor socket, and an 8-lead TO-5 metal can op-amp socket. The single pin sockets accept leads from 0.012 inch (0.305 mm) to 0.023 inch (0.584 mm) in diameter.

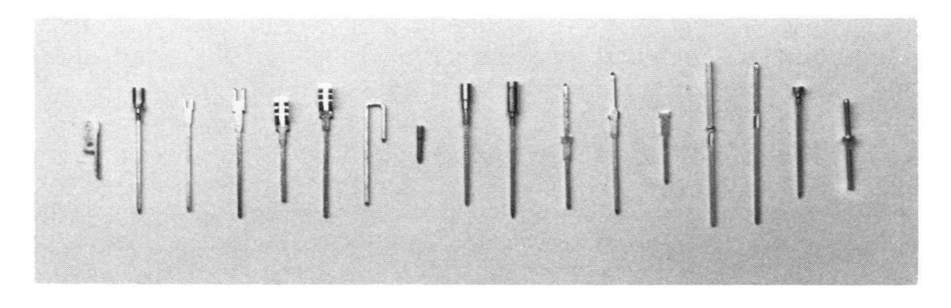

(A) Assortment of types from Vector Electronic Co. and OK Machine & Tool Corporation.

(B) Pins installed in perforated board.

Fig. 1-22. Terminals and pins used for breadboarding and wire-wrapping.

(c) Fig. 1-24C shows how versatile "J" wire-wrap pins can be used to attach large component leads and hookup wire, which is then soldered.

(d) Finally, Fig. 1-24D illustrates the use of a 16-pin DIP socket "header" for mounting small discrete components; after mounting, the header is plugged into a socket as if it were an IC. This method of component attachment is compact and popular for use on prototyping boards which are fully socketed for ICs and have no extra space for discrete components.

Connectors • If the reader is unfamiliar with electronic connectors, a catalog from a general parts supplier such as Newark Electronics provides good drawings and descriptions of products from many manufacturers. The Newark "connector" section covers more than 100 pages! Most of these products are intended for use in modern printed circuit assemblies, but many can be installed in wire-wrap boards. The use of appropriate external connectors in a wire-wrapped assembly cannot be overemphasized. Careful planning is needed to determine what signals will enter or leave the board, and through what type of connection system. The following questions should be answered to avoid problems in design, construction, testing, and final performance:

1. Can the board be easily mounted or removed from its enclosure? These operations may occur more often than planned!

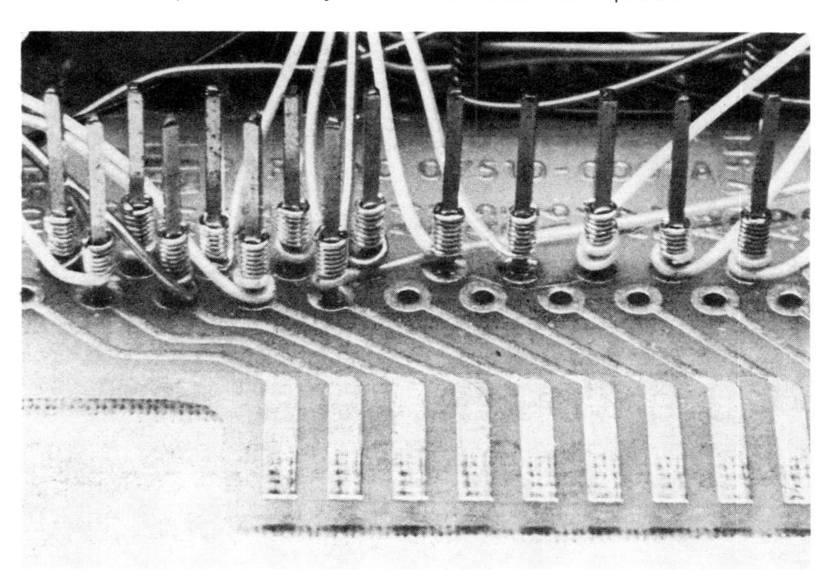

Fig. 1-23. Wrap posts terminating edge-connector fingers.

2. Are all signals accessible for troubleshooting when the board is connected and powered? If not, long connecting cables may be necessary so that the board can be removed from its case while still operating.
3. Are connectors reliable, and can they be disconnected/reconnected repeatedly? The most frequent problems encountered in electronic maintenance involve poor connections where a mechanical contact is present.
4. Are cables and connectors neat? A maze of tangled wires will lead to frustration and difficult troubleshooting.
5. Are high-voltage lines, such as 117-V ac power, protected to avoid shock hazard when you reach to make an adjustment?
6. Can current measurements be made easily on dc power supply lines? An in-line fuse holder provides good protection, as well as an accessible means of breaking lines to measure current loads.
7. Can all connectors be physically attached to the wire-wrap board? How will the terminals be wired to on-board signals?

(A) Forked and clip terminals with square-post bottom pins.

(B) Single-lead socket wire-wrap pins arranged to form custom sockets.

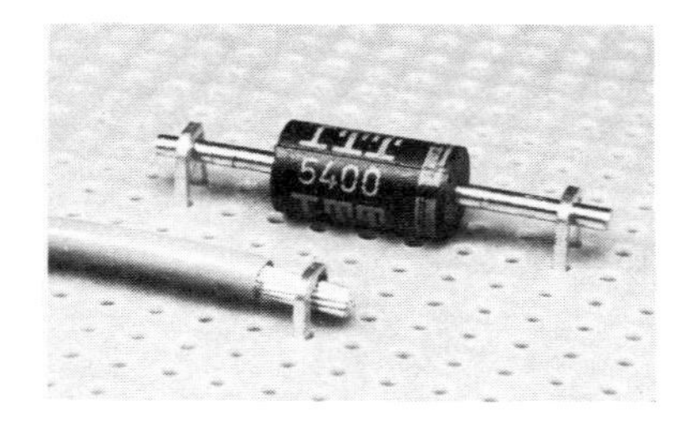

(C) "J" wire-wrap pins for securing large leads.

(D) DIP "header" for mounting small components.

Fig. 1-24. Methods for mounting components on various Vector terminals.

8. Has proper consideration been given to noise problems? These include improper grounding, ground loops, excessive cable lengths, improper shielding, improper line terminations, radiation, and crosstalk. *Crosstalk* occurs when the signal in one wire becomes coupled by electrostatic and/or magnetic fields into an adjacent wire, as with multiwire cable and groups of wires bundled together for some distance.

Noise is probably the greatest plague to electronic circuit designers and hobbyists. Typically, noise problems vary with physical construction methods, particularly connector hardware, and become more troublesome at higher frequencies. Thus, the same circuit may perform differently on a solderless breadboard[2], a wire-wrap board, and a PCB. The author once assembled a computer interface for a drum plotter on a solderless breadboard, and connected it to the I/O bus pins of the computer with individual 22 AWG insulated hookup wires. The circuit performed satisfactorily after some minor changes. A prototype was then assembled on a wire-wrap board, with a 3-foot ribbon cable as the I/O bus connector. This new arrangement failed to function properly due to excessive crosstalk in the cable, and the interconnection system required modification so that a larger ribbon cable could be used with alternating ground and signal lines.

Some of the most popular connectors for prototyping will be discussed next; they are available from hobbyist suppliers such as Digi-Key, Vector, Jameco, and Radio Shack.

Edge Connectors — Edge-card connectors are the most widely used system for bringing signals into and out of circuit boards, since they accommodate a large number of signals as well as provide physical support for the board after it is plugged into the mating connector. Contacts are generally gold-plated both on the board fingers and in the receptacles to avoid poor connections and corrosion. Edge-card female receptacles are available with the following three basic types of protruding pins:

1. Wire-wrap, with standard 0.025 inch (0.635 mm) square posts 0.6 inch (15.24 mm) long (see Fig. 1-25).

2. Solder eye, for attaching hookup wire by soldering.

3. Solder tail, for soldering the receptacle into another assembly. In computer systems, a master assembly of edge connectors is known as the *backplane,* and generally includes a mechanical cage which allows boards to be guided into place on rails. If backplane connections are made with a printed circuit board, this is known as the *motherboard.*

Ribbon Cable Connectors — Ribbon cable is a flat assembly of typically 10 to 50 wires of size 28 AWG, encapsulated in plastic insulation. Wires are often color coded and give the cable a rainbow appearance, thereby making individual signals traceable by eye. Ribbon cables are popular in computer and logic systems because they provide a neat, manageable way to route and terminate a large number of parallel signals. Fig. 1-26 illustrates the three simplest ways to terminate a ribbon cable on a wire-wrap board, using cable components from AP Products, Inc., and described as follows:

1. Edge-card connector — The 4609 Apple II plugboard from Vector includes a 20-contact finger for this exact purpose, to the right of the picture.
2. DIP socket — The cable is terminated in a package that resembles an IC, and can be handled exactly as you would an IC — plugged into a socket, or soldered into place. The cable is shown in the middle of the picture, plugged into a wire-wrap socket.
3. Pins spaced 0.1 inch (2.54 mm) apart — To the left of the picture, a ribbon cable terminates in a female connector that mates with a dual row of 0.025 inch (0.635 mm) square pins. The pins may be individual wire-wrap posts soldered into place, or more conveniently, a one-piece "header" assembly, such as the connector shown, which has wire-wrap pins extending below the board.

Fig. 1-25. PCB edge-card connector.

Terminal Blocks — Three varieties of connector blocks are seen in Fig. 1-27. These blocks are available in many lengths and numbers of contacts, and are generally used to bring signals or power leads in and out using stranded hookup wire. (Wire connections which will be frequently disconnected should be made with stranded rather than solid wire for strength and flexibility.) The Fig. 1-27A terminal strip from OK Machine & Tool is very useful for wire-wrap applications, with 0.04 inch (1 mm) pins that insert snugly into 0.042 inch (1.07 mm) holes on 0.2 inch (5.08 mm) centers. These pins are round and should be soldered following pencil wired connection to board signals. External connection is made by inserting wire leads into the terminal holes, and then screwing them down tight with screw clamps that are built into the strip.

The larger terminal block in Fig. 1-27B requires some drilling to mount on circuit boards. Feedthrough pins are too large to make it through 0.042 inch (1.07 mm) holes, and a screw-hole at each end must be drilled out for attachment. These strips make very solid, reliable connectors in situations where wires must be repeatably detached and reconnected; wire ends are soldered or crimped to push-on lugs for easy removal.

The connector blocks shown in Fig. 1-27C have a male header that is attached to the board, and a female strip containing wires that plugs into the header. Cable or individual wires are attached to the female strip with removable socket inserts. This is the most widely used con-

Fig. 1-26. Three types of ribbon cable terminations on a wire-wrap board: edge-card connector, DIP socket, and header pin rows.

nector system for joining a small number of wires to circuit boards, because it provides for quick-disconnection and can use almost any type of wire or cable.

Quick-Disconnect Connectors — Incoming wires can be soldered directly into a board, but these connections are not easily broken apart. A *quick-disconnect* connector can be installed in series with cables at strategic points to allow for fast insertion and removal of boards from larger assemblies. Some of the most popular plug/socket combinations are shown in Chapter 11, Fig. 11-15 from Molex, together with pins that are installed in the nylon casings for wire attach-

(A) Terminal strip.

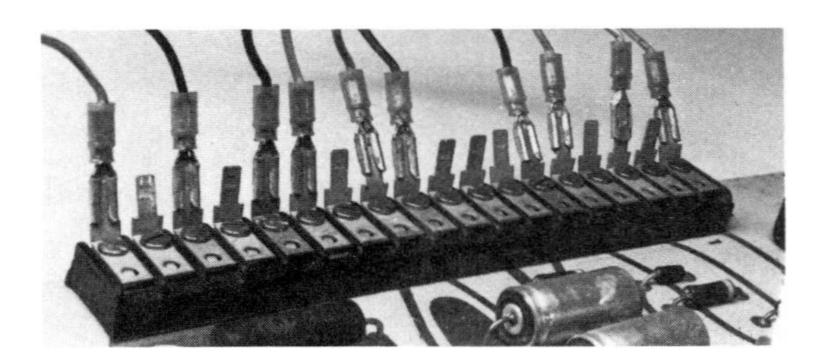

(B) Terminal block.

(C) Header strips.

Fig. 1-27. Three types of external wire connector blocks.

ment. The Molex series connectors are available in many shapes, sizes, and contact numbers, ranging from 1 to 36 circuits. Plugs are polarized mechanically to prevent improper connection, and most housings can also be panel mounted. The pins can be easily removed from their housings for circuit changes. If printed circuit pins are used, plug or socket connectors may be soldered into a board after pins are installed, producing a connection system similar to that of Fig. 1-27C.

Another useful way to temporarily attach hookup wire to wire-wrap boards is to use single-wire socket terminals that push down directly on 0.025 inch square wire-wrap posts.

METHODS FOR MOUNTING BOARDS •

Screw Mounting • Individual boards may be fastened down to a chassis base with screws and standoffs attached to each corner. The standoffs are usually aluminum spacers, sometimes threaded, with sufficient length to allow wire-wrap pins to clear the mounting base. This is the simplest method for securing boards. Spacer and standoff hardware is available from E.F. Johnson, a well-known product line distributed by Digi-Key.

Card Guides • Card guides and brackets may be used to install a circuit board in an enclosure, as seen in Fig. 1-16. The guides shown are K169 from OK Machine & Tool, and have plastic clips to grip the board tightly after it is pushed between the mounting brackets. After the board is inserted, a card edge connector may be pushed down on the top edge finger. (Other card mounts are designed to guide the board into an edge connector mounted at the base.)

Commercial Enclosures • Commercial enclosures are available with rows of card guides and edge connectors already installed. This is the standard system used for computer and logic assemblies, where a number of identical-size boards are inserted in parallel into a master interconnection system. Enclosure hardware for multiple boards is available for all major styles of circuit board connectors; for example, Fig. 1-28 illustrates the Vector CCK100 rack-mount cage which holds up to 21 S-100 microcomputer cards. Similar enclosures also include room to mount a power supply.

Unoccupied Slots • Most computer systems are supplied with extra unoccupied card slots for adding additional circuit boards and custom interface circuitry. If a standard microcomputer "plugboard" of correct dimensions is used for wire-wrap, installation consists of simply plugging the board into an appropriate blank slot in the computer chassis. In the construction example at the end of this chapter, Fig. 1-37 shows a prototype wire-wrap board plugged into the Apple II microcomputer.

POWER SUPPLIES • Unless it is absolutely necessary to incorporate power supplies into a wire-wrap board, modular power supply kits are recommended, such as the ones shown in Fig. 1-29 from Jameco Electronics. These low-cost power supplies include a 5 V/ 1-Amp regulated model (JE 200), an adjustable 5–15 V/0.5–1.5 Amp regulated model (JE 210A), and also a handy, inexpensive multivoltage adapter unit (JE 205A) that uses a switching transformer to convert 5-V dc to negative voltages and other positive voltages. Unregulated outputs of −5 V, ±9 V, and ±12 V are produced simultaneously at current ratings of 250 mA, 200 mA, and 160 mA, respectively. The need for multiple power supplies is thus eliminated. The outputs will shut down if overloaded, and an extra 13 turns of wire on each side of the toroidal transformer's secondary winding will bring output voltages up to ±15 V at a current rating of 100 mA.

Some microcomputer bus systems assume that raw power will be distributed to voltage regulation circuitry present on each card, such as in the S-100 system. Unregulated +8-V dc is supplied to each board for the +5 V supply, and 1.5 amp 7805 ICs are generally used as the 5 V series regulators. The Vector S-100 prototyping cards, such as the 8800V previously pictured in Fig. 1-18, have space reserved for 2 heat

Fig. 1-28. Rack-mount card cage for S-100 boards. ***(Courtesy Vector Electronic Co.)***

sinks and 4 TO-220 regulators because of these S-100 power requirements.

If power supplies must be constructed, it is best to locate the heavy, heat-producing components such as transformers, filter capacitors, and rectifier diodes away from circuit boards and on a sturdier support. These items can be chassis-mounted, or thick perforated board can be used such as Vector's "A" pattern (0.265 inch (6.73 mm) grid pattern, holes 0.093 inch (2.36 mm) in diameter, three-thirty-seconds inch thick). Connections are made with point-to-point wiring on solder lugs or push-in terminals available for the larger hole size.

The use of commonly available pocket calculator power supply modules should not be overlooked as an inexpensive source of *unregulated* low-voltage power of up to a few hundred milliamperes.

TOOLS AND ACCESSORIES • In addition to standard hand tools, such as small electronic pliers, cutters, needle files, X-Acto knives, screwdrivers, nutdrivers, wrenches, wire strippers, etc., some tools and accessories are particularly useful for wire-wrap applications.

Lighting • A well-lit work area is mandatory for assembling compact electronic boards. Adjustable lamps which clamp on the edge of a workbench and swing out over the table are ideal for this type of work; some models have a built-in magnifying lens for viewing small objects. Consult your local hobby store, electronic parts retailer, and office supply house to choose the most suitable lighting equipment.

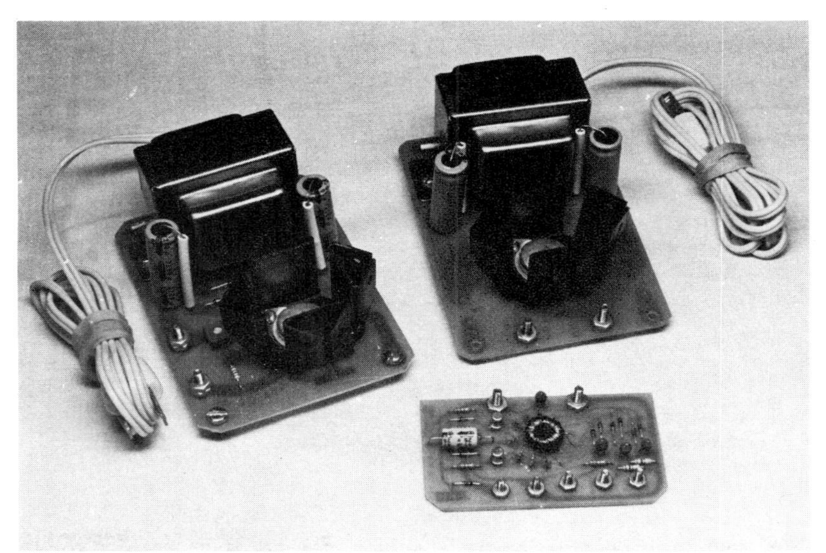

Fig. 1-29. Modular power supply kits from Jameco Electronics.

Desoldering • Removing small soldered items such as IC sockets and wire-wrap pins is not impossible when the right tools and techniques are used. The prime concern when working with printed circuit soldering is to avoid excessive time and heat that causes solder pads to "lift" off the board due to adhesive decomposition. A 25-watt pencil soldering iron is generally safe and sufficient for most wire-wrap work, and small enough in size (Fig. 1-31) to maneuver in tight spots. Even better is a soldering iron with a controlled-temperature tip, such as the popular Weller WTCPN shown in Chapter 9, Fig. 9-9. This iron will supply heating power in the range of 15–100 watts for many sizes of interchangeable tips available in 600°, 700°, and 800°F temperatures. The pencil itself is lightweight and handles well.

Desoldering is accomplished by vacuum suction of molten solder, or by absorbing solder with fine, braided copper wire. A typical suction device is shown in Fig. 1-30, the Soldapullt from Edsyn. This spring-loaded cylinder is placed over a joint, solder melted with an iron, and a button releases the tool's handle to suck molten solder away from the joint and into a chamber inside the tool. Care must be exercised when using this tool to desolder printed circuit pads; excessive heat combined with suction will tear pads away from the board.

Braided copper wire is another good method for desoldering. Fig. 1-30 pictures a roll of Tech-Wick brand copper braid, and a plastic holder which dispenses the braided wire from a guide to avoid burning fingers. To perform properly, braided wire must be made up of fine strands, it must be clean, and it must be well-coated with rosin flux. The braid end is placed directly over the pad to be desoldered, and the soldering iron tip is pressed down firmly on the braid until the wire and

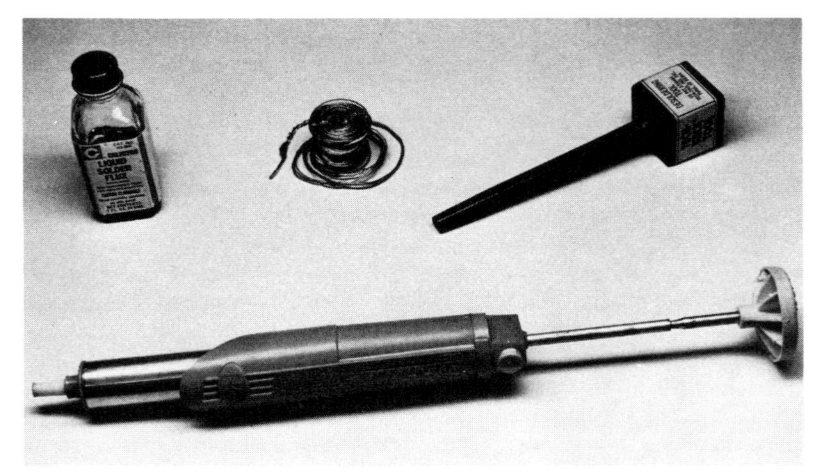

Fig. 1-30. Basic desoldering tools: solder sucker (Edsyn), braided copper wire with holder (Tech-Wick), and liquid rosin flux (GC Electronics).

pad are both hot enough to melt the solder, which quickly "wicks" up into the copper braid. The spent solder-coated end is snipped off before proceeding to the next joint. A properly wicked connection can be desoldered in a few seconds to leave a joint almost completely free of solder; wiggling the protruding wire end or lead with a small pair of needle nose pliers or tweezers may help to break the connection loose from any remaining thin solder film. Although braided wire for desoldering comes precoated with flux, for best results the wire should be dipped in liquid rosin flux just prior to use. A suitable isopropanol rosin solution from GC Electronics is also seen in Fig. 1-30.

One of the most difficult desoldering operations is the removal of an IC or IC socket from a PCB with plated-through holes, which tend to retain solder by their own wicking action. If you are willing to sacrifice the component, snip off its leads from the body, or crush it with pliers to separate each lead; the pins can then be individually removed from their holes, and each hole can be cleaned out with braided wire or a solder sucker. Large flat soldering iron tips are available to simultaneously heat all of the pins on an IC, but this is a risky approach because of the amount of heat generated. In any case, it is best to prevent IC desoldering problems from occurring in your own boards by always installing IC sockets.

Vises • An electronic vise with a circuit board holder is shown in Fig. 1-31. The PanaVise vises with interchangeable heads and bases allow work to be rotated and held at almost any angle, being well-suited for wire-wrap work where both sides of the board need to be simultaneously accessible. Circuit board holders are particularly helpful during desoldering, when it is necessary to heat one side of the board and pull a component lead away from the other side.

Drills • Wire-wrap boards are supplied predrilled or perforated, making assembly neat and less tedious. However, it is often necessary to drill extra holes, enlarge existing holes, or perform routing and cutting operations with a small electric drill. A miniature drill press from GC Electronics is shown in Fig. 1-32, with adjustable speed control and an assortment of 30 bits. Although the mounting bracket does not have enough depth for drilling wide circuit boards, the motor can be removed for hand-held operation. Dremel and Unimat are two other suppliers of hobby drilling equipment, and Vector supplies miniature drills and accessory bits to accomplish circuit board operations, such as routing, shaping, and line-cutting.

Also shown in Fig. 1-32 is a pin-vise for manually drilling small holes, and a drill index box containing small circuit board bits ranging from No. 61 to No. 80 — 0.039 inch (0.99 mm) to 0.0135 inch (0.34 mm) — in size. Small drill bits are carried by industrial hardware outlets.

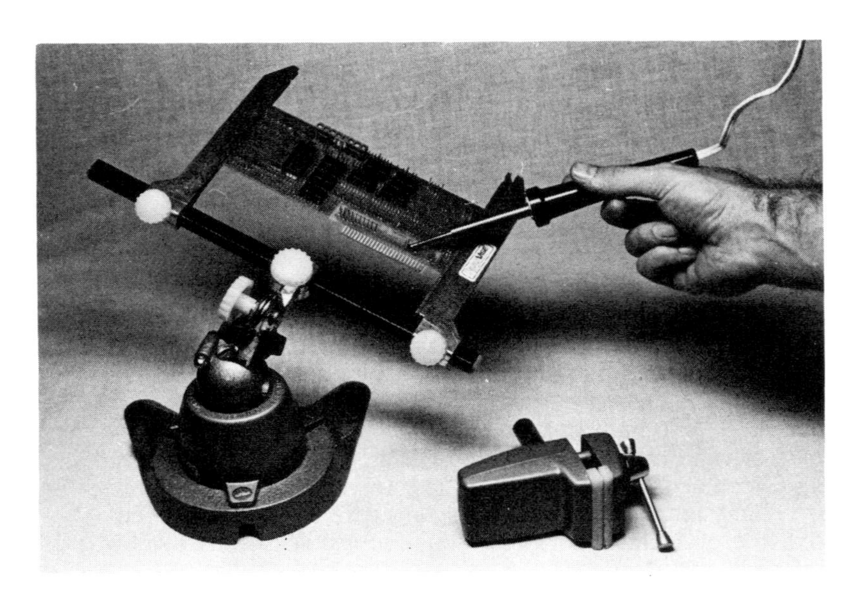

Fig. 1-31. Electronic vise with circuit board holder attachment.

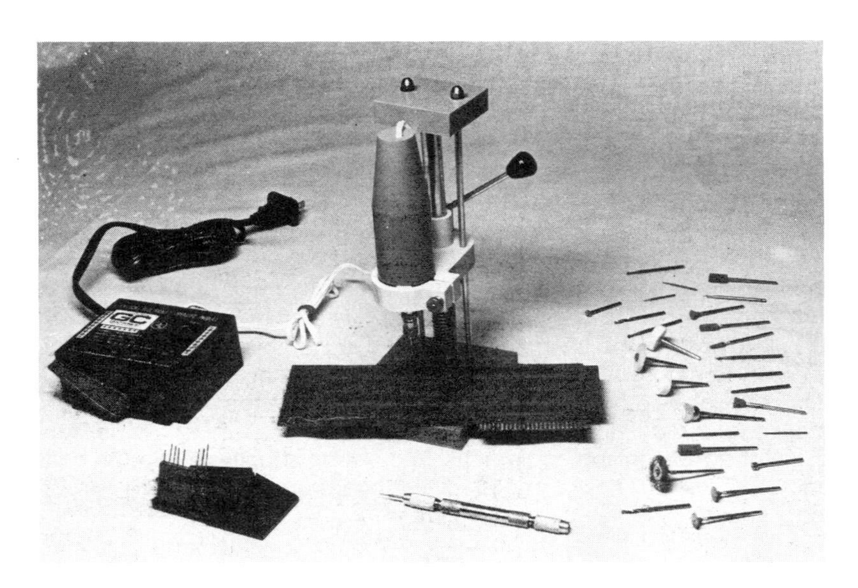

Fig. 1-32. Miniature drilling equipment, including drill press, manual pin-vise, and assorted bits.

Continuity Tester • After wire-wrapping a large number of connections, it is advisable to turn the board with its component side up and probe each contact with a resistance meter to see if the schematic was followed and connections were properly made. This check procedure is performed most efficiently if your eyes do not have to watch the meter. Audible continuity testers are available commercially, but you can easily modify any multimeter to perform this valuable time-saving test function which allows both hands and eyes to operate together in probing connections and following a schematic diagram. The author mounted a small dc buzzer inside a Radio Shack multitester and wired it to a jack input that was seldom used (high-voltage ac). By plugging the common meter lead into this jack, the leads are placed in series with the meter battery and the buzzer, and continuity can be verified by listening for the proper-pitched buzz of a completed circuit. Piezo buzzers are ideal for this purpose with their low current and voltage demands and high frequency audio output.

DESIGN AND CONSTRUCTION •

Design Considerations • Once it has been decided to construct a prototype with wire-wrap methods, a number of decisions must be made before actual assembly begins. The following considerations are basic to wire-wrap design:

1. Enclosure — What type of case will be needed, how much space is available, and how will the wire-wrap board be mounted? How will components external to the board be mounted in the enclosure?
2. Board — What size and type of board will form the basis for the assembly? Many possibilities exist for perforated boards, commercial plugboards, various hole spacings and etched patterns, and various power supply and ground configurations for noise reduction. How much space is needed for possible later circuit additions?
3. Power supply — How will the assembly be powered? Should the supply be constructed as part of the circuit board, or should an external modular supply be used? Is the power supply fused and connected in such a way as to make troubleshooting simple?
4. Wrap method — Which wiring method and what tools will be required? Can the same gauge wire be used for all connections? Frequently a certain amount of soldering is unavoidable for wiring and proper attachment of pins and components. Will this hinder the ability to make easy circuit changes?
5. Components — How will components be attached to the board? What type of sockets, pins, connectors, and mounting devices will

be necessary? How will odd-sized devices such as relays, switches, and large capacitors be mounted?

6. Layout — How will components be physically arranged on the board?
7. External connectors — What signals must enter or leave the board, and through what type of connection scheme? Connector problems were discussed earlier in some detail.
8. Testing and troubleshooting — Is the board accessible for test measurements, and are all signals available? Can the board be easily removed to make physical changes and circuit alterations? Remember that a prototype is not expected to be free of problems, so assemble it with the idea that some modifications will be necessary!

Assembly Example • A wire-wrap construction example will be presented in this section, going from schematic to final hardware. The device to be constructed is an 8-bit Digital-to-Analog Converter (DAC) for the Apple II computer. It was desired to use this interface module as a general purpose controller with four channels of command, and a digital input and digital output line were, therefore, included with each of the four multiplexed DAC outputs for control purposes. The complete schematic of the DAC/controller is shown in Fig. 1-33 for reference. However, its electronic characteristics will not be discussed further since we are interested primarily in physical construction techniques.

Most of the circuitry in Fig. 1-33 was tested on a solderless breadboard before it was deemed ready for wire-wrap construction. However, no connections were actually made to the Apple II bus; these input signals were artificially simulated with switches and single pulses. The reader should be forewarned that bringing microprocessor bus signals out to a solderless breadboard with long wires can change the electrical characteristics of the entire bus and adversely affect the operation of a computer. Interestingly, this is a design situation where wire-wrap plugboards are more suitable than solderless breadboards for testing new ideas, because signal length can be controlled with a plugboard. In any case, the various IC register and buffer configurations connecting the DAC interface to the Apple II bus had been previously proven in other projects and were considered sound.

Enclosure, board, and power supply were all fairly rigid variables. The enclosure was the Apple II cabinet, which contains eight I/O edge-connectors for interface cards. The board was limited to a plugboard of appropriate dimensions to fit into this system. The Apple II computer supplies +5-, +12-, and −12-V dc power to reach edge connector; as seen in the Fig. 1-33 schematic, power requirements for the inter-

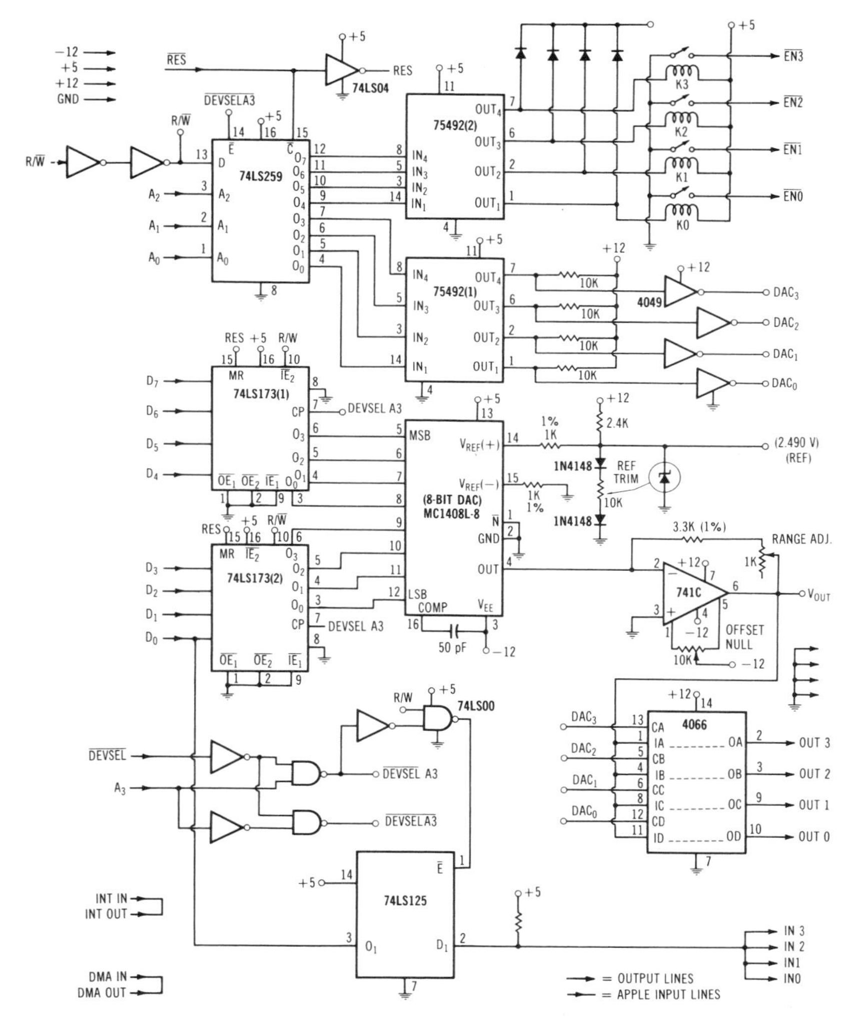

Fig 1-33. Schematic of a Digital-To-Analog Converter/controller for the Apple II computer.

face board were +5 for 74LS-series logic, relay coils, and the DAC IC, +12 to generate analog outputs up to 10 volts, and −12 to supply the DAC and op-amp ICs on their negative ends. Low current demands of this circuitry were not expected to tax the computer power supply which is protected against shorts and excessive currents. Bypass capacitors were included for each IC power supply input as a noise-reduction measure.

The wire-wrap plugboard chosen for this project was the CCS 7510A prototyping board previously shown in Fig. 1-17. The etched pattern of this board makes it well-suited for reverse wrapping and solder mounting of all components, an approach that results in a low-profile assembly well-suited for closely spaced Apple II boards.

The final product was expected to be somewhat inflexible as far as moving or changing components, but very flexible for making wiring changes. All components had leads that could be directly soldered on 0.1 inch (2.54 mm) centers, making the mounting straightforward; these parts included 4 reed relays, two 8-contact terminal blocks, 13 ICs, 6 diodes, 15 capacitors, and 12 resistors, three of which were variable trimpots. Screw-down terminal blocks of the type shown in Fig. 1-27A were chosen as the simplest method of bringing wires into the board, and no noise problems were expected with the low-frequency D/A output lines. Each D/A control channel required a 4-wire cable: D/A output line, digital input, digital output, and common ground return. Holes in the 7510A board required some drilling to enlarge the 0.038 inch (0.965 mm) holes for terminal block feedthroughs, but no other board alterations were necessary.

Assembly consisted of the following consecutive steps:

1. Layout — Components were physically moved around on the blank 7510A board until a neat, orderly arrangement of parts was obtained. The CMOS integrated circuits requiring static protection were not handled during this process, but were replaced with dummy ICs of the same size. Special attention was given to finding a layout which would leave space for the installation of reverse wrap pins connecting to each component lead.

2. Documentation — The layout was transferred to 0.1 inch (2.54 mm) grid graph paper on a 2× scale, as shown in Fig. 1-34. This task was made easier with the help of electronic design templates from Bishop Graphics. (The Chapter 3 sections covering PCB design and layout describe these templates in more detail.) The layout diagram provides a proper record and guide for soldering in pins and leads during the next assembly step. As a visual aid, wire-wrap posts were colored in with a green pencil, and components or socket terminals were colored in red. Each edge-connector signal used from the Apple II bus was also labeled with its name, and shaded to indicate the need for a wire-wrap post.

3. Reverse Wrap Posts — Wire-wrap posts were soldered into place, as shown in Fig. 1-35. These posts extend 0.4 inch (10.16 mm) above the board for 2 levels of wrap; they are shorter than the standard 0.6 inch (15.24 mm) 3-level posts. Pins were installed by first soldering a post of the proper length in each corner. The board was then supported on these corner posts while the

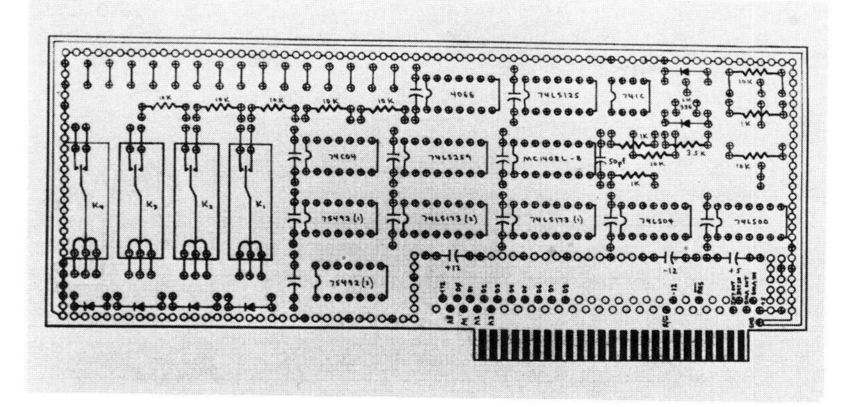

Fig. 1-34. Layout is recorded on 0.1 inch (2.54 mm) grid graph paper as a record and guide for assembly.

remaining pins were dropped in from the back side and soldered into place, producing a uniform set of posts protruding equidistant on the component side. Pins were cut flush on the back side. It might be of interest to note that the large number of reverse wrap posts needed for this assembly were obtained by cannibalizing a junked computer backplane found in a salvage yard. Obsolete backplanes are fairly common and provide thousands of gold-plated wire-wrap posts.

4. Components — Components and sockets were soldered into place as seen in Fig. 1-35. Note that only two channels of D/A were actually implemented at this point; space was set aside for two more reed relays and a second 8-pin connector block when needed.

5. Wiring — Following the schematic diagram, all power supply connections were made with a daisy chain tool and 30 AWG Kynar wire. Red insulation was used for (+) lines, black for (−). The signal connections were then made with conventional modified wraps using various colored prestripped wires. (As many random colors as possible were included to make signal tracing easier.) An orderly way of accomplishing the tedious task of wiring is to list every component on a sheet of paper, and proceed to connect all leads on each component successively to their appropriate destinations, ignoring any connections previously made. A complete "wire-run" list can also be constructed before beginning to wrap. During wrapping, the routing of wires is accomplished by following natural pathways around sockets and pins, using a small screwdriver to press-fit wires into neat channels with a mini-

mum of slack. After wiring was complete, the board was examined to make sure that at least one wrap was installed on every post. The final wrapped board is shown in Fig. 1-36. The reverse side is flat, having only solder connections.

6. Continuity Testing — The board was tested for wiring errors using an audible continuity tester to follow the schematic. Several problems were found and corrected: connections left out, two identical 75492 ICs partially cross-wired, and an omitted solder joint. Particular attention was paid to making sure that wires connecting to the Apple II bus were correctly installed; a mistake on these lines could result in costly computer repairs!

7. Electrical Testing — Power supply inputs were first checked with a resistance measurement between themselves and ground to de-

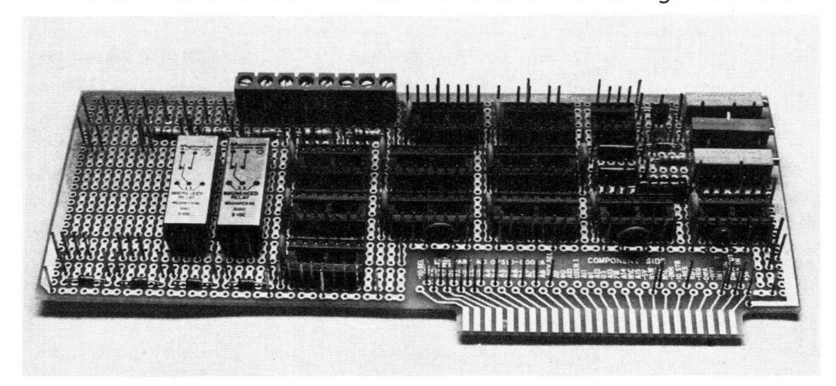

Fig. 1-35. Components, sockets, and pins are soldered into place.

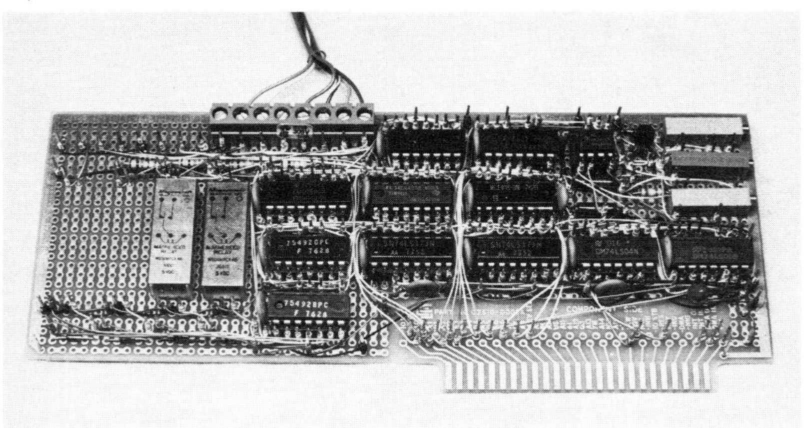

Fig. 1-36. Final board with wire-wrapping complete.

Fig. 1-37. Prototype board is plugged into the Apple II computer for testing.

tect the presence of any dead shorts. No such problems were found. Current measurements were then made on the +5, +12, and −12 power inputs in succession by connecting them through a meter to a lab power supply. No excessive currents were seen; as expected, the +5 supply drew the most power, less than 100 mA with relays off. Permanent power connections were made. Each chip was then checked for excessive heat by touching and smelling all areas of the board. The experienced prototyper will learn to quickly recognize the odor of burnt semiconductors and plastic! In fact, the LM336 voltage reference IC was destroyed because of a small piece of loose wire that came into intermittent contact with the input line, shorting it to +12.

8. Calibration Adjustments — Several potentiometer adjustments were made with the aid of a digital voltmeter to accurately calibrate the analog output signals. It was necessary to replace two fixed resistors with slightly different values in order to obtain proper adjustment ranges for the variable resistors.

9. Computer Testing — The completed board was plugged into the Apple II computer in series with an *extender board* (California Computer Systems 7520A) to make for accessible testing (see Fig. 1-37). The extender board maintains electrical contacts while raising the prototype out into open space for convenient troubleshooting under power. All possible computer operations were tested by issuing appropriate BASIC programming commands and observing the results with a meter.

Chapter 2
Printed Circuit Technology

INTRODUCTION • Since its initial development during World War II, the printed circuit board (PCB) has evolved into the prime basis for modern electronic assembly. The original objective for printed wiring was to develop a lightweight interconnection system that was amenable to mass production, with military applications in mind; however, a number of other advantages have since become apparent for the printed circuit and are as follows:

1. Support — The rigid printed board provides support, protection, and heat dissipation for electronic components.
2. Versatility — Modern miniaturized components are particularly well-suited for printed circuit mounting, with few restrictions placed on the packaging design.
3. Electrical Characteristics — These factors can be designed into a PCB, with repeatable results obtained from one assembly to the next.
4. Production — The neatness and simplicity of this technologically complex product result in low cost, fast assembly, and straightforward testing procedures. The personnel involved require little training and few technical skills.
5. Maintenance — Troubleshooting of equipment is vastly simplified with flat, modular PCBs containing easily located components. The modern electronic technician exchanges plug-in boards to correct a problem, and ships the defective unit back to the factory where specialized equipment and personnel are available to efficiently make repairs.

The characteristics that make PCB construction ideal for mass production of electronic assemblies also have some inherent disadvantages: complex design, and high cost for low quantity production. Repair may also be difficult for highly miniaturized, dense, multilayer boards; however, a steady decrease in electronic hardware costs is leading to "throw-away" boards whose actual cost is determined largely by design/overhead expenses and the technical support provided by the manufacturer. This situation is obviously desirable for mass production where fixed expenses can be averaged out over a large number of product items.

Despite the technological complexities behind conventional PCB manufacture, the hobbyist or electronic designer has well within his reach the ability to assemble prototype circuitry with printed wiring methods, provided he is willing to substitute his own time for expensive equipment. Happily, the compromises that must be made in fabrication do not have a significant effect on product quality; it is possible to use manual design and construction methods that are not amenable to mass production, but that do permit the assembly of a prototype PCB in just a few days time, within a small laboratory or workshop setting.

OBJECTIVES • The main objective of this chapter is to give the reader an overview of current industrial PCB practices. It is felt that this knowledge will be helpful before proceeding to other chapters detailing simplified manual approaches that are feasible for prototype construction. Advances in the electronic packaging industry have led to a variety of new processes, automated tools, and product styles; we will not attempt to cover all of the important offshoots such as *flexible* boards, *multilayer* boards, and *multiwire* (bonded wiring). The model for our discussion here will be the industry-standard double-sided PCB of the type used in most modern computer and digital logic circuits. Since single-sided construction methods are essentially a subset of the double-sided PCB methods, they will not be considered separately. However, the reader should be aware that the single-sided PCB is still extremely important in industry for producing low-cost electronic equipment having relatively simple circuitry.

A second objective of this chapter is to outline a simplified approach for producing low-volume PCB prototypes with limited facilities; the actual procedures will be discussed in subsequent chapters. With the background provided by an initial discussion of conventional industrial methods, the reader will have a better perspective as he considers the compromises needed to design a low-cost laboratory system capable of producing quality results.

For more extensive information on printed circuit design and construction, the reader is referred to reference books 3 through 7 in Appendix A. Several trade journals (8, 9, 10) are also recommended to the reader interested in current PCB manufacturing trends and up-to-date sources for materials and equipment.

GENERAL INDUSTRIAL PROCESS •

Standard Double-Sided PCB • The conventional industrial process to be described in this chapter is for the design and construction of standard double-sided PCBs with *plated-through-holes.* A typical board is shown in Fig. 2-1. This product is usually formed on a one-sixteenth inch *laminated epoxy-glass* substrate, with copper foil patterns that are chemically etched on the two surfaces to form conductive pads and lines. The interconnecting lines join integrated circuit (IC) and discrete component leads which are typically mounted on 0.1 inch (2.54 mm) centered holes, approximately 0.030 inch (0.762 mm) in diameter. The hole walls are plated inside with copper and solder to provide connections from one side of the board to the other, leading to the descriptor "plated-through-holes."

Fig. 2-1. Typical commercial double-sided printed circuit board.

After components have been inserted in their proper mounting holes, component leads are soldered into place, providing electrical connection and mechanical support. A principal advantage of the plated-through-hole is the result that only one side of the board must be soldered; if components are all mounted from the opposite side, this design leads to feasible methods for automated solder assembly, and thus low-cost mass production.

During fabrication, copper patterns are coated with solder to serve as an *etch resist,* as well as to provide protection against oxidation and to preserve solderability. Edge connector fingers are plated with gold and nickel to make a durable surface for reliable connection into the final electronic package. Before automated solder assembly of components, the entire board is coated with *solder mask resist* to direct molten solder to proper areas without "bridging" closely spaced conductors. The solder mask is usually a transparent green coating that also serves as a general protective coating for the entire board. Another final step before component assembly is the screen-printing of information on the component side of the board, such as component locations, polarities, and troubleshooting test points; this information is usually printed with white epoxy-base paint. Following solder assembly, the entire board is cleaned with solvent to remove flux and other process residues. For this reason, almost all components used in modern printed circuit boards are encapsulated in solvent-resistant packages. Solder flux which is compatible with aqueous cleaning solutions is coming into use, but components must still be sealed to protect against moisture entrapment.

Manufacturing Steps • The principal design and construction steps for a double-sided PCB with plated-through-holes proceed sequentially as follows:

1. Design PCB layout and connections.
2. Prepare artwork.
3. Photograph and prepare positive photo masks.
4. Drill holes.
5. Sensitize hole walls and apply thin copper "flash" coating.
6. Transfer negative image to both sides of board.
7. Plate patterns and through-holes with copper.
8. Plate patterns and holes with solder.
9. Etch copper leaving desired patterns.
10. Reflow (smooth) solder coating.
11. Plate connector fingers with nickel/gold.
12. Screen print solder mask resist.
13. Screen print labels.
14. Machine board to final dimensions.
15. Insert components and assemble by wave-soldering.
16. Calibrate and test final PCB.

Numerous intermediate steps are also necessary for proper cleaning and preparation of surfaces. The important chemical plating and etching steps used to form conductive patterns are illustrated with diagrams in Fig. 2-2. It should be noted that there are several different manufacturing approaches to the formation of PCB patterns, and the one illustrated in Fig. 2-2 is known as *pattern plating;* this approach involves plating through-holes and conductive patterns with copper *before* etching away the undesired surrounding copper foil. The PCB fabrication processes are generally classified as being *subtractive* or *additive*, depending upon whether copper is *deposited* on the board, or *removed* by etching to form patterns. However, the pattern-plating approach uses a combination of both methods and is probably the most prevalent large-scale manufacturing process seen today.

Individual manufacturing steps will now be described as they proceed in conventional PCB production operations.

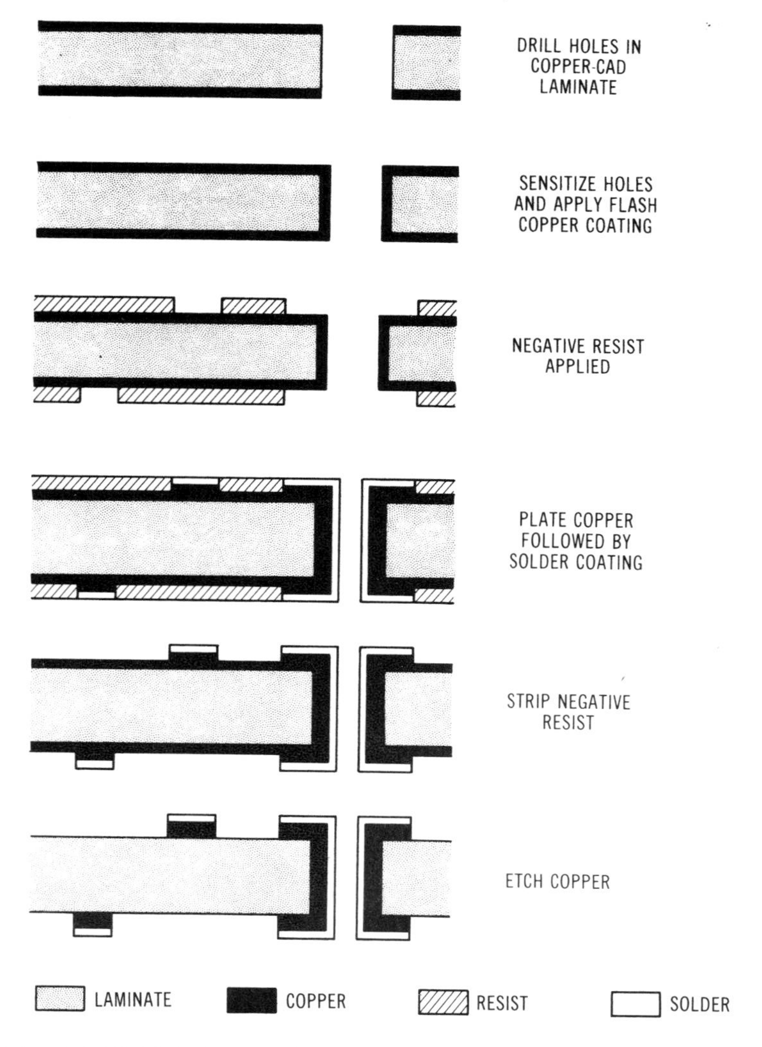

Fig. 2-2. PCB fabrication steps for pattern plating.

Design and Layout • Design of a printed circuit board starts with a careful study of the electrical schematic, which should be drawn as neatly as possible. The PCB designer's first attempt at component layout will probably look very similar to the schematic, so it is important that this drawing be properly constructed. The most important characteristic of a good schematic is that electrically related components should be grouped together to organize the drawing for smooth signal flow; this approach will also avoid unnecessarily long signal paths.

Before components and their interconnections can actually be laid out, a number of other considerations must be worked out between the PCB designer and the original electronic designer. These considerations can be roughly divided into three areas that affect the final design: mechanical, electrical, and manufacturing. By taking all three factors into account early, problems with assembly, performance, and maintenance can be avoided.

Mechanical — The main mechanical consideration is usually board size. The designer's job becomes more difficult as component density increases, and signal routing becomes more restricted. Typical densities that can be achieved with digital circuits on double-sided PCBs are in the neighborhood of one 16-pin integrated circuit per square inch of board space, although this figure varies with the number of discrete components accompanying ICs in a circuit. The example board shown in Fig. 2-1 has a density of about one IC per square inch. A better general estimate of packaging density is the total number of component lead holes per square inch of board space, which is typically 10–20 for commercial double-sided boards with plated-through-holes. In establishing board size, the designer must consider the total electronic package. For example, cabinet geometry may dictate that a particular circuit be broken up into several equal-sized cards for ease of mounting. The general trend in electronic packaging is towards modularity, because this approach simplifies all aspects of design, assembly, mounting, and service; for this reason, separate functions such as power supply, logic, and displays are usually found on separate circuit boards.

Once the general variables of board size and circuitry content have been decided, some other mechanical factors must be defined before the PCB designer can proceed with layout. He needs to be aware of heat dissipation problems, high current signals, shielding requirements, and the complete external connector scheme. He also needs to obtain packaging specifications for all unfamiliar components so that their physical size and lead patterns are known. Finally, hardware such as heat sinks and card guides must be considered to allow for component clearance.

Electrical — The schematic diagram defines all electronic components and their interconnections. However, other information must also be communicated to the layout designer for planning sound electrical performance on a printed circuit board. The following factors should be considered:

1. Conductor Size — The current carrying capability of etched copper lines is limited. Copper foil is typically specified to be one ounce per square foot of area for double-sided PCBs, giving an average foil thickness of 0.0014 inch (0.0356 mm). To avoid voltage drops in critical circuitry (such as power and ground distribution networks) the width of conductors is chosen to maintain low resistance over the path of each connection.
2. Conductor Spacings — The high resistance of epoxy-glass laminates is such that voltage arcing or leakage from closely spaced PCB conductors is not usually a problem in low-voltage logic circuitry. However, conductor spacings can have a critical effect in high-frequency circuits because of capacitance and inductive coupling; popular G-10 epoxy-glass laminates have a dielectric constant of 5–6 at frequencies from 25 MHz to 1 KHz. It is possible to control the characteristic impedance of parallel conductors to some extent by introducing a ground plane of copper foil on the opposite side of the board. Ground planes are also used to shield radio frequency circuits in combination with mechanical shields around critical components.
3. Conductor Lengths — A good general PCB design approach keeps conductor lengths as short as possible. This becomes especially important in high-frequency circuits and in high-gain amplifiers where radio frequency interference, signal coupling, and signal delays must be avoided. The original circuit designer should be careful to isolate and define critical circuitry where PCB layout might affect electrical performance, and in some cases he may need to directly assist in the layout.

Manufacturing — The actual manufacturing procedures to be followed may have an effect on the initial PCB design, so they must be considered early. For example, if automatic component insertion machinery is used for assembly, it is important to maintain standard component lead spacings, minimize the number of wire jumpers, and organize the components in consistent horizontal and vertical patterns. It may often appear desirable to orient components so that routing of their interconnections is simplified; however, the effect of random IC orientations can prevent the cost-effective use of an IC insertion device for automatic assembly later on. Another important design consideration is the proper generation of documentation and alignment patterns to guide automatic drilling and assembly tools during later fabrication steps. Even the type of solder mask coating to be applied to final PCBs may affect the initial design, because of the change it might produce in surface dielectric properties.

Layout — The actual layout of printed circuit patterns is accomplished by trial and error on a background grid using drafting vellum (large sheets of semitransparent erasable paper). The standardization of modern component packages to leads spaced on 0.1 inch (2.54 mm) centers makes 0.1 inch (2.54 mm) grid vellum ideal for the layout; if drawings are made 2× larger than final dimensions, this gives an effective 0.05 inch (1.27 mm) cross-hatch pattern for design. Component placement is worked out by moving around *paper dolls* within the outlines of the board until a suitable arrangement is found to maintain electrical and mechanical requirements while preserving space and producing a neat, orderly appearance. (''Paper dolls'' are cardboard scale cut-outs of the various devices to be mounted on the PCB.)

Once a suitable layout is obtained, component outline patterns are transferred to the vellum with pencil and electronic *templates* to speed the drawing process. The designer is now ready to work out signal connections using colored pencils to represent patterns on the two opposite sides of the board. General principles for layout of double-sided PCBs involve using one side of the board exclusively for vertical runs, and the other side for horizontal runs; through-holes are established to bring signals from one side to the other. Strict adherence to this X-Y system will theoretically always allow a new connection to be added between any two points on the surface of the board. A vertical run is first made from one point to the proper Y coordinate of the second point; a through-hole is established, and the connection is completed with a horizontal run to the final X coordinate of the second point. In practice, component density and board space may limit the success of this approach, particularly when through-holes start blocking up thoroughfares. Examination of a finished commercial PCB will show that although the vertical/horizontal approach is usually followed in principle, a certain number of short vertical runs are made on the horizontal side (and vice-versa) to limit the number of through-holes. Also, many runs are not accomplished with two simple X and Y motions; rather, they are broken up into a series of X and Y zig-zags to use routing space where it is available.

The layout designer will probably go through three or four drafts of the PCB connections before he arrives at a final design. He will also probably rearrange some components along the way. A proper layout will have a neat, orderly arrangement of devices with the conductors fairly evenly spaced, or *balanced,* throughout the surface area of the board. In addition, the original electrical, mechanical, and manufacturing constraints are successfully incorporated into the layout. It is clear from this discussion that a considerable amount of time and experience is needed to design a double-sided PCB, particularly as the parts count and density increase.

Computer-Aided Design — Electronic manufacturers are now commonly using computer equipment to aid in the PCB design process; these systems are known as CAD (computer-aided design) systems. Commercial CAD systems for PCB design include a large crt (cathode ray tube) screen for displaying patterns, a digitizer/electronic pen/keyboard for entering instructions, a computer for calculations and control, a plotter to generate hard-copy drawings, and a magnetic storage system for filing results. A typical CAD system is shown in Fig. 2-3, the PC-800 Model 3 from the Gerber Scientific Instrument Company. The

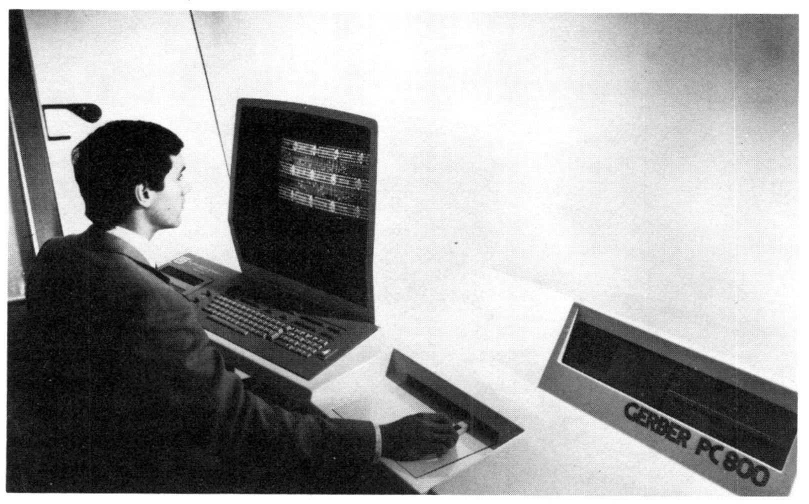

Fig. 2-3. Computer-aided design (CAD) system. *(Courtesy Gerber Scientific Instrument Co.)*

crt screen in such a system often has a light pen or other digitizing device attached to allow the designer to enter information directly on the screen, as if it were a sketch pad.

It is interesting to note that CAD systems have in fact not been completely successful to date for making complex design decisions, or for producing quality PCB layouts from scratch. The number of initial considerations are simply too voluminous to take into simultaneous account using a computer program which runs in any reasonable length of time. However, the CAD system is extremely valuable when used as a *tool* by an experienced PCB designer, in much the same way that a secretary saves time with a word-processing system. The designer uses a CAD system for the nitty-gritty work of drawing lines, moving patterns, redrawing a complete layout following a minor component move, and producing finished drawings in record time. The actual design is accomplished by directing component placement, letting the computer run connections in short controlled bursts, and helping it out when it backs itself into a corner. Typical patterns produced on a CAD crt screen during layout of PCB connections are shown in Fig. 2-4.

Fig. 2-4. CAD patterns on a crt screen. *(Courtesy Gerber Scientific Instrument Co.)*

By relieving design engineers of their tedious, repetitious tasks, by preventing simple errors, and by streamlining drafting steps, the CAD system provides designers with the time they need to be creative and apply the sound judgment and experience needed for a good design. It is expected that faster computers and more sophisticated software will steadily improve the capabilities of CAD systems; in any case, the inherent power of a large, high-resolution crt display/digitizer linked to an accurate plotter is already changing the entire engineering world of design and drafting considerably.

Artwork and Documentation • Following completion of design, master artwork must be prepared, from which photo masks will be made for PCB fabrication. For double-sided PCBs, the artwork will usually consist of the following four accurately scaled patterns:

1. Front view of the PCB showing all pads, interconnections, and labels to be plated and etched into the board.
2. Back view of the PCB showing all final patterns.
3. Pads to be soldered. This artwork will be used to produce a screen printed solder mask.
4. Labels to be screen printed on the final board with ink.

In addition to the actual PCB patterns, these four pieces of artwork may include additional markings to aid in tooling alignment during fabrication.

If the artwork is prepared by hand, it is generally taped up on dimensionally stable polyester film, such as 0.005 inch (0.127 mm) thick Mylar sheets. During preparation, the sheets are usually attached to a transparent "light table" with a precision grid background to serve as a reference for the artist's scale work. Dry transfer patterns, letters, and precision circuit tapes are used to form a high-contrast image on a scale of 2× or 4× larger than the final PCB; this image will be photographically reduced to form the final, accurate photo mask. Fig. 2-5 shows an example of PCB artwork being prepared manually on a light table.

If artwork is generated by a computer system, a companion digital plotter may have sufficient accuracy to produce camera-ready output on a 1:1 scale. In addition to providing accuracy, an automatic artwork generation system can speed up the PCB design sequence considerably if the CAD system used for layout can also produce data that will directly feed into the artwork plotter. Other by-products of CAD equipment include master tapes for numerically controlled (NC) drilling machines, documentation for manufacture, and input data for equipment used to automatically test the final assembled PCB.

Formal manufacturing documentation for double-sided PCBs typically consists of the following items:

1. Schematic — The original basis for the printed circuit design.
2. Layout — The final diagram produced in the design step.
3. Artwork copies — Reduced copies of the master artwork for a reference file.
4. Master (Fabrication) drawing — Complete fabrication instructions

for PCB production. These instructions include engineering information for drilling, plating, etching, coating, screening, and routing, as well as all dimensional information and tolerances.

5. Assembly drawing — Complete drawing for locating components and assembling the PCB. This item usually includes a diagram of the finished assembly, and a parts list.
6. Calibration and test procedures.

Fig. 2-5. Manual tape-up of PCB artwork. (Courtesy Bishop Graphics, Inc.)

Photography • Large *process cameras* are used to produce positives or negatives from the master artwork. These cameras have copy boards that can hold artwork up to 40 by 48 inches (1.02 by 1.22 m) in size, and a lens system that can typically make reductions down to one-eighth or enlargements up to 2×, with a final accuracy of ±0.001 inch to 0.005 inch (0.025 to 0.127 mm). A typical process camera is shown in Fig. 2-6, the Mercury Model LPC-2024SR from R. W. Borrowdale Company. The sheet film that is used is low-speed, high-contrast, high-resolution "line" film of the type used for lithographic processes, and is generally on a polyester base of 0.004 to 0.007 inch (0.102 to 0.178 mm) for dimensional stability. A good copy camera will have a ground-glass focusing plate which allows the operator to precisely focus and set the reduction dimensions before inserting and exposing a piece of film. Because of the low speed of the film, the actual exposure is often

Fig. 2-6. Graphic arts process camera. ***(Courtesy R. W. Borrowdale Co.)***

accomplished with high output photoflood lamps. Film is processed with standard darkroom equipment, although a great deal of attention is paid to maintaining a dust-free environment for avoiding pin-holes and scratches in the film emulsion.

If commonly available negative-acting film is used, contact prints are made onto a second sheet of film to produce positives. Positives and negatives are the final photographic tools (masks) needed for screen-printing and image transfer operations in PCB manufacture. Photo masks must be carefully inspected on a light table to reveal defects. Corrections are made by applying liquid opaque with a fine brush, or by eradicating the emulsion where dark areas must be removed. As the final step before releasing masks to production, alignment holes are punched in photographic films to match with tooling pins inserted during the image transfer operations.

A fairly recent innovation in photographic reproduction processes involving precision, expanded-scale artwork is the use of *wash-off* films (for example, Kodagraph wash-off Autopositive films WA1 and WA2). These films are available in widths up to 42 inches (1.07 m) and can be used to make positive contact reproductions from master artwork onto a translucent base. The resultant high-contrast photographic lines are wet-erasable with a soft moistened eraser, allowing revisions to be made to master artwork without disturbing it. Wash-off film is used extensively to copy and preserve original taped-up PCB artwork, which may stretch, creep, or otherwise deform during lengthy storage. The resultant copy is more dimensionally stable than the original, and permits easy revisions with the erasable feature. Kodagraph films and papers are also available with negative-acting emulsions, or with permanent, nonerasable final images.

Base Material • Printed circuit board *laminates* are flat, insulating boards that have been formed under heat and pressure by joining together layers of various materials with bonding resins. Copper foil is then attached to the laminate surfaces using adhesive and pressure. A typical laminating machine is shown in Fig. 2-7, the Model PC125-2424-8TM Hydraulic Press from Wabash Metal Products, Inc. A wide variety of laminates are available with various electrical and mechanical

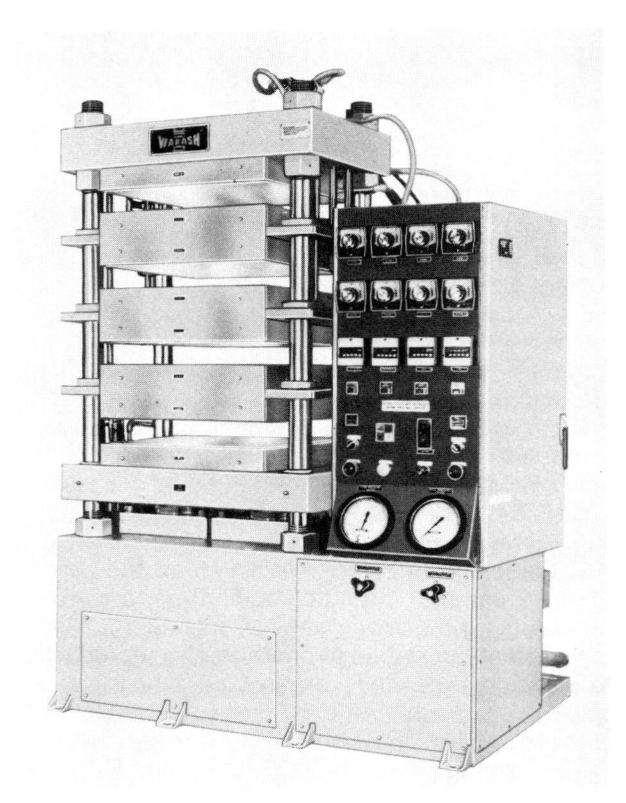

Fig. 2-7. PCB laminating machine. *(Courtesy Wabash Metal Products, Inc.)*

properties; however, the most widely used product for double-sided PCBs with plated-through-holes is the G-10 *epoxy-glass* laminate. This material is composed of glass cloth fibers with epoxy thermosetting plastic resin, and epoxy adhesive is used to bond the copper foil. Circuit boards made from G-10 material are a familiar translucent green in color, and exhibit excellent electrical characteristics such as insulation resistance and dielectric constant. Mechanical strength and flexibility are also excellent, and G-10 ranks high in hot solder resistance, which is a measure of the tendency for copper foil to lift during soldering. Perhaps the worst characteristic of G-10 epoxy-glass laminate is its machinability; the inherent hardness of glass fiber makes it fairly difficult to drill, and almost unacceptable for punching. For this reason, and for reasons of cost, many noncritical PCB applications use the cheaper *paper-phenolic* laminates such as XXP-grade, which exhibit generally lower mechanical and electrical properties than G-10, but are easy to punch and machine. It should be noted that paper-phenolic materials have thermal expansion characteristics that make them undesirable for use in PCB designs requiring plated-through-holes; the plated hole walls tend to develop cracks from temperature changes in this material.

For very high-frequency applications (above 40 MHz) G-10 laminates may have unsuitable electrical properties, and more expensive *silicone-glass* or *teflon-glass* laminates are available with lower dielectric constants.

Unless there is a need for greater mechanical support, G-10 circuit boards are usually constructed using standard one-sixteenth inch thick material. Another fairly standard parameter is a final copper foil thickness of one ounce copper per square foot of board, which is suitable for most circuit paths not exceeding 3 amperes in current demand. One ounce copper foil has an approximate thickness of 0.0014 inch (0.0356 mm) and is made from 99.5% pure metal to ensure high electrical conductivity. Other thicknesses of copper foil and laminates are available for PCBs that carry large currents or support heavy components. If an *additive* PCB fabrication process is used, the initial stock laminate may have copper foil density much less than one ounce, but the copper thickness is built up later during plating.

Machining • The mechanical processes used to shape double-sided PCBs include shearing, sawing, drilling, and routing. Punching of epoxy-glass laminates is seldom seen as an alternative to drilling holes; this is due to the abrasiveness of the material, and because of the resulting finish inside the holes which may result in poor-quality plated-through-holes. However, die-punching is a standard method for forming holes in phenolic-base laminates.

As the first step in PCB fabrication, blank copper-clad stock material is cut to individual panel size using a shear or a circular saw; saw blades are made with carbide teeth or diamond-steel-bonded teeth to cut abrasive glass-base laminates. Each panel may eventually include several circuit board patterns to maximize plating throughput. The panels are then drilled to specification using high-speed drilling machines with carbide bits.

The drilling operation is extremely critical for the manufacture of PCBs with plated-through-holes. Because of the eventual build-up of copper and solder in these holes, drilling specifications must call for slightly oversized holes to arrive at proper final dimensions. The accuracy required of the drilling operation is evident when one considers that modern dense double-sided boards have pads only 0.040 inch (1.02 mm) in diameter with 0.030 inch (0.762 mm) holes. This leaves only 0.005 inch (0.127 mm) borders around the holes!

Burrs must naturally be avoided on entry and exit of the drill bit, but in addition to this the walls of the holes must exhibit proper characteristics. If the drill feed rate is too fast, holes will be roughened by the punch-through action, allowing glass fibers to protrude and interfere with smooth plating. However, if the feed rate is too slow, or if the drill bit runs too hot, hole surfaces can become so smooth that plating does not adhere properly; heat can also cause epoxy resin to become smeared inside the hole walls. Resin smear is a common problem in plated-through-holes and its end effect is plating discontinuity, a difficult defect to repair. Modern PCB manufacturing shops use precision computer-controlled drilling machines that can position holes to within ±0.001 inch (0.025 mm), and drill through a stack of boards at spindle speeds of 90,000 rpm and at feed rates faster than 300 holes per minute. A computerized multiple spindle machine is seen in Fig. 2-8,

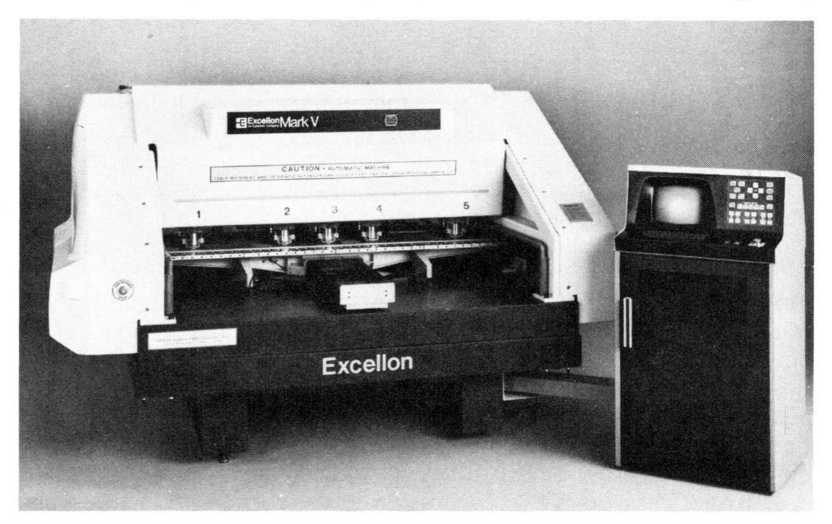

Fig. 2-8. Computerized numerically controlled (CNC) drill. ***(Courtesy Excellon Automation)***

the Excellon Automation Mark V. Because of the fragile nature of small carbide drill bits, the work must be held rigidly and securely to the table as the spindle makes a vibration-free entry into the stack; any slight binding or seizing will break the bit. To prevent burrs, the board stack is clamped tightly together and entry/backup materials, such as aluminum-coated boards, are used. Bits are examined frequently with a microscope to discern wear and improper cutting edges.

Following the fabrication processes of image transfer, plating, etching, and coating, finished boards are cut to their final shape with routing machines. Using automated equipment, router bits can cut through a stack of four circuit boards at feed rates faster than 100 inches per minute, while maintaining closer tolerances and smoother edges than older methods, such as shearing, sawing, or die blanking. Close tolerance is important when signal traces come close to the board edges, and when precision edge connector fingers must be formed; registration holes may be located within the PCB pattern during the original design stage to accommodate the final routing operation.

Copper Deposition • Because electroplating methods cannot be used to coat nonconductive surfaces, such as through-holes drilled in epoxy-glass, these surfaces must first be coated with a thin metal film by autocatalytic, or *electroless* chemical methods. A typical procedure to form a thin, 10–20 μin (0.00001–0.00002 inch) copper deposit involves seeding or *sensitizing* the board surfaces with a wet suspension of palladium metal. The precious metal sites that are created on the laminate then serve as catalysts to initiate the chemical reduction of copper from a basic formaldehyde solution. The touchy electroless copper process must be carefully controlled because of many critical variables such as cleanliness, solution impurities, reagent concentrations, temperature, and surface preparation. Electroless copper deposition is almost always carried out *before* image transfer to ensure perfectly clean surfaces. An example of a typical problem encountered in the electroless process is the presence of epoxy-resin smear in holes following drilling, which may lead to poorly formed or nonadherent deposits. To avoid this problem, boards are frequently treated briefly with a powerful epoxy etchant to remove residues from the holes just prior to seeding and surface activation. Chemical etching creates safety and disposal handling problems, so it is being replaced with a more recently developed hole-cleaning treatment known as *plasma etching* (see Fig. 2-9) which subjects PCB holes to intense electric fields prior to copper deposition.

Once formed, the electroless copper deposit inside holes is so thin that it is very sensitive to air oxidation, and so it may be immediately followed by a "flash" coating of plated copper to build up the metal thickness to several ten-thousandths of an inch in hole walls prior to image transfer.

Fig. 2-9. Plasma etcher for cleaning through-holes. *(Courtesy Branson International Plasma Corp.)*

Image Transfer • The patterns on the surface of a printed circuit board are initially established through an *image transfer* process that produces a coating in appropriate areas. The coating that is formed uniquely defines the pad shapes and lines from the master artwork, and this coating is termed *resist* because it must be chemically and mechanically resistant to highly corrosive plating and etching solutions. Another important functional property of protective resist is that it must be able to form coatings with the proper *resolution* and *line definition* required for a particular circuit pattern. Smallest line size used in typical double-sided boards of medium component density is 0.015 inch (0.381 mm); some typical etched lines are shown in Fig. 2-10. Two principal image transfer materials are used to form resist images in commercial PCB production: *photo resists,* and *screened resists.*

Photo Resist — The most common types of photo resists are organic compounds that are sensitive to light in the ultraviolet range, causing chemical polymerization (increase in molecular weight) upon exposure. This process is the basis for forming images through a photographic film mask that selectively allows light to reach the coating. After exposure, unpolymerized resist is washed away, or *developed* with a suitable solvent, leaving only the polymerized, insoluble areas behind as a durable coating. Photo resists are capable of producing extremely fine line definition for image transfer, and 0.005 inch (0.127

mm) lines can be easily plated or etched in one ounce copper boards using this type of coating.

The five basic steps in forming PCB photo resist patterns are surface preparation, coating, exposure, developing, and post-treatment:

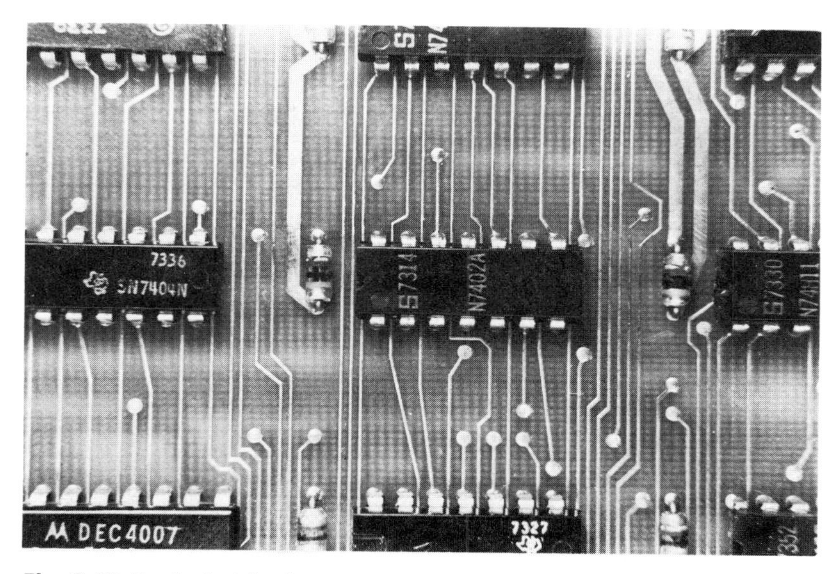

Fig. 2-10. Typical etched patterns in a commercial double-sided PCB.

1. Proper *preparation* of copper surfaces is necessary to ensure adhesion of photo resist during later harsh plating or etching steps. Routine board preparation may include solvent degreasing, surface roughening with an abrasive/cleaner, washing, acid-dip, and a final water rinse followed by thorough drying.

2. Photo-sensitive resists are *coated* in two forms: *liquid solution* and *dry films.* The liquid form is coated on copper boards by spraying, dipping (see Fig. 2-11), or roller-coating, followed by drying. Dry film resist is coated by bonding a thin pre-prepared sheet to the copper surface with heat and pressure (*lamination*). Liquid resists are more difficult to apply in uniform density, particularly for thicker coatings. The coating thickness is easier to control with pre-prepared dry film sheets, so this method is coming into wider use; it is particularly desirable to use thick resist coatings when boards must be exposed to a series of electrochemical plating baths, as is done to form plated-through-holes. Typical final coating thicknesses are 0.0001 inch (0.0025 mm) for dipped liquid photo resist, and 0.0005 to 0.0030 inch (0.127 to 0.076 mm) for dry films.

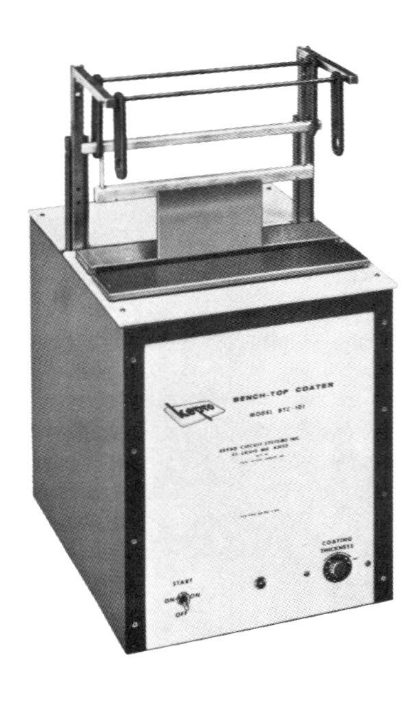

Fig. 2-11. Device for coating photo resist by controlled dipping/withdrawal. *(Courtesy Kepro Circuit Systems, Inc.)*

3. Images are formed in photo resist coatings by *exposing* the board to a high intensity, ultraviolet-rich light source through a photographic film mask; the mask is held in intimate contact with the surface of the board during exposure. The Douthitt Model DMP vacuum exposure system is shown in Fig. 2-12. Double-sided PCBs are coated on both sides, so a registration system must be used to clamp the two film masks in perfect alignment during exposure of these boards.

4. Images are *developed* by washing the board in a solvent that removes unexposed resist. Finest patterns are obtained by spray-developing with nozzles positioned uniformly around the board. It is important that solvent be forced through all of the drilled holes to completely remove any traces of resist.

5. *Final treatment* of the photo resist coating includes dyeing (to make patterns visible for examination), washing, and drying/baking to cure the coating. Boards are then inspected for defects. Touch-up is accomplished with liquid resist ink and an artist's brush, or by scraping away resist with a fine knife blade.

Most photo resists are *negative-acting,* which means that a *positive* photo mask (patterns black, background clear) produces a *negative* coating on the copper (background coated with resist, patterns are clean exposed copper surface). This is the desirable result for a commer-

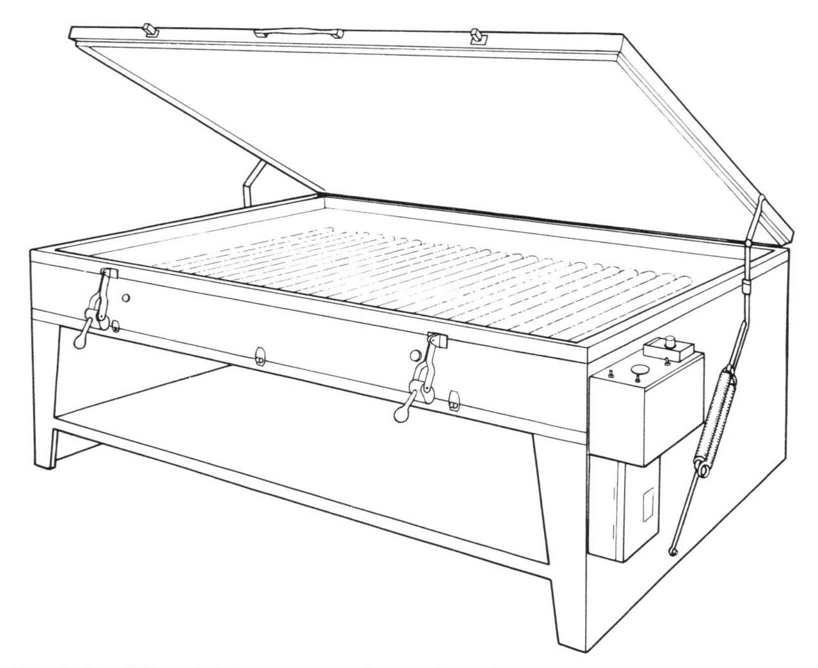

Fig. 2-12. Ultraviolet exposure frame for photo resist image transfer or photographic screen printing. *(Courtesy Cincinnati Screen Process Supplies, Inc.)*

cial plated-through-hole process. Subsequent steps will plate the exposed pattern and holes with copper followed by solder. Resist is then stripped, exposing the background copper to be removed with a chemical etching bath that attacks copper but does not react with solder.

Screened Resists — Screen printing techniques are commonly used to directly print liquid ink on PCB laminate surfaces as a plating and etch resist. A semiautomatic high precision (tolerance ±0.01 mm) screen printing machine for PCBs is shown in Fig. 2-13, the Argon Modular with electropneumatic controls. While photo resist is an organic polymer, *screened resist* is generally a vinyl-base solvent-soluble ink having the necessary properties of chemical resistance, surface adhesion, and good screen processing characteristics, such as a grease-like consistency. Solvent-free inks which can be cured with ultraviolet radiation and stripped with aqueous alkali are also coming into widespread use. The techniques used to screen print electronic circuits are essentially identical to all modern photographic screen processes. A film mask is formed photographically from a light-sensitive emulsion;

the mask (stencil) is imbedded in a tightly stretched finely woven screen made of silk, polyester, or stainless steel; and ink is forced through the screen to form patterns on the underlying surface, which in the case of PCBs is a clean, blank circuit board.

The screening variables of specific importance to PCB image transfer are mainly the type of ink used, and the registration system for forming two-sided board patterns in proper alignment. Although screen printing processes do not produce the extremely high-resolution images that can be obtained with photo resists, conventional screening techniques are capable of forming the 0.015 inch (0.381 mm) lines common to double-sided circuit boards. Screen printing is widely used for producing PCBs with plated-through-holes because of two other advantages relative to photo resist: greater physical resistance to electrochemical plating action, and less tendency to deposit resist in the predrilled holes.

The decision to use photo resist or screened resist in commercial PCB production depends on economic factors, production volume, and image resolution requirements. For the finest line definitions and

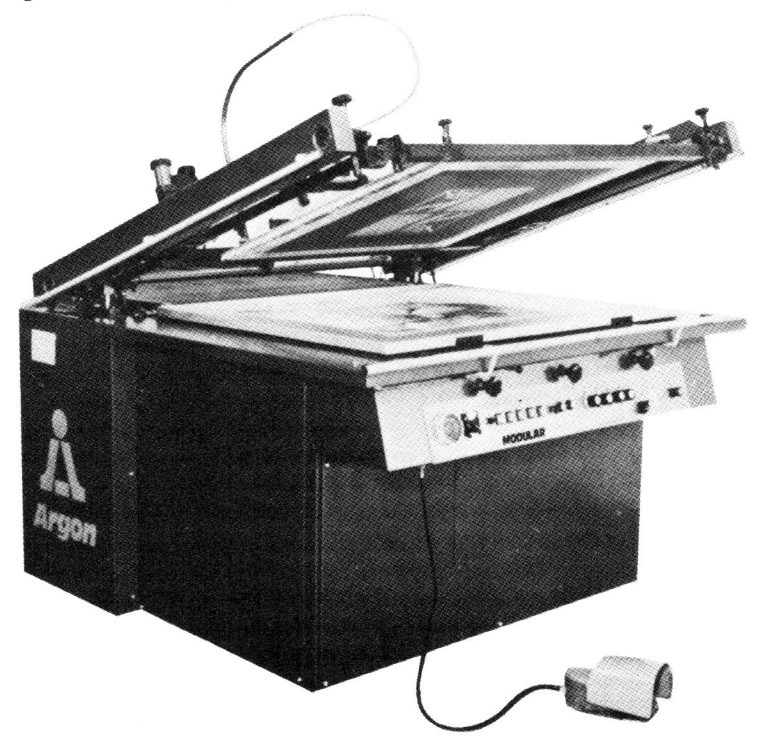

Fig. 2-13. PCB screen printing machine. ***(Courtesy International Printing Machines Corp.)***

sharpest pattern contours, photo resist is the preferred method. Screened resist tends to be superior as far as chemical resistance, but it is more difficult to set up for production. Photo resist is preferred for short production runs, but screen printing is an overall cheaper process for volume manufacture. Finally, side-to-side registration is usually more accurate with photo resist exposure techniques. Both methods of image transfer are in widespread use for fabrication of double-sided PCBs with plated-through-holes.

Plating • Three plating operations are required to form plated-through-holes and an etch-resistant surface coating for conventional double-sided PCBs. The first plating step is carried out before image transfer and has already been discussed: electroless deposition of a thin layer of copper inside hole walls to produce a conductive surface for further plating. The other two steps are *copper plating* and *solder plating.*

***Copper Plating* —** Electroless copper coatings are not usually thick enough to form reliable through-hole connections, but they do serve as a conductive surface for the subsequent *electroplating* of a more durable copper coating at least 0.001 inch (0.025 mm) thick. A light etch step is often included between image transfer and copper plating to clean the surfaces and remove resist residues; the presence of resist (screened or photo resist) in holes is a common plating problem which must be avoided during image transfer.

Copper electroplating of printed circuit patterns is accomplished by immersing the resist-coated panels in an appropriate solution (see Fig. 2-14), and attaching them to the negative side of a dc power supply. Anodes of pure copper are also hung in the solution and attached to the positive dc source. As current is passed through the solution at proper temperature, flow rate, and agitation, copper ions are reduced and deposited at the negative surfaces of the board, and copper metal is oxidized into solution at the positive anode. Copper plating is carried out with a variety of electrochemical baths, generally purchased as proprietary formulations. One of the most common baths used today is a copper pyrophosphate formulation. Various additives are used to control deposit characteristics such as brightness, hardness, smoothness, and uniformity of coating (*leveling*).

In addition to controlling such variables as surface cleanliness, bath composition, impurities, and plating parameters, a major problem in modern high-volume electroplating operations is *waste disposal.* Environmental controls are particularly stringent on effluents containing metal salts, so the printed circuit manufacturer must give waste considerations high priority when choosing plating and etching processes.

It should be noted in conjunction with this copper plating discussion that the negative resist image applied in earlier image transfer

Fig. 2-14. Copper electroplating operation. ***(Courtesy PEC Industries and the Shipley Company, Inc.)***

steps exposes the PCB *patterns* as well as the *holes.* This means that copper is plated on foil traces as well as in the through-holes where it is desired. For this reason, thin ½ ounce foil (0.0007 inch or 0.0178 mm thick) laminates are often used in pattern plating. If 0.001 inch (0.025 mm) of copper is then plated in holes and on traces, the final patterns will be only slightly thicker than standard one ounce foil (0.0014 inch or 0.0356 mm). It is only recently that very thin copper foil laminates have become available to the PCB industry; one-fourth and one-eighth ounce foils are also being used today. The big advantage of the thinner foils in pattern plating is that the subsequent etching step is fast and consumes very little copper, thus minimizing waste disposal and pollution problems.

Solder Plating — Solder is normally plated following copper to serve as an etch-resist during the subsequent step to etch the final PCB patterns. Solder can be electroplated as the common 60% tin, 40% lead alloy used in electronic soldering, and this coating serves a triple purpose:

1. Etch-resist.
2. Protective coating for oxidizable copper.
3. Preservation of surface solderability for later assembly of components. The only other practical coating to provide these functions is gold, but its use has declined because of the escalating cost of precious metals. An average solder coating thickness considered adequate for final surface protection is about 0.0005 inch (0.0127 mm).

Solder is plated from baths containing fluoboric acid, with tin and lead as their fluoborate salts. Tin and lead are used as the metal anodes. Normal plating considerations are temperature, agitation, current density, additives and brighteners, carbon treatment, contamination, and concentration of bath constituents. The maintenance of solder plating baths is a continuous problem because of the air oxidation of stannous tin, and organic contaminants such as photo resist can result in poor deposits.

To improve the general characteristics of electroplated solder coatings, PCBs are usually *reflowed* (heated above the melting point of solder) to fuse solder to the base copper metal and produce an even, bright appearance. This operation is carried out following the etching step, and the results generally reveal any problems up to that point in the PCB fabrication processes. Three methods of solder reflow are in common use: IR (infrared) fusing, hot oil bath immersion, and hot air leveling. Equipment for the hot air process is shown in Fig. 2-15, the Levelair product from Electrovert which is also capable of producing a uniform, dipped solder coating on PCBs without prior tin-lead plating.

Etching • With the exception of special coatings, etching is the final step in forming printed circuit patterns by subtractive processes, or by additive-subtractive processes such as pattern plating. These are processes where a uniform layer of copper is selectively removed by chemical action to form the final pads and traces. Of the steps leading up to etching, image transfer is most critical because the resist coating for plating must uniquely define and protect final copper surfaces several steps before etching. With a durable, adherent, and well-defined initial image, solder outlines are well-formed during plating, and etching becomes one of the most straightforward operations in PCB fabrication.

The most critical parameter in PCB etching is the degree of *undercutting* in the horizontal direction underneath a resist coating. This undesirable action is measured with a value known as the *etch factor,* which is simply the ratio of vertical to horizontal etch rates (see Fig. 2-16). Undercutting must be minimized because when out of control, it leads to unpredictable changes in line widths, and may cause foil traces to become less adherent. An etch factor $\geqslant 3$ is usually considered

Fig. 2-15. Solder coating and hot air leveling system. ***(Courtesy AM&P Services, Inc. and Electrovert Ltd.)***

satisfactory. The master artwork must take the etch factor into account when close tolerances are required for final pattern dimensions.

The overall rate of etching is another important variable for volume PCB production. In most cases a faster rate will also give less horizontal etching action, so this is generally desirable. Most commercial PCB equipment can now etch one ounce copper foil in less than 2 minutes, and speed provides an important advantage — it lessens the chance for resist breakdown with shorter chemical contact time. Etching processes are, therefore, designed to remove copper in the least amount of time, with the least amount of undercutting, and with the least destruction of resist, whether it be solder, ink, or photopolymer.

Variables in the various approaches for achieving optimum etch results include the etchant's chemical composition, the mechanical action used to achieve even, fast etching, and the process temperature. General considerations for large-scale manufacture demand a *continuous*

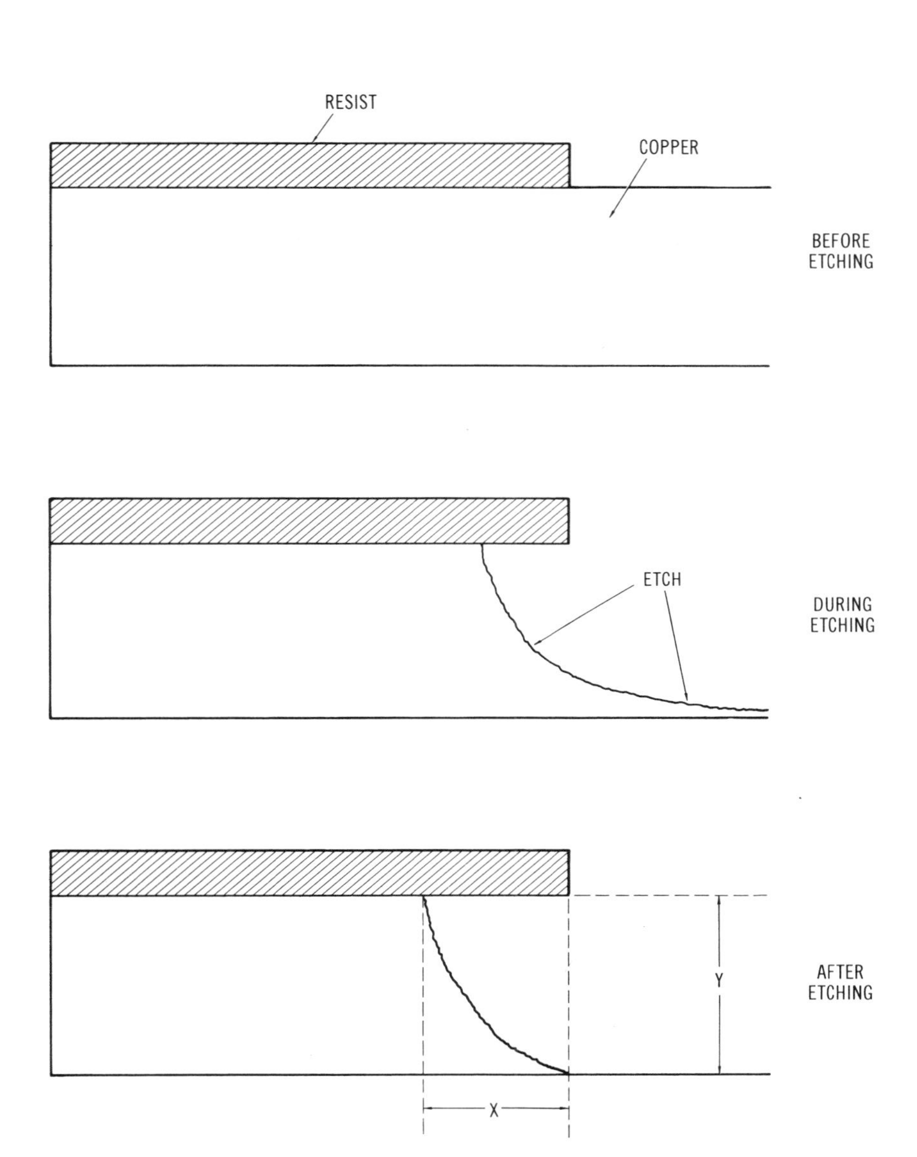

Fig. 2-16. The etch factor and etch undercut.

system that allows boards to proceed through a conveyorized treatment with chemicals whose composition can be constantly adjusted and replenished to give a constant etch rate. The Hunt/DEA system shown in Fig. 2-17 is based on the chemically regenerative nature of cupric chloride etchant, whose chemistry is detailed in Chapter 6.

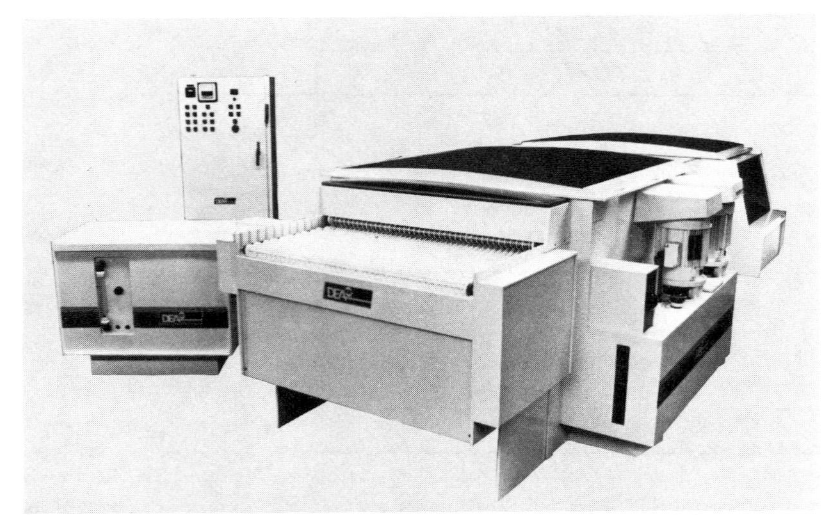

Fig. 2-17. Continuous PCB etching system. ***(Courtesy Philip A. Hunt Chemical Corp., DEA Equipment Div.)***

Because of the pollution problems associated with metal-containing wastes, development of regenerative chemical etch systems has been a major influence in the industry. For example, at one time ferric chloride was the most widely used copper etchant due to its low cost and high capacity for metal dissolution. However, ferric chloride cannot be easily regenerated, and its disposal is currently a legal problem. In addition to severe corrosion to metal plumbing and sewers, ferric chloride etchant is extremely harsh on ecological systems due to a high final content of copper salts. The reader should be aware that it is destructive and may be illegal to pour metal etching solutions down the household drain.

The actual chemical formulations used to etch copper include both acidic and basic solutions: ferric chloride, cupric chloride, chromic-sulfuric acid, peroxide-sulfuric acid, alkaline ammonia, and ammonium/sodium persulfate. Of these, the last four are applicable to resist coatings formed from solder, the process we have discussed for a double-sided plated-through-hole PCB. A complete etching process following solder plating consists of the following sequential steps:

1. Resist stripping — The photo resist or screened resist ink used to define solder electroplating patterns must be removed with a solvent to expose the bare copper to be etched. Solvent-soaking and light mechanical scrubbing are needed.

2. Examination, touch-up, and cleaning — Resist and excess solder must be removed from copper background areas, and damaged circuit patterns must be hand-painted with ink resist. Alkaline,

acid, and water rinses are used to expose a clean copper surface for etching.

3. Etching — Common methods include spraying, splashing, and air-bubbling to achieve the intense mechanical action needed for fast, even etching. Pressurized spray nozzles inside a sump chamber provide the best combination of etching speed, uniformity, and etch factor for treating double-sided circuit boards in one controlled operation.
4. Neutralization and cleaning of etch residues.
5. Solder reflow — In addition to producing a clean, protective surface, the fusion of molten solder to underlying copper eliminates the solder plate *overhang* which results after etch undercutting.

Of the etching chemistries with which solder coatings are compatible, ammonium formulations offer the best possibilities for pollution control and high copper capacity. The alkaline ammonia bath has fume problems, but with proper ventilation, it is an easily controlled system that is often used for batch operation. Ammonium persulfate is in widespread use for volume production with continuous etching systems and chemical recycling. Unfortunately, it is a fairly unstable formulation whose composition is difficult to control for low-volume batch etching over an extended time period. Sodium persulfate is a variation of ammonium persulfate used for batch operation, since it offers greater chemical stability and fewer disposal problems as well. Peroxide-sulfuric acid etchant is another promising system for both regenerative and batchwise etching processes, and is under considerable development.

Coatings • Four coatings are used routinely on double-sided printed circuit boards for protective and functional purposes: solder, solder resist, gold/nickel plating, and labels. We have already discussed the use of solder plating for etch resist, copper protection, and preserving solderability. The other coatings are applied after etching but prior to assembly and are discussed as follows:

Solder Resist — This resist is usually a transparent green epoxy ink that is applied using screen printing techniques. Solder resist may be used on both sides of a board for general protection, but its main function is on the wiring side of the board during assembly with hot dipping or wave soldering methods. By exposing only the terminal pads where component leads must be soldered, this high-temperature coating directs molten solder to the areas where it is needed and keeps it away from interconnection traces; a broad smear of thick solder across the board is thus avoided, preventing *bridging* between closely spaced lines.

Gold / Nickel Plating — Most PCBs use edge connector fingers for mounting and connecting the board into a larger electronic assembly. For durability and reliable electrical contacts, the fingers are usually electroplated with hard nickel (0.0005 inch or 0.0127 mm) followed by a thin coating of gold (0.0001 inch or 0.0025 mm), using equipment such as that illustrated in Fig. 2-18. Since the fingers are selectively coated, the rest of the board is temporarily masked during plating. During component assembly, another temporary mask is used on the gold surfaces to prevent hot solder from coating the fingers. Both of these masks are screen printed and stripped later with a suitable solvent.

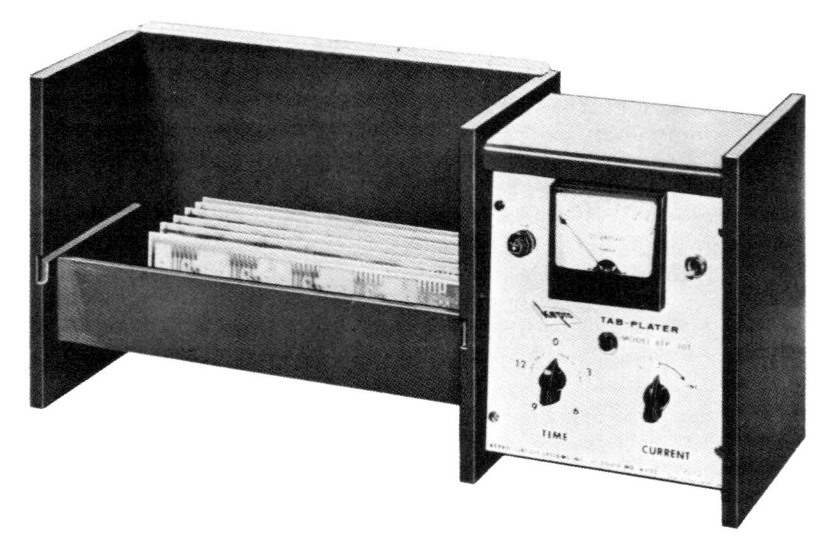

Fig. 2-18. Electroplating edge connector fingers with gold. ***(Courtesy Kepro Circuit Systems, Inc.)***

Labels — Labels are often printed on the component side of a PCB as a visual aid during assembly, testing, and troubleshooting. Although a certain amount of information can be included in etch patterns, the use of a nonconductive ink allows a great deal more information to be printed on the board without regard to bridging etched patterns. A permanent, solvent-resistant white epoxy ink is the preferred coating for screen-printing this informational coating; typical PCB labels are seen in Fig. 2-19.

Assembly • Once all of the desired coatings have been applied and the board is machined to its final dimensions, components must be mounted in place and soldered. Although this can be done manually, conventional high-volume production uses an automatic machine to

Fig. 2-19. Screen-printed labels on the PCB component side.

bend and cut component leads to their proper shapes, insert them in the board holes, and clinch the leads on the opposite side to hold components in place. A microprocessor-controlled automatic axial component insertion machine is illustrated in Fig. 2-20, the VCD-F from Dyna/Pert. This machine is capable of inserting up to 27,500 components per hour! A closeup view of the insertion head is shown in Fig. 2-21 feeding resistors into a crowded PCB assembly; the head is guided by digital instructions, and optical corrections assist its movements.

After all components are mounted, the back side of the board is coated with a solder aid such as rosin flux, and processed through a wave soldering machine which completes the assembly. Wave soldering is an interesting technique which involves moving the PCB on a conveyor belt across a pool of molten solder in which waves or ripples have been created to pass across the undersurface of the board (see Fig. 2-22). Soldering time and temperature are carefully controlled so that the molten solder is properly "wicked" up around the leads and into through-holes. This wicking action is encouraged by the plating inside hole walls. A properly wave-soldered board will appear as if it had been soldered on both sides. The engineering combination of plated-through-holes and wave soldering is a unique approach that allows the simultaneous soldering of hundreds of component leads without ever exposing the PCB's top component side to hot solder and excessive heat.

Following the wave solder operation, flux residues are removed by dipping PCBs in solvents such as xylene, alcohols, and aqueous alkali. A final dip-cleaning operation demands that all board coatings be solvent resistant, and that components themselves be hermetically sealed. If it is necessary to include components that cannot be wave-soldered or solvent-washed, then these parts must be assembled manually after cleanup. The final board is then ready for calibration adjustments and testing.

Fig. 2-20. Automatic PCB component insertion system. *(Courtesy Dyna/Pert Div., Emhart Corp.)*

PROTOTYPE PCB CONSIDERATIONS • Unfortunately, the disadvantages of conventional PCB construction methods are particularly distressing to the electronic hobbyist or designer building a prototype. If the initial tasks of learning techniques in seven separate fields (electronics, drafting, photography, electroplating, chemistry, machining, printing) seem difficult, the following drawbacks may appear overwhelming:

1. Flexibility — A minor PCB circuit change may require the initiation of a major design/construction sequence to construct a new board.
2. Equipment — Complex equipment is used to produce dense PCBs with plated-through-holes. Devices such as a process camera, drilling machine, spray etcher, and plating bath represent major

Fig. 2-21. Resistors in process of being inserted automatically. ***(Courtesy Dyna/Pert Div., Emhart Corp.)***

investments and are expensive to use at the low-volume end of PCB production.

If the prototyper desires to use printed circuit construction, there is obviously no substitute for the time and effort needed to design boards and prepare artwork.

However, the fabrication of dense, double-sided PCBs is in fact *feasible* and *practical* for the prototype designer with limited budget and facilities, if certain *compromises* can be incorporated into the fabrication steps. The two principal compromises that must be reached are generally time versus equipment, and hand-assembly versus plated-through-holes. It is the intent of this book to describe a practical, low-cost process that satisfies the quality and density requirements of a PCB prototype, while eliminating the need for costly equipment and complex process controls.

Plated-Through-Holes • Review of the conventional commercial fabrication process for PCBs described in this chapter will show that

Fig. 2-22. Wave soldering operation. ***(Courtesy Hollis Engineering, Inc.)***

most of the complexities and equipment demands are directly related to the need for plated-through-holes which simplify assembly. The difficult construction steps related to through-holes include precision drilling with carbide bits, electroless copper deposition, copper plating, solder plating, and etching with ammonium bath formulations. For proper results, all of these steps require precision equipment, a variety of chemical baths, and careful process controls. Therefore, the first major compromise in PCB prototyping is to eliminate the plated-through-hole.

Surprisingly enough, this has little effect on the final product quality; it does have a significant effect on initial design considerations and on assembly time and methods. A little thought will reveal that there are many ways to complete a through-hole connection without plating the hole walls. For example, dedicated through-holes which connect signal traces on opposite sides of the board can be easily soldered-through with a short piece of wire. For discrete components like resistors, diodes, and capacitors, the component leads themselves can be soldered on both sides of the board to complete through-connections. If leads are not accessible on the component side, as with some miniature PCB potentiometers, extra through-holes are simply created during the design stage so that all connecting traces originate from the reverse side of the board.

It is usually a good idea to socket all multiple-lead components (like ICs) in a prototype design. These sockets should be chosen so that their pins are accessible for soldering on both sides of the board, completing any through-hole connections needed. An alternative for ICs and IC sockets is again to *design in* extra through-holes so that traces do not connect to IC pads on the component side. The author has found that this approach of creating extra through-holes has little effect on the final board density, although it does require some careful thought during design and some extra soldering of pins during assembly.

Interestingly, elimination of plated-through-holes in PCB prototypes is desirable for reasons other than simplified construction. A frequent problem in perfecting circuit performance is the need to change component values after the board has already been assembled. Because of the solder-wicking action produced by plated holes, removal of soldered parts can be difficult without destruction of components and board; however, desoldering becomes much easier without the presence of metal plating in holes.

In PCB fabrication, elimination of plated-through-holes generally results in a trade-off between the need for complex processing/equipment, and the prototyper's own time. However, for one or two PCBs the time factor is not significant when compared to the delays involved in obtaining PCBs from outside prototyping shops, and the convenience of being able to construct in-house PCB test assemblies.

Registration and Double-Sided Processing • A major problem in producing quality double-sided PCB prototypes without expensive machinery is the need for precise side-to-side registration of the dual printed circuit patterns. Even if photographic masks are properly made, specialized equipment is needed to ensure alignment during image transfer. Further processing of dual images creates other equipment demands: resist must be coated evenly on both sides of the board, and double-sided developing and etching tanks are required.

In general, the simultaneous even processing of two opposite board sides becomes very difficult without using equipment specifically designed for that purpose. In addition to initial registration difficulties, typical problems that will occur include smearing of patterns during image transfer, and uneven etching which results in one side being over-etched before the other is completed. When these problems occur, a great deal of time and supplies may be wasted if the prototype board must then be thrown away at the very end of fabrication, and a new one processed from scratch.

To eliminate the need for double-sided processing equipment and precise registration methods, the PCB prototyping approach to be described in this book makes a second major compromise — only *single-sided* fabrication methods will be used! This may seem impossible at first thought, but it is in fact a very straightforward approach that substitutes time for equipment. The approach can be simply outlined as Print-Etch-Drill-Print-Etch, which means that each side is processed separately while the other side is protected and ignored. The sequential fabrication steps are illustrated as a diagram in Fig. 2-23. Pattern registration is accomplished by drilling some strategic holes after the first side has been etched; this allows the second image transfer step to have reference line-up holes for correct exposure. If any problems are seen in the image transfer steps, the unsatisfactory resist pattern is simply removed and reapplied without affecting the other side of the board. With this approach, little material and a minimum amount of time are wasted; by paying close attention to each side separately, the processing is simple and produces few rejects. One consequence of taking each side individually is that the prototyper must learn to coat photo resist on blank circuit boards; presensitized double-sided boards are of little use except in the initial image transfer step.

Commercial Prototyping Equipment • It should be mentioned that all of the PCB prototyping methods to be discussed in this book can be refined if double-sided processing equipment is available. Therefore, some commercial equipment will be presented at various points to illustrate some things that are available to the low-volume PCB manufacturer who is willing to improve his fabrication procedures and throughput at some investment. However, the need for plated-through-holes will not be considered; the reader should simply be aware that low-volume equipment for this purpose is also on the market. Unfortunately, the move to a plated-through-process demands more than increased capital outlay; there is also a big jump in process complexity and chemical controls.

Artwork and Photography • One area where little compromise is possible is in the preparation of artwork and photo masks. The final quality of a PCB can be no better than the initial artwork, and we are demanding dense circuitry with fine line definition. Although a certain amount of fuzziness may be permitted in the patterns for a prototype board on a negative, the masks must be in perfect overlay *alignment* to allow the use of small through-hole pads. We would also like to be able to pass signal traces between IC pins spaced on 0.1 inch (2.54 mm) centers, and so these pads must also be small; improper side-to-side registration in the negative masks would make hole drilling difficult, destroying pads. Methods for producing artwork that lead to well-registered photographic masks will be described in detail, and the actual photography can be done with an outside professional process camera. Camera service is available through printers and architects, particularly printers who perform color separation work. Typical charge for a pair of 6 by 9 inch photo masks is $20.00 to $30.00.

It is also possible to produce quality double-sided negatives with good registration using only a 35 mm SLR (single lens reflex) camera and standard black-and-white darkroom equipment. A unique process

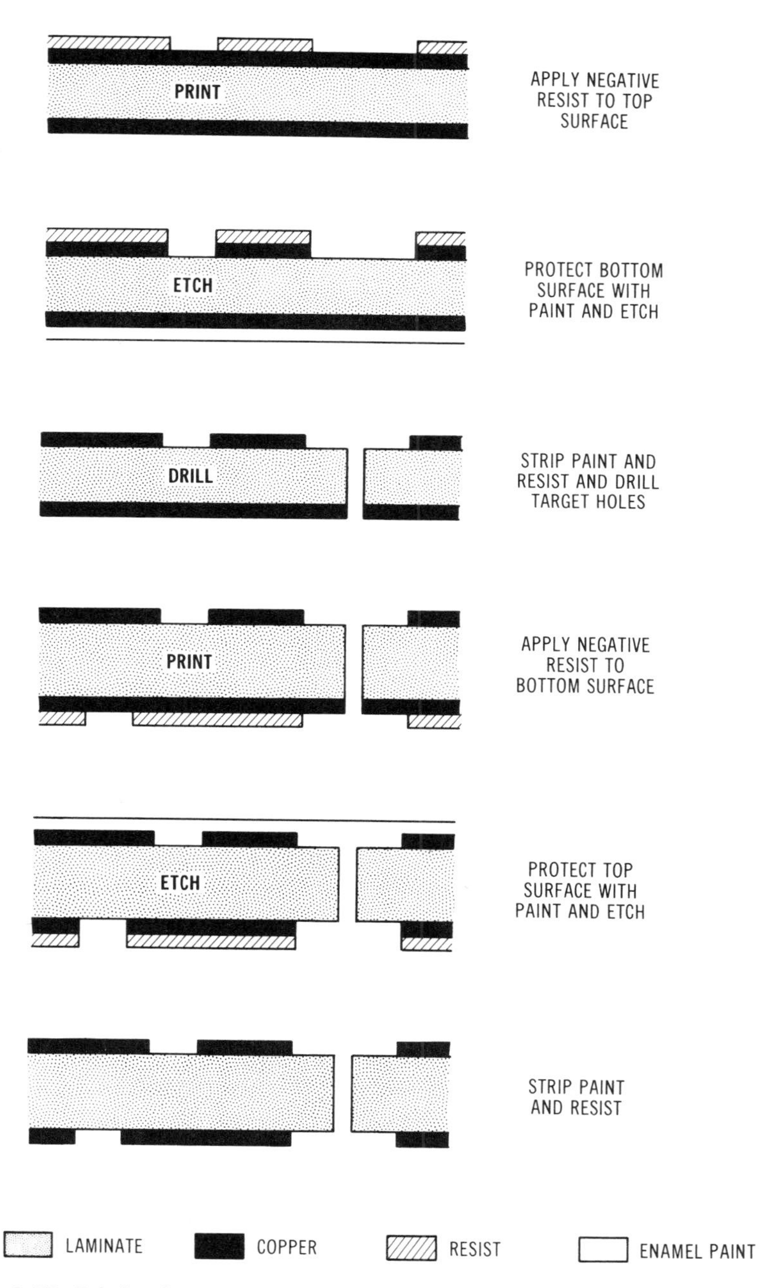

Fig. 2-23. Fabrication steps for PCB prototypes using the author's Print-Etch-Drill-Print-Etch approach.

will be described in Chapter 4 that allows any amateur photographer with an enlarger to obtain good PCB photo masks up to 8 by 10 inches in size, with little expense and in a minimum amount of time.

Coatings • Final board coatings such as gold/nickel for edge connectors, labels and component outlines, and solder masks are *optional* for prototypes. However, since these coatings are often desirable and require little equipment expense, plating and screen printing methods will be discussed in detail in later chapters. Screen printing requires a certain amount of time for setup and is not often used for low-volume work, but it is an important tool since it can also be used for image transfer of etch-resist ink, and for labeling chassis and electronic enclosures.

One coating which should *not* be considered optional is a thin layer of solder on all etched copper patterns. Solder serves as a protective coating for copper and prevents oxidation (discoloration, corrosion) but an even more important function is the preservation of surface solderability for ease of assembly. Since extensive hand soldering is necessary to assemble a prototype PCB, it is crucial that this be accomplished with a minimum amount of heating that could destroy components and lift copper traces from the laminate base. For establishing through-hole connections, component leads will often be soldered close to the component itself with little heatsink protection, so a readily soldered PCB surface coating allows quick, efficient hand soldering with only 2 to 3 seconds heat-up time per connection.

Solder coating was previously described with a *plating process* followed by *fused reflow.* However, the same results can be obtained manually without using a plating bath. Fresh, clean PCB copper traces can be easily coated manually with solder using a small 25-watt iron and a minimal amount of solder. The board is first coated with liquid rosin flux, and the final solder coating can be smoothed out in a hot oil bath if desired.

Alternatives to solder coating for providing a readily soldered surface include immersion tin and bare copper (cleaned just before assembly). Immersion tin is a chemical solution of tin salts that plates out autocatalytically on copper surfaces when the clean PCB is dipped for several minutes. Proprietary formulations for immersion tin are available from various plating suppliers and are not expensive. However, the author has found immersion tin coatings to be short-lived in their ability to preserve solderability; the coating can become very difficult to "wet" with solder in just a few days following its application, and the reader may not find this approach satisfactory.

Bare clean copper is a better surface for hand-solder assembly with rosin flux, and it can be used for prototype boards that are assembled soon after etching. However, this approach does not provide any long-term protection for copper surfaces that remain exposed.

PROTOTYPE PROCESS OUTLINE • Based on previous considerations, and following the fabrication sequence illustrated in Fig. 2-23, the following steps will constitute a simple laboratory or workshop approach for constructing double-sided PCBs without the use of expensive equipment:

1. Design PCB.
2. Prepare artwork.
3. Photograph and generate negative photo masks.
4. Coat photo resist on one side of board, and transfer image.
5. Dye resist coating, touch-up, protect reverse side with spray enamel, and bake.
6. Etch first side and remove residues.
7. Remove protective enamel and clean.
8. Drill 3–4 strategic holes in various areas of the board.
9. Coat photo resist on second side of board, and transfer image.
10. Dye resist coating, touch-up, protect first (etched) side with spray enamel, and bake.
11. Etch second side and remove residues.
12. Remove protective enamel, dye, and photoresist from all surfaces and clean copper thoroughly.
13. Electroplate connector tabs with nickel/gold.
14. Coat solder and reflow in hot oil bath.
15. Add optional screened coatings, such as label information on component side.
16. Drill all holes and deburr.
17. Assemble and solder all components and through-hole connectors.

Individual steps will be described in detail in subsequent chapters. This fabrication approach is time-consuming and impractical for producing a large number of PCBs; however, it is ideal for making one or two prototypes, and provides careful control at each step to prevent rejects and wasted time. Once the negatives are prepared, complete fabrication can be accomplished in one or two days, giving the electronic designer a relatively quick route for producing custom PCBs in his or her own laboratory.

CHAPTER 3
PCB Design, Layout, and Artwork

INTRODUCTION • In this chapter we will consider the design and layout steps necessary to generate final PCB artwork for photofabrication. Although the chapter was written primarily with double-sided PCBs in mind, certain considerations unique to single-sided PCBs will be mentioned where applicable; the same general design principles apply, but signal routing becomes more restricted for single-sided boards since only one plane of wiring is available for interconnections.

The design of a printed circuit board proceeds through some distinct stages. The first step is to prepare a well-drawn schematic, and the designer must know exactly what circuitry he or she wants to include on the board. He must also have a pretty good idea of the packaging requirements for the total electronic system; for example, the board mounting method and external connector hardware will play an important role in the total PCB design.

The next task is to establish board dimensions, and lay out parts and signal connections, keeping in mind the total electrical and mechanical requirements of the circuit. The PCB layout requires a fundamental knowledge of the fabrication processes to be used, including photography, etching, and machining. The designer must know what component density can be achieved, what tolerances can be expected for side-to-side registration, and what size pads, lines, and holes can be photographed, etched, drilled, and soldered satisfactorily. An important constraint in prototype fabrication is the lack of plated-through-holes, which are technically unnecessary for low-volume PCB production; it is possible to make a high density board without plating holes, but special considerations are necessary during design to ensure that the through-hole connections can be established with solder ioints.

The final step is artwork preparation. Although other methods are frequently recommended in the hobby literature for etching one-of-a-kind circuit boards, the only *good* method for making quality double-sided PCBs with sharp patterns and proper side-to-side registration involves the use of photographic transparencies (masks) which are prepared through the photographic reduction of expanded-scale artwork patterns. Master artwork must, therefore, be generated following PCB layout. With a well-drawn layout, the completion of artwork should be straightforward; as will be seen, the preparation of this artwork is not difficult or expensive. In Chapter 4, the reader will find that even the photography can be accomplished with a simple amateur black-and-white darkroom.

OBJECTIVES ● The objectives of this chapter are as follows:

- Explain the importance and characteristics of a well-drawn *schematic* as the original basis for a PCB design.
- Show how the PCB *dimensions* are established.
- Detail the considerations that go into *parts layout* and *line layout,* including their interaction with prototype PCB fabrication constraints.
- Discuss the materials and methods for generation of a *final layout drawing.*
- Explain the various approaches to converting a PCB layout drawing into *master artwork.*
- Detail the recommended *three-layer* artwork approach for double-sided PCBs: materials, procedures, and techniques.
- Discuss optional artwork patterns for *screen printing* and *solder masking.*

DESIGN ● Development of a good printed circuit board design proceeds through the following general steps:

1. Breadboard / wire-wrap testing — The electronic design should be assembled in breadboard fashion to test circuit operation and to determine final component values. The printed circuit board is an inflexible medium for modifying circuitry, since only minor alterations are practical following construction. For example, it may be possible to change resistor and capacitor values after PCB assembly, but a complete board re-design would be needed to allow substitution of a new IC with different pinout connections. In general, a PCB design should not be attempted before the electronic circuit has been physically tested in some form.
2. Prepare a final schematic — A well-organized schematic with smooth signal flow and complete information is the first key to effective PCB design.
3. Determine PCB dimensions — The board must be large enough to contain its own components, but small enough to fit into the overall packaging scheme.

4. Layout of components and external connectors — The parts layout should result in a balanced, uncrowded organization of components and hardware to be mounted on the board.
5. Design on-board interconnections — The routing of connecting lines must be accomplished within the constraints of available space, while satisfying electrical requirements such as signal length, resistance, and protection from interference. Through-hole connections must be established, and some layout changes may be necessary to achieve satisfactory routing patterns.
6. Final design drawing — The final design drawing for a PCB prototype should furnish all necessary information for artwork generation, board fabrication, and assembly of parts. To minimize documentation, the electronic designer should need only three pieces of information while working with an assembled prototype: the schematic, the design drawing, and the artwork patterns.

Schematic Drawing • The traditional approach in PCB design is for an electronic circuit designer to prepare a final schematic for use by a PCB designer as the basic reference. This drawing must transfer a great deal more information than simply the electronic circuit; it must also give the designer the necessary ideas and specifications he or she needs to design reliable electrical characteristics into the final assembly.

However, in the case of a one-of-a-kind prototype, the circuit designer may also be doing final construction, so the designer's complete knowledge helps bridge the gap between schematic and PCB design. There may, therefore, be a tendency to start laying out a PCB from rough schematics as soon as the circuit is functioning by breadboard testing. Impatience should be avoided, because the creation of a complete, final schematic is the most effective way of organizing a successful approach to PCB design.

Digital and Analog — Some examples of finished schematics are shown in Fig. 3-1 (digital circuit) and in Fig. 3-2 (analog circuit). These types of circuits are fundamentally different, although they may be combined in an electronic design. *Digital logic* uses two-state electronics in the form of integrated circuits to make decisions through logical "reasoning." A whole group of symbols has been created to aid in drawing logic schematics that represent digital electronic functions, and so the schematic appears quite different from *analog circuitry* which contains many more discrete components. Another difference is seen in the interconnection patterns. Digital electronic circuits tend to have many common signals that occur throughout the schematic, resulting in many long signal lines traversing the length and width of the drawing; examples are the RESET signal and the eight DATA bits

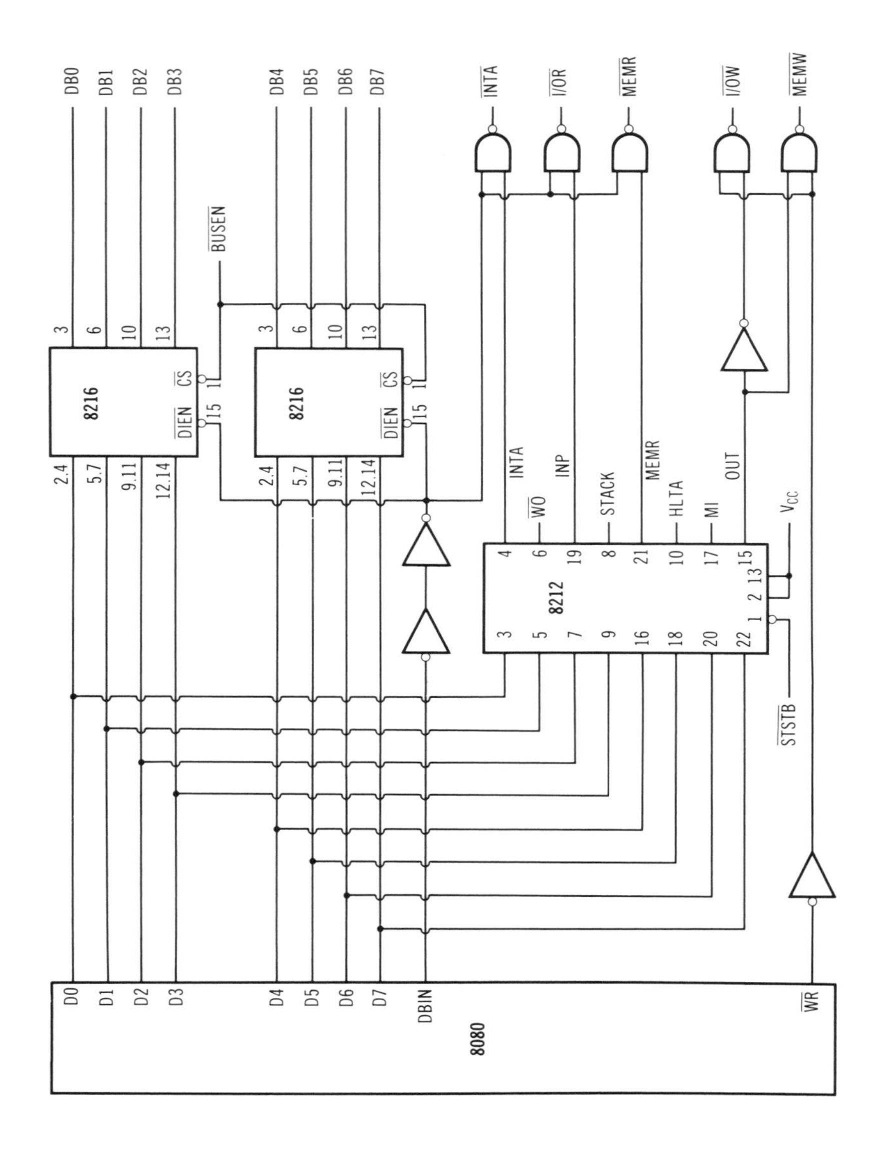

Fig. 3-1. Typical digital electronic schematic.

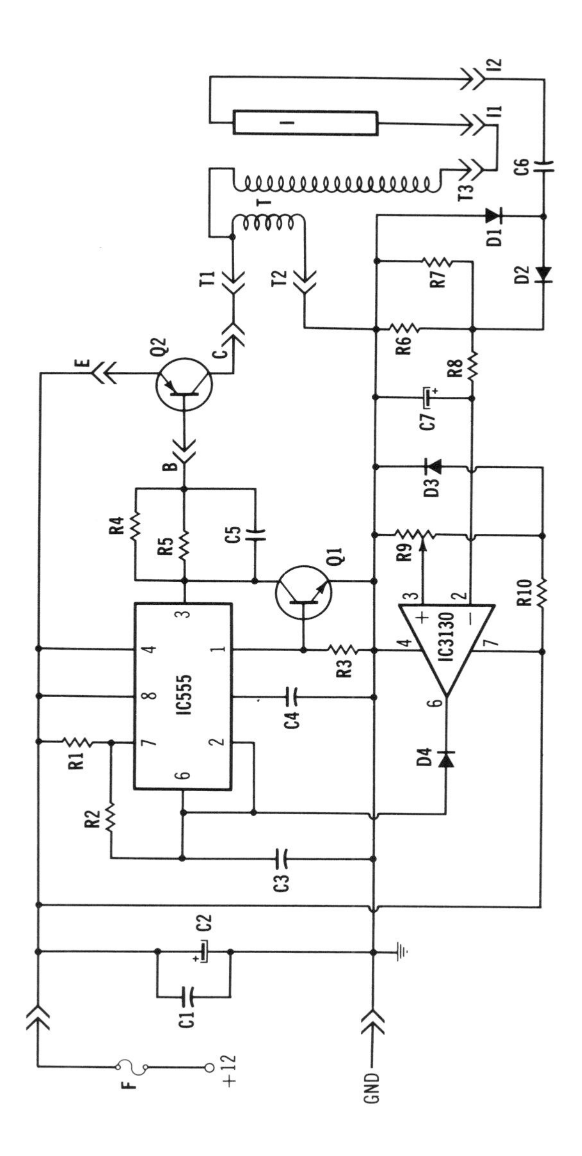

Fig. 3-2. Typical analog electronic schematic.

that are used extensively in 8-bit microprocessor designs. To avoid a mess, logic schematics are often simplified by *labeling* common signals rather than actually drawing every connection as a line.

Analog circuitry is the more traditional electronic approach that uses *discrete* components such as resistors, capacitors, and transistors to achieve the desired results. However, *linear* integrated circuits which perform analog functions have come into widespread use, such as the common op-amp and phase-locked loop ICs. The fundamental difference between analog and digital electronics is not in the type of components involved, but rather in the type of signals which are being treated. Analog circuitry is concerned with *continuously variable* signal levels, including voltage, current, and frequency; digital signals are primarily two-state voltage levels whose occurrences with respect to time and each other determine logical conditions. (Of increasing importance are *mixed* ICs that use logic to control analog signal levels, and vice-versa.)

Analog schematics have a fairly different appearance than digital ones because the interconnections are mostly confined to component blocks. If one block represents a stage of signal amplification, for example, the internal component connections may be quite complex, but only one signal enters and only one signal leaves the area to proceed to the next functional block. Discrete analog components may require close physical positioning on the PCB surface to maintain short signal paths and to prevent coupling or delays between stages; tight component grouping is, therefore, important in the schematic so that PCB layout can follow this drawing.

Schematic Rules — A proper schematic should make the entire circuit visible in as few drawings as possible, and should be constructed in the following ways:

1. Functional groups and subcircuits are apparent, and their components are grouped together in the drawing.
2. Overall signal flow (in/out) should proceed from left to right for smooth progression. This approach cannot be strictly followed for every single signal line, but the general principle will make schematics easier to read.
3. Signal lines should cross as little as possible. In addition to being hard to read, a schematic with tangled lines is harder to convert into PCB patterns, which can never cross on the same side of the board. Digital logic circuits frequently have a number of parallel signals which must be connected in numerous spots throughout the schematic; in this situation, it is better to simply label the connections with their signal abbreviations, rather than draw a complete maze of lines.

4. Critical leads should be apparent or specifically labeled. These are connections which must be kept short or isolated from other signals. A typical example is a low-level analog signal which is shielded as it enters the board, and must be terminated as soon as possible. It may also be desirable to keep a resultant high-level signal routed away from the input to avoid feedback interference in the amplification circuit.
5. All external connectors and components should be clearly indicated. For example, if the power supply line shows a fuse, it should be clear whether this fuse will be chassis-mounted, or mounted on the PCB in an in-line fuseholder. If parallel logic signals enter or leave the board, their connection scheme must be apparent from plug/jack labels, or edge connector numbers.
6. Logic gates should be combined and grouped together before drawing the final schematic. The circuit may include 18 inverters, so a little thought should precede the decision on how to organize these gates into 3 separate 7404 IC packages with 6 inverters on each chip. Similarly, if 19 inverters are needed, it may be desirable to substitute an unused NAND gate for one of the inverters, rather than require the use of an extra 7404 IC chip which would consume extra board space and would result in a number of excess, unused gates.
7. All integrated circuit pinout connections should normally be labeled, including power supply inputs. However, it may be useful to delay the assignment of logic gate connections when a number of equivalent gates exist in the same IC package. By deferring the identification of gate leads until PCB interconnections are actually being worked out, a more efficient design can be achieved, as will be illustrated later.
8. Any unused logic gates or extra subcircuits should be tabulated. These could include NAND gates, flip-flops, and operational amplifiers existing on multiple-unit IC packages. Two benefits result from keeping track of these "extras": their inputs can be properly tied to power supply or ground to prevent IC instability, and their locations are known whenever future circuit modifications must be made on a prototype PCB. It is often possible to correct a circuit problem without redesigning and constructing a new PCB, if extra logic gates are available.
9. Extra components that are created during PCB construction should be added to the schematic. For example, it is good general practice to use extra *bypass capacitors* across the power supply inputs near all IC packages which send or receive signals having high-frequency characteristics; this practice avoids noise problems

on the power lines connecting many different ICs. Another example is the use of multiple resistors in place of a single non-standard value to "trim" a circuit. For example, the revised schematic might indicate two parallel 10K and 36K resistors instead of a single odd value of 7.8K.

***Design Restrictions* —** Some electrical parameters are not apparent from the schematic drawing, although their contribution to the PCB design may be critical. The prototype designer who creates an electronic circuit and then lays out the first test circuit board has a distinct advantage when design restrictions must be heeded in the layout. Typical PCB electrical design considerations that are not obvious from schematics are the following:

1. Etched line widths and thicknesses — These affect resistance, current carrying capability, and voltage drop for each connection. Signal traces can usually be made much narrower than power supply or ground traces.
2. Line spacing — Spacing affects capacitance and cross-coupling between adjacent traces. Line spacing may need to be increased significantly if high-voltage differences exist between conductors. Inductive signal paths from critical circuit elements may require right-angle crossing to prevent coupling.
3. Isolation — Ground planes and shielding may be needed to prevent coupling from inductive elements such as transformers and coils.
4. Heat effects — Components generating heat may be mounted on heat sinks, or on wide foil patterns to dissipate the excess heat. Hot devices should be spaced evenly about the board, and away from any heat-sensitive components. For example, the oscillation frequency of a clock circuit may be directly affected by the temperature of critical circuit elements, so the critical parts should be positioned away from heat-producing components on the same board.
5. Component placement — Amplifiers and high-frequency circuits may require specific component orientations and spacings to avoid coupling, feedback, and radio frequency interference (rfi).

***Drawing Materials* —** For preparation of a final schematic drawing, the use of templates and drafting vellum (semi-transparent drafting paper) is recommended. Templates are available to reproduce most electronic/logic symbols (see Fig. 3-3) with the aid of a thin lead mechanical pencil, such as the popular 0.5 mm Pentel P205. Vellum with a 0.1 inch (2.54 mm) grid pattern is suitable for most work, and erasable rolls of this gridded drafting paper are available in large 42

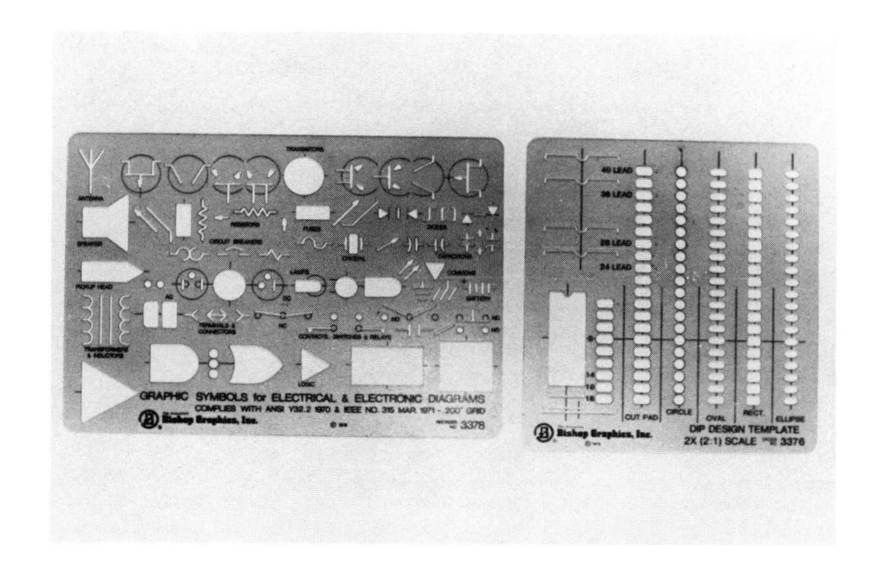

Fig. 3-3. Templates for schematic drawings — electronic/logic symbols and DIP patterns.

inch (1.07 m) widths from graphic arts suppliers. The grid background pattern helps keep symbols, lines, and labels uniform, neat, and well-organized. As stated before, it is advantageous in PCB design to have the entire schematic drawn on as few pieces of paper as possible. A good discussion of standard electronic components and their symbols is found in a publication[11] from Radio Shack, including examples of well-drawn schematics. Other good examples of modern schematic drawings are found in the monthly hobby magazines such as *Radio Electronics, Computers and Electronics,* and *Byte.*

Board Dimensions • The first step in actually laying out PCB components is to establish the outline and final dimensions of the board. The design process becomes easier as board area increases, so the largest practical size should be used. Board size may be limited by the enclosure and mounting hardware, or it may be rigidly fixed, as in the case of a microcomputer interface card to be plugged into a specific bus system. Because of limitations in the PCB construction methods described in this book (photography, image transfer, etching, drilling, etc.) the largest practical size for a double-sided prototype PCB is about 8 by 10 inches (203 by 254 mm). Dimensions larger than this will present problems at various points in the fabrication process, and will almost certainly require professional photographic assistance in preparing accurate photo masks from artwork.

Once fabrication and packaging constraints have been determined, the major factor in establishing PCB dimensions will be component density. For digital electronic prototype designs having relatively few discrete components, the typical double-sided density that can be readily achieved is about two square inches per IC, or about 10 component lead holes per square inch of board space (see Fig. 3-4). However, there is no reason that the experienced prototyper cannot achieve the common commercial double-sided density of one IC or 20 lead holes per square inch; construction techniques to be described in later chapters are compatible with this objective.

For single-sided boards, at least twice as much area is normally required as compared to a double-sided design; even more space may be consumed if a significant number of wire jumpers are necessary.

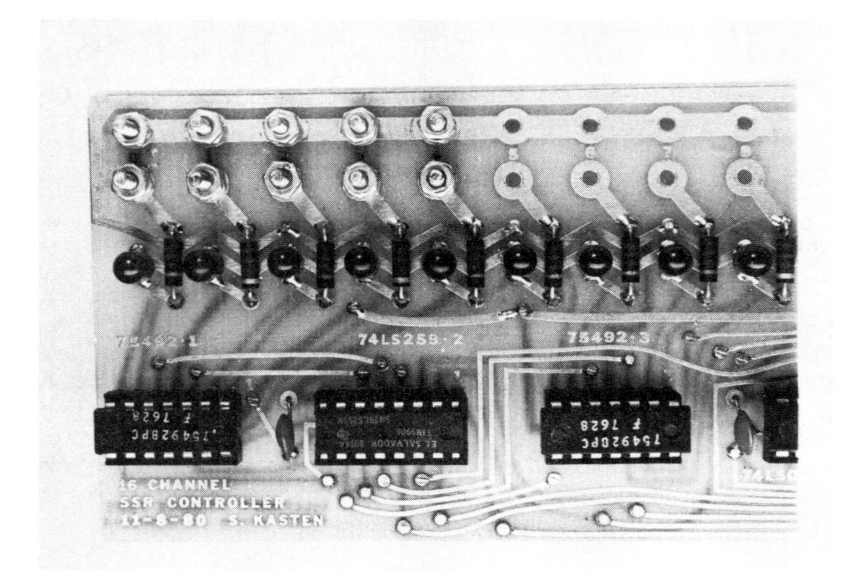

Fig. 3-4. Typical double-sided prototype PCB density.

Parts Layout • The design is begun by taping a suitable-size piece of 0.l inch (2.54 mm) grid paper to a desk top and drawing the proposed PCB outline in pencil. It is recommended that all design drawings be made on an expanded 2× scale to magnify small components and connecting lines; the 0.1 inch (2.54 mm) grid background then represents an effective 0.05 inch (1.27 mm) crosshatch pattern for the final board. This should provide plenty of resolution for positioning any component lead holes or connecting traces.

Once the outline has been established, all parts to be mounted on the board should be considered in review. These parts will include ICs, discrete components, connectors, terminals, and mounting hardware such as flanges and heat sinks. If possible, they should be gathered together to examine sizes, shapes, and mounting configurations.

The next step in parts layout is to select external connectors and determine their positions along the board edges. The positioning of I/O (input/output) connectors, including edge fingers, is crucial because it affects PCB mounting, it ensures compatibility with cable attachments, and it determines the manner in which power, ground, and I/O signals enter and leave the board. Once these connectors have been established, their geometry affects the total layout. An edgeboard finger connector will contain many signals that must be terminated as soon as possible; the layout of associated circuitry is thus dependent on leaving clear routes open to I/O connectors.

Layout of electronic components is accomplished with the help of paper dolls (cardboard cutouts) which represent the space consumed by each individual item, and have proper lead spacings based on the 2× layout scale. Dolls can be easily moved around within the PCB drawing outline to visually evaluate their effect on the total layout (see Fig. 3-5). (At the professional level, reusable layout dolls are available from Bishop Graphics, Inc. as their Puppets vinyl adherent symbols for both schematic and PCB layout.) The general objective of the component layout operation is to achieve a neat, balanced organization of parts; components should be uncrowded, evenly distributed about the board area, and positioned so as to allow convenient interconnection routes during line layout. To achieve these results, the following approaches are useful:

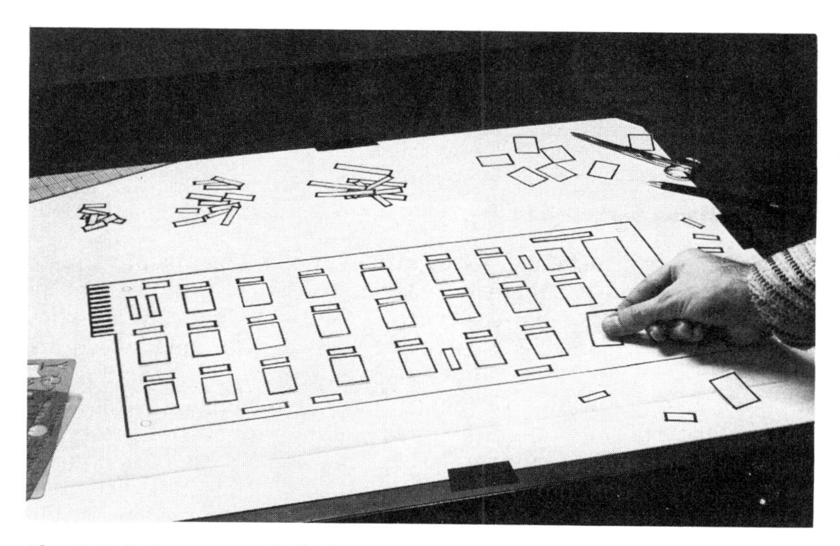

Fig. 3-5. Using paper dolls for component layout.

1. Schematic viewpoint — A well-organized schematic drawing provides the first approximation for circuit layout. Arrange components in the same *functional groups* as they appear in the schematic, and you will achieve the shortest possible interconnection patterns. Although signal flow may proceed in a rectangular path around the board, rather than in a straight line, the overall flow will be smooth if schematic groups are preserved intact. The principle of PCB layout based on functional schematic subcircuits is most important for conserving board space, minimizing signal lengths, improving electrical performance, and producing an assembly with easily located parts.
2. Central components — Examination of the schematic will generally show some key components within larger groups. These key devices should be centrally positioned without support components oriented around them. An example of such analog layout is found in Fig. 3-6, which shows central 555 and 3130 IC packages supporting numerous discrete resistors and capacitors around them. Compare this layout with the corresponding schematic in Fig. 3-2.

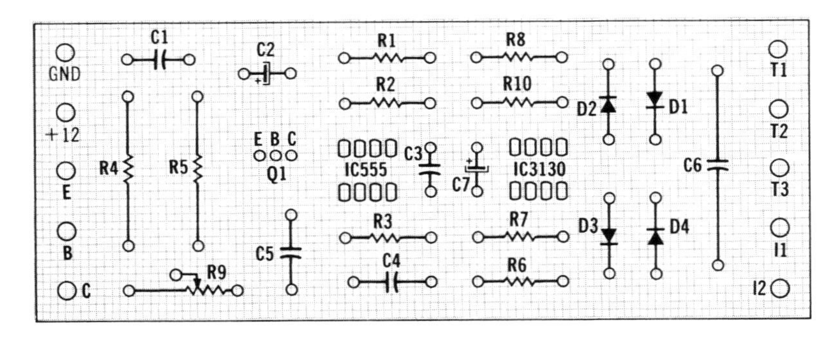

Fig. 3-6. Analog parts layout.

3. Squares and rectangles — By grouping components into square and rectangular subareas, space is conserved and the smaller areas will fit together easier to fill the total allotted space. This approach is particularly successful with IC packages, which are themselves rectangular in shape. By organizing ICs into rows and columns, the rectangular principle is maintained, and clear interconnection routes are established in parallel directions across the board. Discrete components accompanying ICs are positioned in vertical and horizontal segments so as to least obstruct open signal pathways. These IC layout principles are evident in the digital layout being developed in Fig. 3-5.

4. Key areas — Key board areas are evident where a large number of signals converge in a small space. Typical examples are the areas around edge connectors with many I/O signals, and the areas around large scale integration (LSI) IC packages, such as microprocessor chips and serial communication UARTs (Universal Asynchronous Receiver Transmitter). Circuit components with many connections into these areas must be carefully organized around the key space to prevent signal obstructions and insufficient conductor routing space during the subsequent line layout.

5. Component orientation — The important objective in component orientation is to maintain parallel/perpendicular mounting patterns with uniform spacing between devices; this principle will result in easy conductor layout because natural paths will exist in rows and columns across the board. Remember that the overall approach to line layout for a double-sided PCB involves creating primarily vertical traces on one side of the board, and horizontal traces on the other side; any two points on the board can then be connected by a series of traces joined by through-holes.

   The orientation of integrated circuit packages during layout is an interesting problem, because the common practice of mounting ICs in standard, identical orientations across the board does not appear to be efficient from an interconnection standpoint. Ignoring the need for automated insertion equipment (which does not exist for prototype PCBs), it would seem in theory that more efficient use of routing space could be made if each IC were positioned in some optimum configuration to obtain the simplest interconnection patterns with surrounding components. However, experience has shown that little is gained by this approach in digital logic circuitry. The improvements obtained in one area of the board tend to create new problems in adjacent areas, signal lines become blocked, and power traces begin to zigzag around. Troubleshooting/circuit testing also become confusing later when ICs are randomly oriented and their pins become difficult to count and probe with test equipment. For these reasons, it is recommended that pin No. 1 on each IC be oriented in a consistent direction.

6. Space — Two spatial considerations in component placement are (1) the need to make test points and adjustable components *accessible* for calibration and test purposes (for example, trimmer potentiometers — see Fig. 3-7), and (2) the need to establish *space* between components that generate excessive heat, require special mounting hardware/shields, or require physical separation for electrical reasons noted earlier. Space must also be allotted around the board perimeter for routing long lines and for establishing component clearance due to the physical mounting

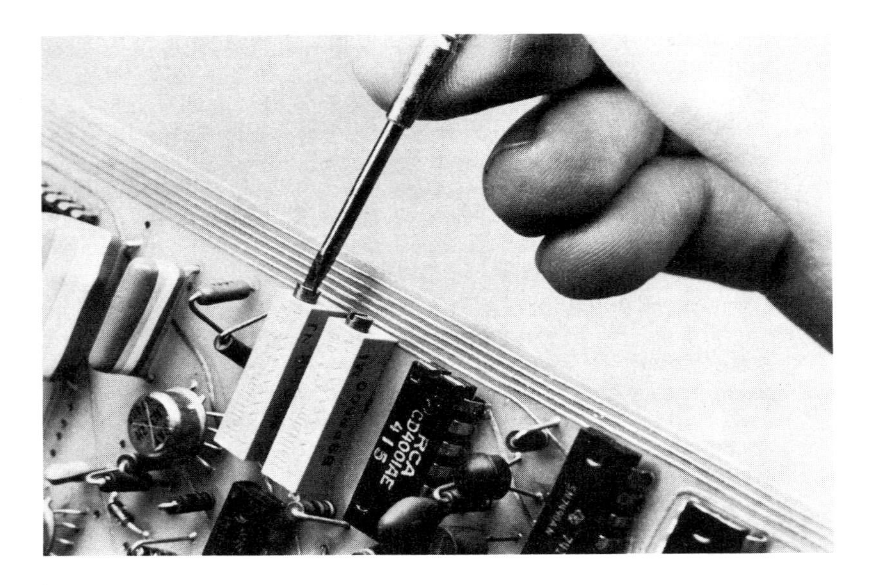

Fig. 3-7. Adjustable components should be accessible.

configuration of the board. Component placement should ideally allow all electrical labels to be visible, and should not hinder the ability to install or remove a single item on the board. Obstruction of components is especially undesirable for prototype PCBs, where it is highly likely that alterations will be needed, such as changes in resistance and capacitance values. Finally, appropriate space must be allotted for wide power and ground traces, and around components in metal packages which might contact other conductive parts.

When a satisfactory arrangement for all components, connectors, and hardware is obtained, this information is transferred in pencil to the PCB outline drawing to complete the parts layout phase of design. *Templates* are useful for this purpose (see Fig. 3-8), since they permit the drawing to be made rapidly with accurate positioning of lead holes. Two of the most useful layout templates include one for 2×-scale DIP-IC patterns, and one for various size circles. These and other templates are available from suppliers such as Bishop Graphics to represent almost any standard electronic component and its specific lead pattern on a scale of 1×, 2×, or 4×. The 2× scale is recommended for double-sided PCB design.

Where possible, all component, connector, and terminal holes should be positioned at the intersection of 0.1 inch (2.54 mm) grid lines on the drafting vellum paper. This standardization of component placement will lead to accurate artwork generation later when the master

Fig. 3-8. Using a template for component layout. ***(Courtesy Bishop Graphics, Inc.)***

tape-up is done on a similar grid background. Working at a 2× design scale, 0.1 inch (2.54 mm) increments should be small enough to accurately position almost any component or lead pattern at the grid intersection points. Some attempt at using final pad diameters should also be made during the layout sketch, because hole clearance problems will then be evident during the line layout phase of design. After labeling each component and terminal, ambiguous lead holes are identified, such as the emitter, base, and collector leads of a transistor, pin No. 1 of each integrated circuit, and polarity of diodes and electrolytic capacitors.

It is expected that most double-sided layout designs will progress through 2–4 revisions, so initial drawings should be made in rough form (including freehand sketching) to save time. There is no reason to produce an overly neat component placement drawing for the first attempt at line layout, because the layout will quickly become a mess of erasures, freehand traces, and smudges where components have been moved around and entire board areas have been redrawn. How closely the final PCB resembles the initial component layout drawing will depend largely on the designer's experience, the original schematic organization, and the completeness of layout considerations. However, it is assumed that some problems experienced during trace routing will always require a certain amount of component positional changes during the line layout.

Line Layout •

Color Code — It has been found useful to develop interconnection drawings with a *color-coded* system to simplify the presentation of double-sided PCB patterns during design. This approach is found in both manual drawings and in CAD systems, which perfect the idea using large color crt display screens. In our case, the use of gridded drafting vellum and colored pencils will give the same final results. A typical color-coded system is as follows:

- Black — Board outline, holes, pads, and hardware.
- Green — Component electrical symbols and outlines.
- Red — Connecting traces on the component side.
- Blue — Connecting traces on the reverse side.
- Yellow — Through-holes.

By following this system, an entire PCB drawing can be made on one sheet of paper, while giving complete information at a glance without confusion. The drawing is usually made as if viewed from the component side of the board.

Rat's Nest — When designing uncomplicated PCBs that are small in size, or have relatively few components filling a large area (low density), it may be possible to begin working out connecting traces as soon as a reasonable component layout is established. However, for denser PCBs approaching 1–2 square inches per IC, an intermediate step is advisable to *test* the proposed layout for major routing problems. This step is known as the *rat's nest* drawing, as presented in the Fig. 3-9 example. The rat's nest drawing is a quick, rough sketch showing all interconnecting lines, but ignoring all routing considerations. Direct lines are drawn between all pads requiring electrical connection, using a red pencil for lines running primarily horizontally (X direction), and blue pencil for lines running mostly in the vertical (Y) direction. The purpose of this drawing is to present the designer with a quick overview of line density requirements across the board. The rat's nest gives a good idea of how well-balanced the final line layout will be using the proposed component layout, and it shows up problem areas where insufficient routing space has been allotted. By revealing potential routing problems before line layout is begun, the designer should be able to rearrange the component layout and avoid any extensive changes during later steps. The rat's nest drawing provides insurance against major time losses that can occur when you design yourself into a corner that allows "no way to get there from here." A solid parts layout will be the result after this step, and the designer will feel more comfortable in proceeding to actual conductor routing.

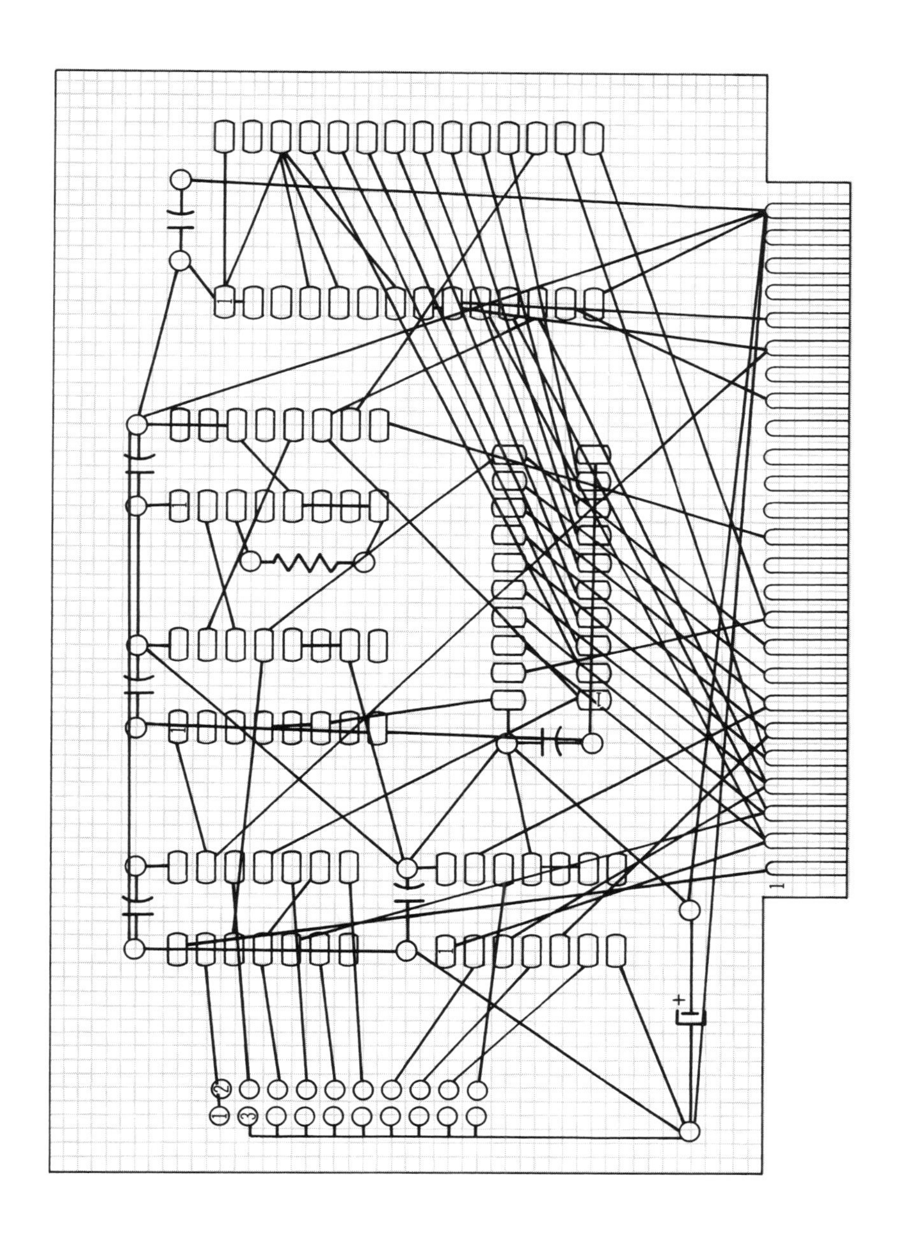

Fig. 3-9. A rat's nest drawing.

Line Widths — The width and spacing of conductor traces is important for reasons of electrical performance, design, and fabrication:

1. Electrical — Line width limits current-carrying capability and determines voltage drop in the etched conductor. Resistance of a 1-ounce copper trace can be estimated from the following equation:

$$R = \frac{0.0005}{W}$$

where,

R is resistance in ohms per linear inch,
W is line width in inches.

This expression is derived from the resistivity constant (ρ) for copper, $\rho = 6.77 \times 10^{-7}$ ohm-inch, which establishes the relation between conductor resistance, length, and cross-sectional area: $R = \rho \times L/A$. (Copper thickness of 0.0014 inch is assumed for 1-ounce copper foil.) If line width is chosen too small, problems with excessive resistance, voltage drop, and conductor heat rise may occur. The chart in Fig. 3-10 takes these factors into general account when specifying minimum conductor widths for various current loads in copper foil circuits.

2. Design — For purposes of design, small conductor widths are desirable because they consume less space in routing.

3. Fabrication — Large conductor widths are desirable from the standpoint of prototype PCB fabrication because greater tolerances and more errors can be permitted during photography and etching. Pinholes and scratches in negatives and photo resist coatings are much less likely to disturb 0.025 inch (0.635 mm) traces as compared to 0.015 inch (0.381 mm) conductors, where careful processing is required to prevent electrical breaks. Another consideration in favor of wider PCB patterns is that soldering heat is less likely to "lift" foil from the laminate surface if the traces and pads are wider.

It is obvious from Fig. 3-10 that conductor width is fairly unimportant in the electrical performance of most modern circuits using low power components, particularly digital logic designs; signal current levels are simply too small to worry about. With the exception of power supply traces, most connections can be made using line widths as small as fabrication technology will allow, typically 0.015 inch (0.381 mm) for prototypes. However, to prevent noise and ground-loop problems in

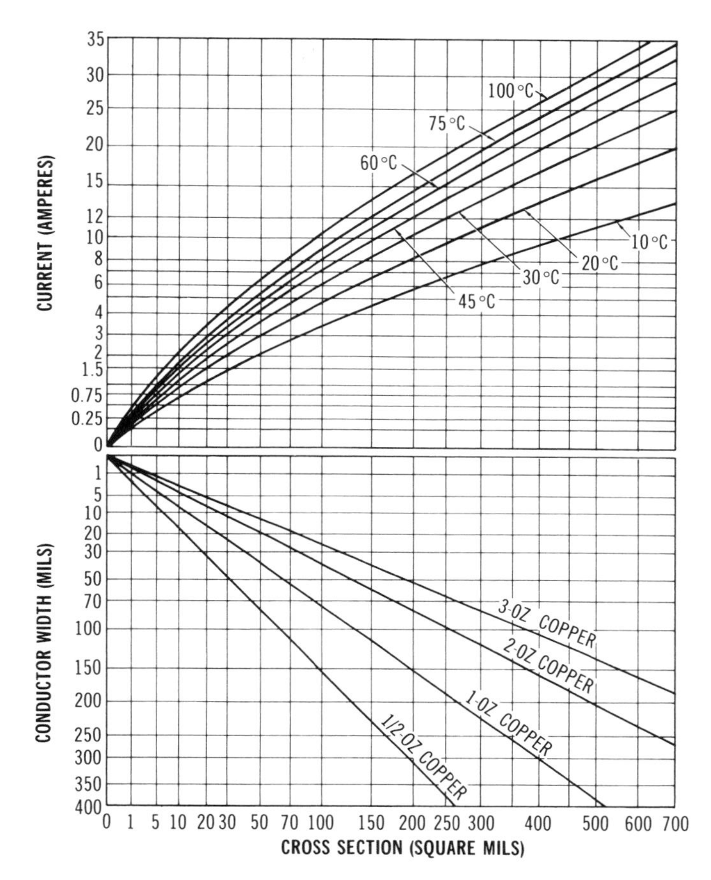

Fig. 3-10. Current carrying capacity versus PCB conductor width. ***(Courtesy Bishop Graphics, Inc.)***

the power connections, most conventional PCB designs use line widths of 0.05 inch (1.27 mm) for power conductors, and wider traces where more current is carried and less noise can be tolerated.

The considerations for design and fabrication lead to directly opposite objectives concerning line widths. Since design is a more important factor, it must take precedence: the board cannot be fabricated if it cannot be designed! A good general approach is, therefore, to use the smallest conductor widths necessary to achieve line layout. Later, during artwork preparation, increase the line widths where space is available, easing the burden on fabrication. Most line layouts do not directly indicate conductor widths when drawn in pencil, so the designer must keep notes and leave space for wider traces in the drawings.

Spacing between conductors is of lesser importance than conductor widths, but certain minimum standards should be followed to prevent problems during the construction processes of photography, etching, and solder assembly. A good general practice is to maintain at least 0.020 inch (0.5 mm) spacing between conductors, and slightly more between conductors and pads that must be soldered close by. High frequency or high impedance analog circuits may require some wider spacing of signals, but in general a 0.020 inch (0.5 mm) spacing standard will satisfy electrical performance demands of most low-voltage circuitry assembled on G-10 epoxy-glass laminates.

Line Patterns — After considering the effects of line width, line spacing, current demands, and fabrication procedures on prototype PCBs, a general guide to line layout drawings can be developed. Some typical approaches are seen in Figs. 3-11 and 3-12, and the corresponding final artwork is also illustrated to show how the etched patterns will eventually appear. These figures show most of the common line patterns encountered in digital electronic PCBs; the reader is encouraged to obtain industrial quality PCBs and carefully examine the various commercial approaches that are used for line layout — they are as follows:

1. Signal traces — Standard signal connection lines are shown in Fig. 3-11. The drawing is prepared on a 2× scale using 0.1 inch (2.54 mm) grid paper, so each grid spacing represents 0.05 inch (1.27 mm) on the final board. It is very convenient to draw each signal trace on a grid line, because this can represent 0.025 inch (0.635 mm)-wide lines with a spacing of 0.025 inch (0.635 mm), typical geometry for prototype PCB traces.
2. Power traces — Standard power supply lines are also shown in Fig. 3-11. These lines will be 0.05 inch (1.27 mm)-wide in the final artwork, so they are spaced wider apart and drawn in somewhat heavier pencil lines for the layout drawing. As seen in the figure, at least two power traces can usually be routed horizontally through a DIP-IC pattern with pins spaced 0.3 inch (7.62 mm) apart; this is the common pattern for 14 or 16 pin IC packages. It is common practice to route power traces through rows of horizontally positioned ICs on the reverse side of the board.
3. High density signal traces — The greatest density of signal lines that can be achieved with prototype PCB fabrication methods described in this book is illustrated in Fig. 3-12. This drawing represents 0.015 inch (0.381 mm) traces separated by 0.020 inch (0.5 mm) spaces, and can be represented in a 2× scale layout on 0.1 inch (2.54 mm) grid paper by spacing signal lines three-fourths of a division apart.
4. High density IC pads — Large component pads are preferred for

prototype construction, as was seen in Fig. 3-11. Large pads make alignment, drilling, and soldering less critical operations. However, for the greatest line density, considerable space may be saved by using *minimum area* pads, such as the oval ones shown in Fig. 3-12. Small pads preserve surface area while still allowing 0.015 inch (0.381 mm) traces to pass between 0.1 inch (2.54 mm)-spaced IC pins in the vertical direction on the component side of the PCB. Caution must be exercised when hand-soldering and particularly drilling small pads; etching these high-density patterns is not difficult, but they can be torn apart with improper

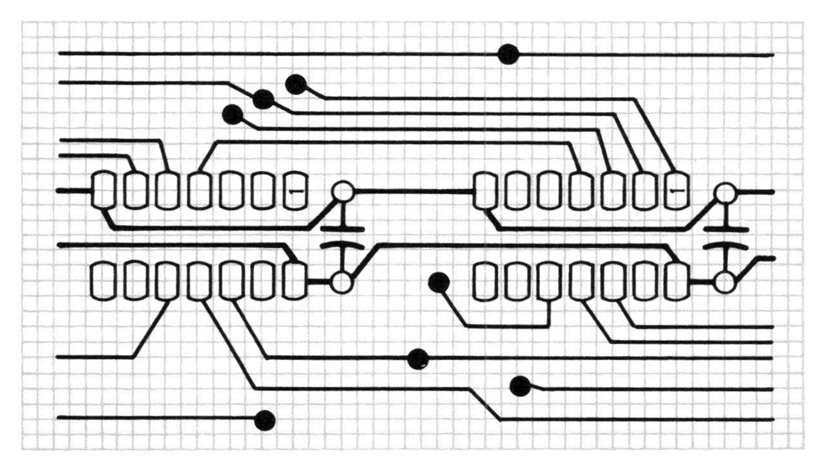

(A) Line layout drawing.

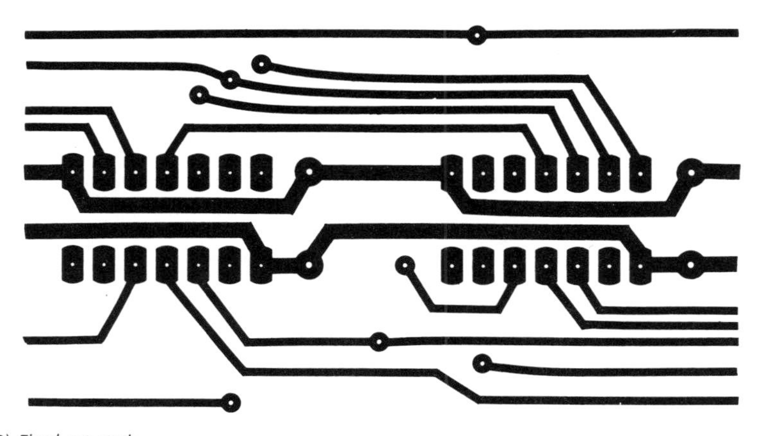

(B) Final artwork.

Fig. 3-11. Standard line patterns for PCB reverse side.

alignment and careless drilling. Small foil patterns are more prone to lifting (delamination) during hand-soldering because of the greater concentration of heat.

Routing Approach — Each PCB designer will develop his or her own system for efficiently routing conductors, but the following guidelines are recommended:

1. Color code — Make use of a color code to represent signals on opposite sides of the board. Easily erasable colored pencils are available, and the extra time needed for drawings is well worth the ability to visually separate conductor patterns and avoid confusion.
2. Balancing — Make use of the available board space. If your designs produce a combination of open areas and high-density conductor areas, there is probably a better way to lay out components. Remember that many connections are not critical as far as length; it may be better to route some signals in a roundabout fashion to avoid congestion, rather than attempt to make each connection in the most direct manner.
3. X/Y system — Use the horizontal/vertical routing system:

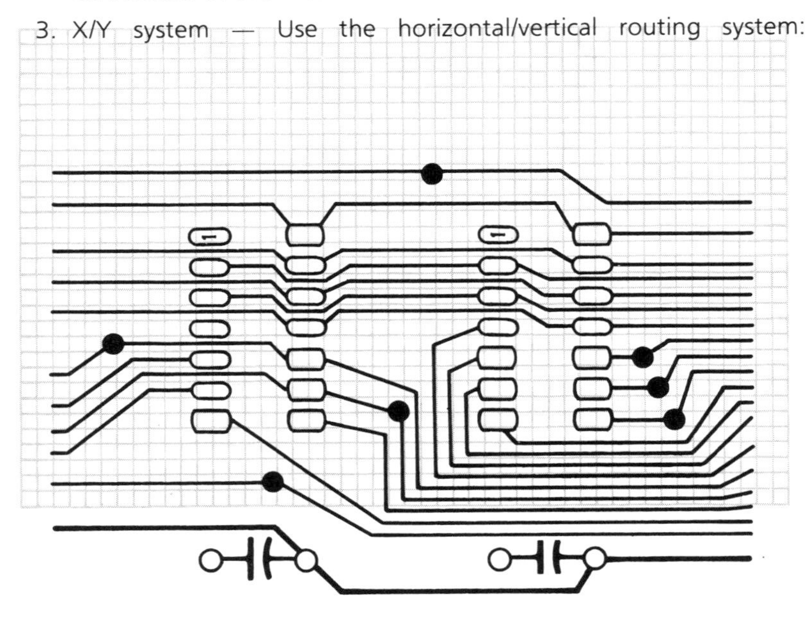

(A) Line layout drawing.

Fig. 3-12. High-density line patterns

horizontal traces on one side of the board, vertical traces on the other. This system is particularly useful in digital circuits where many signals must extend across the entire board length and width. An example of the X/Y system in operation is shown in Fig. 3-13, where horizontal etched traces are routed on the reverse side of a digital PCB; vertical traces on the component side were illustrated previously in Chapter 2, Fig. 2-10. For prototype PCBs, it is preferable to route vertical conductors on the component side, at a 90° angle to IC package rows. Vertical traces are the ones that will have to occasionally travel between 0.1 inch (2.54 mm)-spaced IC pins, and most of the soldering will be done on the opposite side; therefore, use of small pads can be confined to the component side to minimize drilling and soldering problems noted previously.

Analog circuits do not have as many common signals across the board, so the X/Y system is less important for these designs. (For the same reason, it is easier to lay out analog circuits than digital ones on a *single-sided* PCB.) As shown in Fig. 3-14, a double-sided analog PCB tends to have mixed X/Y traces on both sides of the board, minimizing the need for many through-holes. However, during design, it is still best to follow the X/Y layout sys-

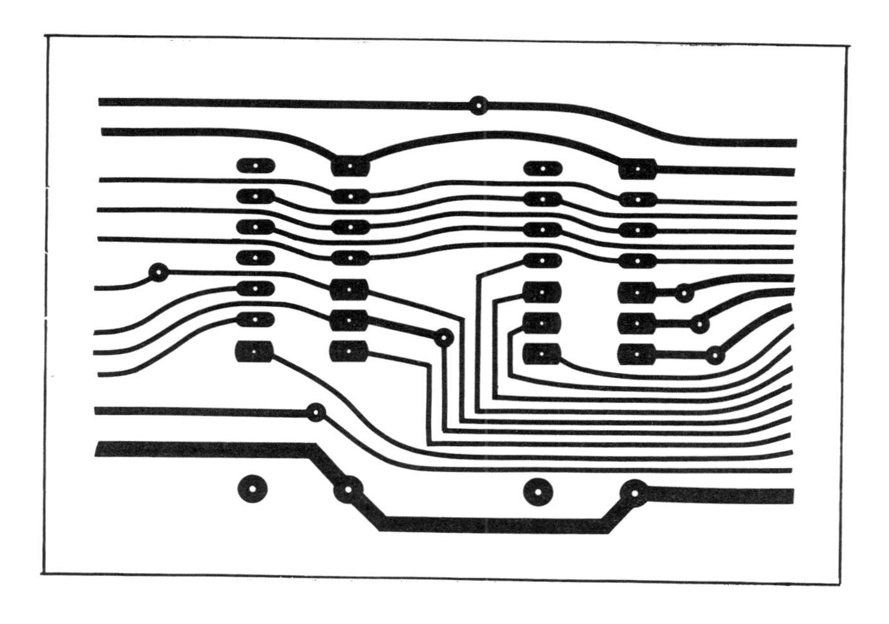

(B) Final artwork.

for PCB component side.

tem for all circuits; it is much easier to reduce and simplify an X/Y layout drawing at the *end* of design, rather than include a mixture of X and Y traces on both sides of the board from the very beginning.

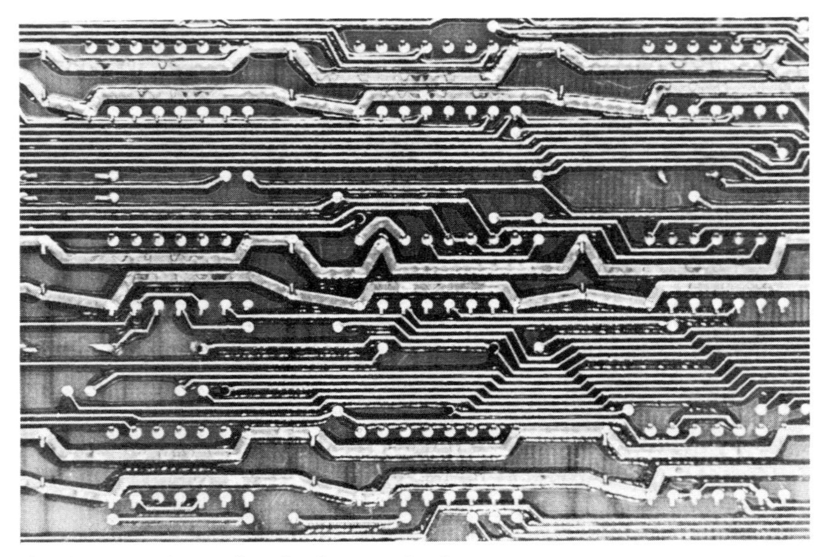

Fig. 3-13. Horizontal etched traces in the X/Y routing system.

Fig. 3-14. Analog PCB patterns with mixed X/Y routing.

4. Through-holes — Make generous use of through-holes. Many of them can be eliminated later during design simplification. Although through-holes take up conductor routing space, they can usually be moved around freely to position them out of the way. Commercial PCBs frequently have 25% through-holes as compared to component lead holes, and little is gained by avoiding their use if board space is available. The design time saved by using extra through-holes in prototypes and maintaining the X/Y routing system intact is well worth the small amount of extra time needed during assembly to solder pins in through-holes.

   Remember to use through-holes to terminate selected signals on the *reverse* side of the board, and thus avoid the need to solder inaccessible leads on the *component* side. This problem is created by the lack of plating in prototype through-holes, but a little thought during design can always circumvent the need to solder in difficult spots.

5. Power supply traces — Lay these out first. They consume a lot of space that will have to be reserved anyway, so go ahead and route them as the initial step. This approach will result in the simplest power distribution network to prevent noise and ground-loop problems later on. If possible, arrange power and ground traces so that the closest common connection to each IC power supply pin is a decoupling capacitor. The simplest routing method for power supply traces is usually to run them horizontally on the reverse side of the board between the legs of IC packages; this brings the power signals very close to every IC pin on the board.

6. Shortest traces first — A good component layout will minimize the need for long signal runs, and most connections will be relatively short. Draw short lines next, particularly if they can be routed in a direct run without the need for through-holes. Typical examples of short traces include power connections, connections between pins on the same IC, and connections between an IC and its discrete support components.

7. Logic substitutions — Leave equivalent logic gate pin assignments undefined until signal routing begins. This approach allows the most efficient layout to be achieved, as illustrated in Fig. 3-15. In this example, neater traces and fewer through-hole connections are the result of exchanging two NAND gates in a 7400 IC package. Note that the output lines from pins 8 and 11 have been interchanged and this must be accounted for elsewhere in the layout.

8. Crowded areas — Bring signals out of crowded areas early to avoid snarl-ups later. It is easier to rearrange several connections

in open areas than to move a single trace in a congested area, such as around a large IC or near an edgeboard connector. The rat's nest drawing is useful for recognizing these problem areas early.

9. Complete connections — After drawing power supply traces, short connections, and crowded patterns, complete all of the remaining connections in a steady sweep across the board. Starting in one corner, work steadily towards the opposite diagonal corner without missing any leads in between.

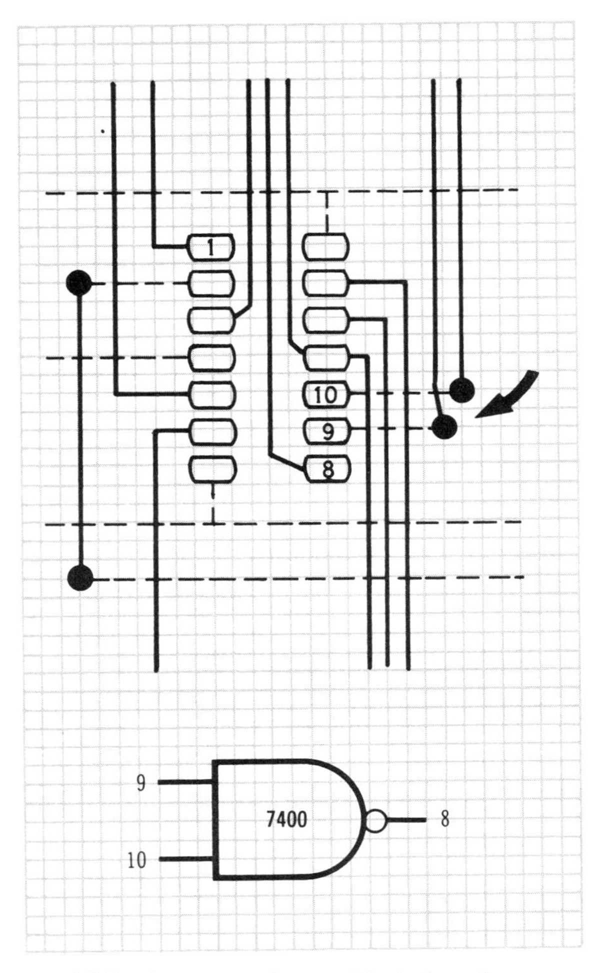

(A) Line layout according to original schematic.

Fig. 3-15. Deferred gate

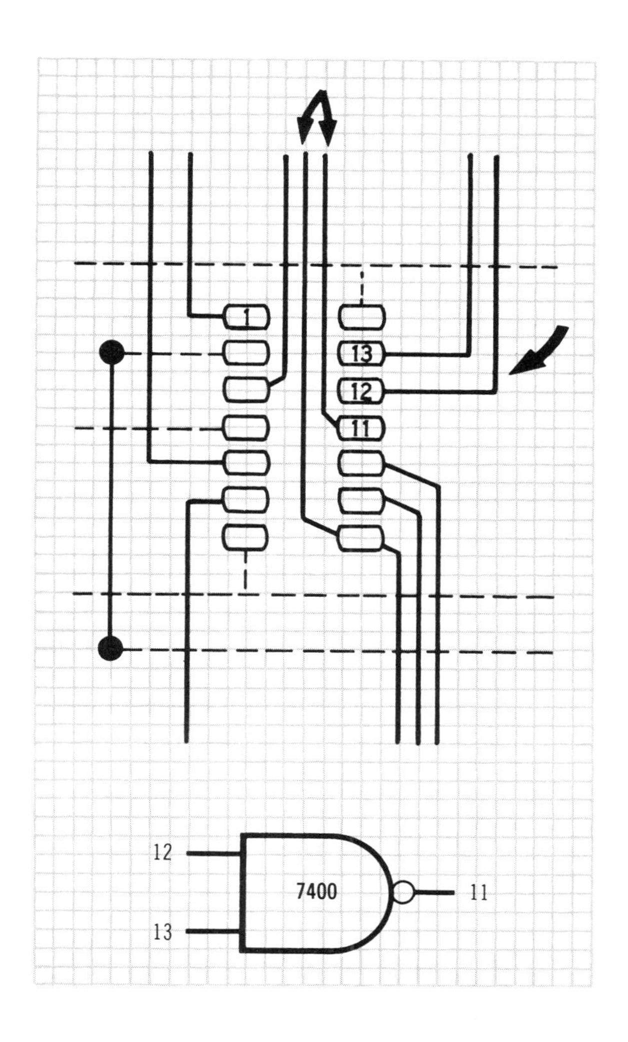

NOTE: COMPONENT-SIDE TRACES = DASHED LINES
REVERSE-SIDE TRACES = DARK LINES

(B) Line layout after switching two NAND gates.

assignments for efficient design.

Optimization — A double-sided PCB line layout will usually proceed through 2–4 design cycles before a final layout is reached. This cyclic process can be called optimization because each revision makes improvements over the previous one. At the conclusion of each line layout cycle, the designer must assess the drawing with respect to the general objectives: balanced layout, efficient routing, minimum trace density, accurate connections, satisfaction of electrical and mechanical requirements, lack of conductor crossovers, elimination of unnecessary through-holes, and well-defined pad positions and line spacings. Each revision should make efforts to correct problems seen in the previous version, often requiring that components be shifted and rearranged before the next cycle begins. Each revision should also become tidier. The initial layout may be drawn freehand with considerable erasures and penciled notes, but the final drawing should contain neatly ruled lines, careful adherence to the color code, and all pads and through-holes positioned at grid intersections. Only in this manner can the final layout be verified and used successfully for artwork preparation. If problems are discovered during tape-up, such as an impossible routing assignment, considerable material and time will be lost in making design changes on the artwork table.

In the final stages of the line layout process, the most important optimization steps are simplification, consideration of electrical and mechanical design factors, and verification of connector accuracy — these steps are as follows:

1. Simplification — After a complete routing system has been worked out for all interconnections, considerable simplification of the line layout can usually be achieved by eliminating through-holes and removing X/Y design restrictions. Since there will be no further lines to route, the designer can combine X and Y runs on either side of the board without worrying about blocking new connections. This process is illustrated in Fig. 3-16. Fig. 3-16A shows the unsimplified layout with strict adherence to the X/Y system, and Fig. 3-16B shows the simplified layout, where nine through-holes have been eliminated, and several XY traces combined on different sides of the board. Note that the lack of red and blue colors in this book has required a different system for line illustrations: — traces on the component side of the board are shown as *dashed* lines, while *solid* lines are used to represent traces on the reverse side.

   A second type of layout simplification is shown in Fig. 3-17, which gives an example of *swapping parallel lines* to eliminate through-holes.

   These two examples of simplification should be studied carefully because they illustrate the most common situations amenable to layout improvement. They are also good examples of refinements to *ignore* early in the design cycle; such simplifications are much easier to spot and correct in the final layout drawing.

2. Electrical/mechanical considerations — These design factors have already been discussed, and include such variables as connector positions, conductor widths, noise, heat-sinking, component mounting, component orientation, and so forth. It is simply stated here that all such important factors must be checked in the final stage of line layout to confirm that they have been satisfactorily designed into the system. One would not want to discover during artwork tape-up that no space was reserved for a power regulator heat sink, for example. Another mechanical consideration to check towards the end of design is the accessibility of pads for soldering on the component side. In prototype PCB assembly, through-hole connections must be soldered on both sides of the board; however, sockets or components themselves may make some leads inaccessible after mounting. This situation is illustrated in Fig. 3-18. The layout in Fig. 3-18A shows original trace routing to an IC pattern; it was decided to mount ICs on sockets whose leads could not be soldered on the component side, and so the layout was changed to permit reverse side soldering, as seen in Fig. 3-18B. Extra through-holes are usually needed to ensure that all IC pads connect to traces terminating on the reverse side.

3. Verification of connections — The designer must be absolutely sure that every single electrical connection has been included in the layout, and that every connection is accurate. A common problem seen with connections occurs when one particular signal joins numerous common points on the board, say six points; three of those points are connected to each other, and the other three are joined together, but the two sets may somehow remain unconnected to each other. Another common problem is that several equivalent IC packages can become partially cross-wired through ambiguous labeling in the schematic or component layout drawing. This situation is particularly troublesome when IC pins are deliberately left unlabeled so that layout can be optimized through the substitution of identical logic gates or functions during signal routing. Be extra careful when establishing schematic changes during layout.

4. Verification of pad and conductor dimensions — The most common problem in artwork preparation is that insufficient space is available for pads, conductors, and clearance. This is often a result of using thin pencil lines rather than actual-size traces in the layout drawing. To avoid this problem, the designer should

draw all pads and holes to their proper dimensions in the layout, and he should standardize drawings with line spacings such as those illustrated previously in Figs. 3-11 and 3-12.

Final Drawing • The final PCB layout should contain all information needed to generate master artwork. This information will include board dimensions, location, and size of all electronic components and mechanical parts, hole sizes and locations, and conductor routing patterns.

Portions of final digital line layout drawings are illustrated in the (B) sections of Figs. 3-15 through 3-18. Typical layout patterns in digital electronic PCBs exhibit even spacing of ICs, close adherence to the X/Y layout system, and strong influence from the position of I/O edge con-

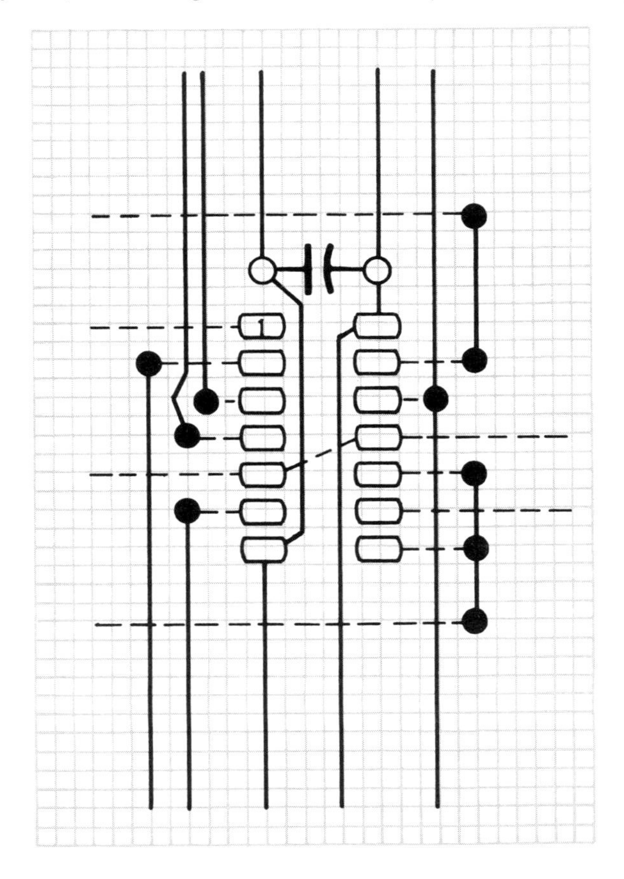

(A) Original line layout.

Fig. 3-16. Layout simplification

nector signals. The digital PCB tends to have long, parallel signal traces broken up into separate X and Y runs on opposite sides of the board, joined with an ample number of through-holes.

A final analog PCB line layout is illustrated in Fig. 3-19. This design shows strong influence from the original schematic (shown in Fig. 3-2) based on component grouping. As compared to digital PCBs, fewer through-holes are needed, and more X/Y runs are combined on the same sides of the board. This is a result of fewer common signals, and more

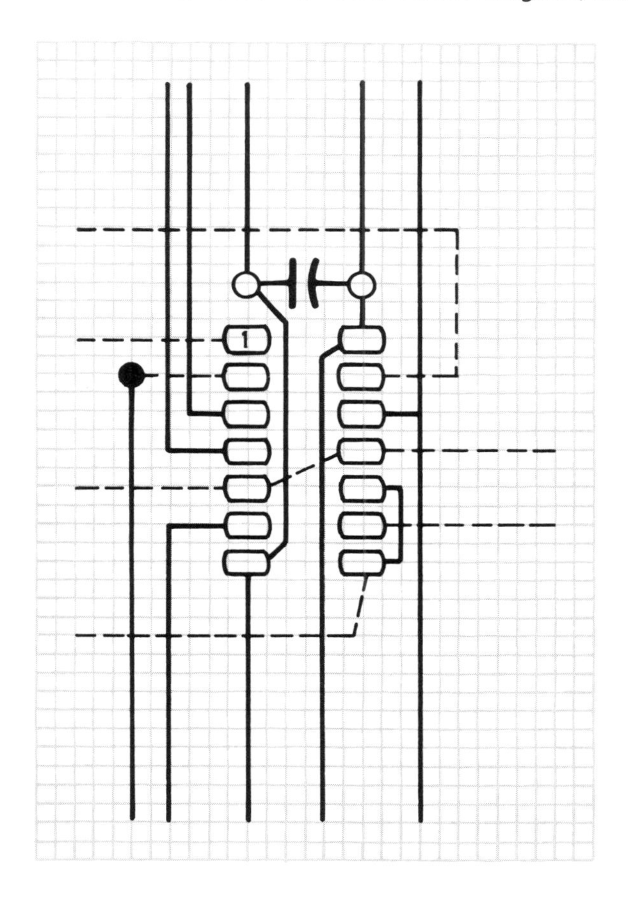

NOTE: COMPONENT-SIDE TRACES = DASHED LINES
REVERSE-SIDE TRACES = DARK LINES

(B) After simplification.

by combining X and Y traces.

discrete components, which can be arranged to optimize conductor layout with little fear of blocking signal paths. The analog PCB tends to have shorter signal connections confined to well-defined component groups, with X and Y runs mixed randomly on both sides of the board.

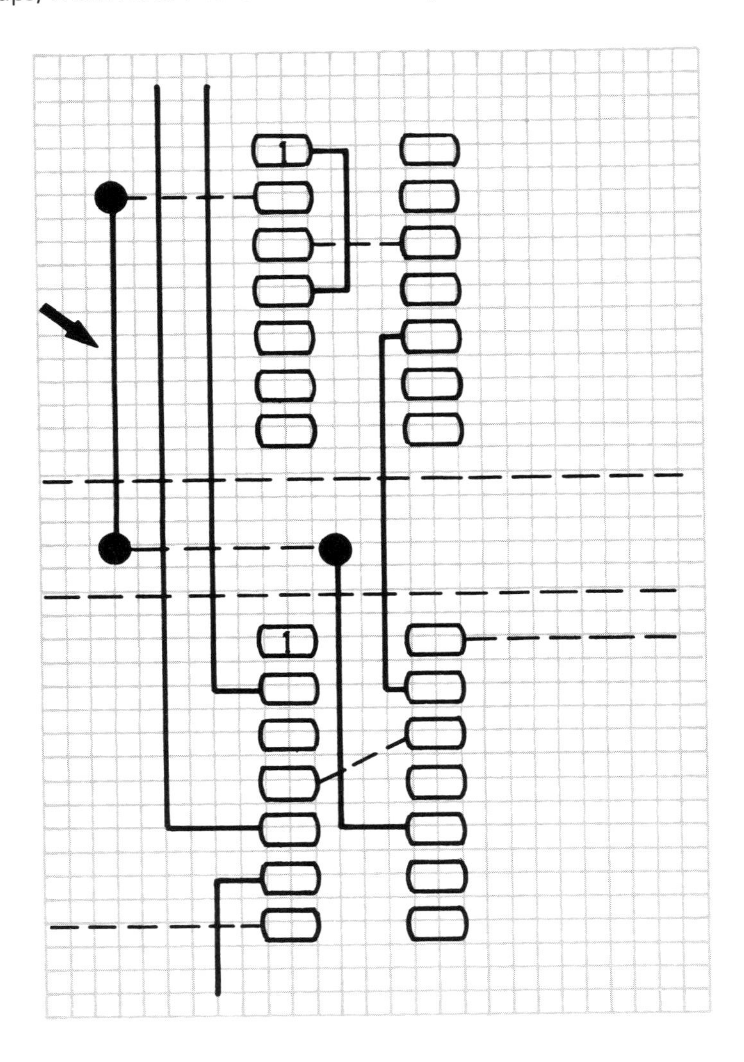

(A) Original line layout.

Fig. 3-17. Elimination of through-holes

In conclusion, the last five figures illustrate good PCB layout practices with neat, well-defined drawings ready to translate into taped-up artwork. The figures also show important PCB design differences between analog and digital circuitry.

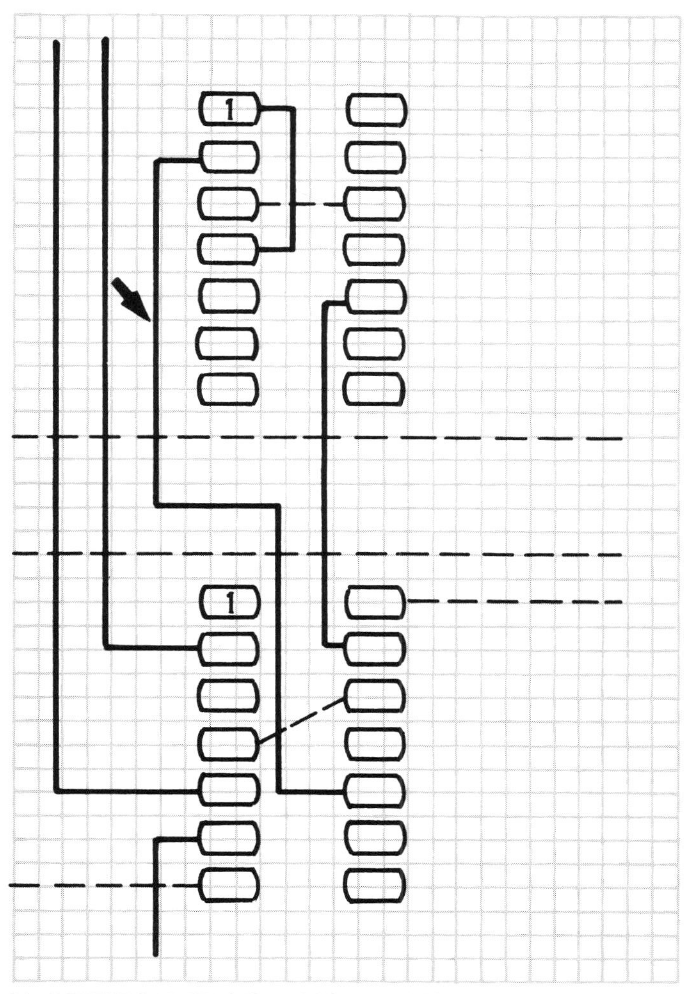

NOTE: COMPONENT-SIDE TRACES = DASHED LINES
REVERSE-SIDE TRACES = DARK LINES

(B) After simplification.

by swapping parallel lines.

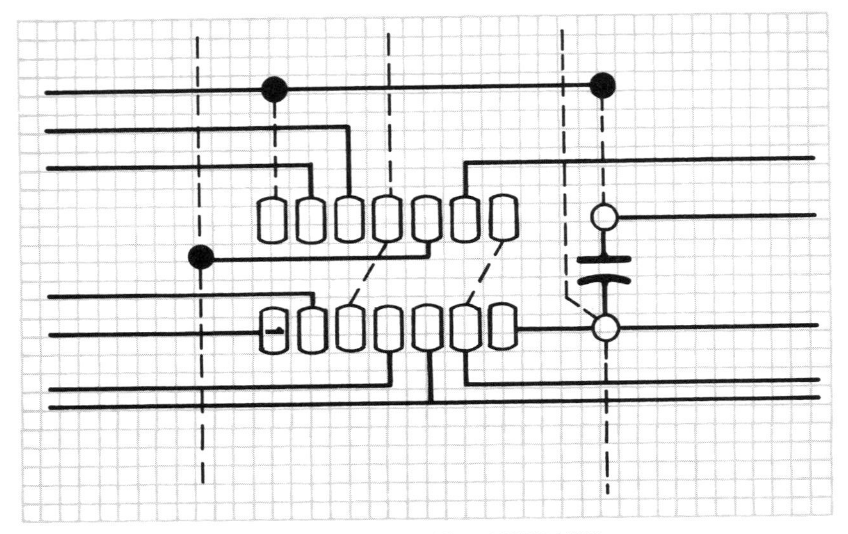

(A) Original line layout.

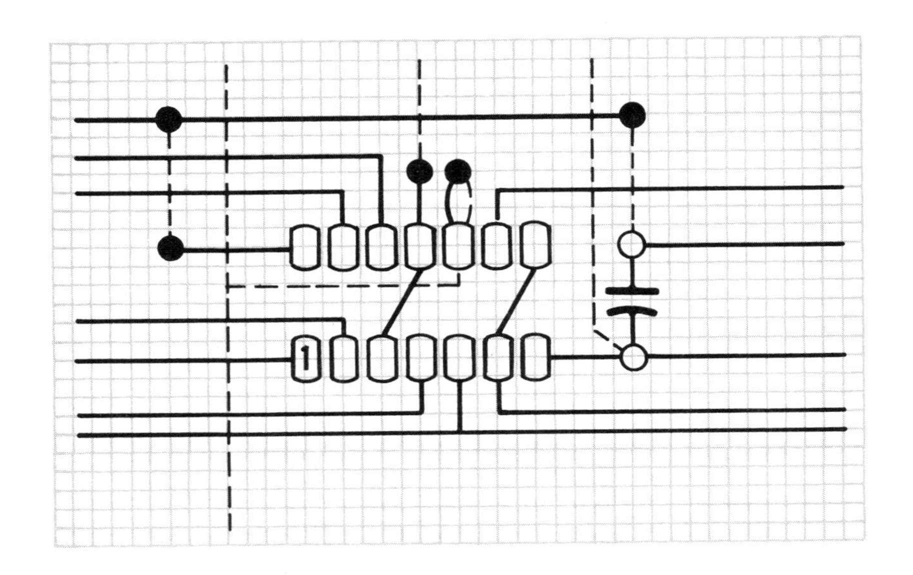

(B) After terminating all connecting traces on the reverse side.

Fig. 3-18. Layout alteration to allow reverse-side soldering of IC pads.

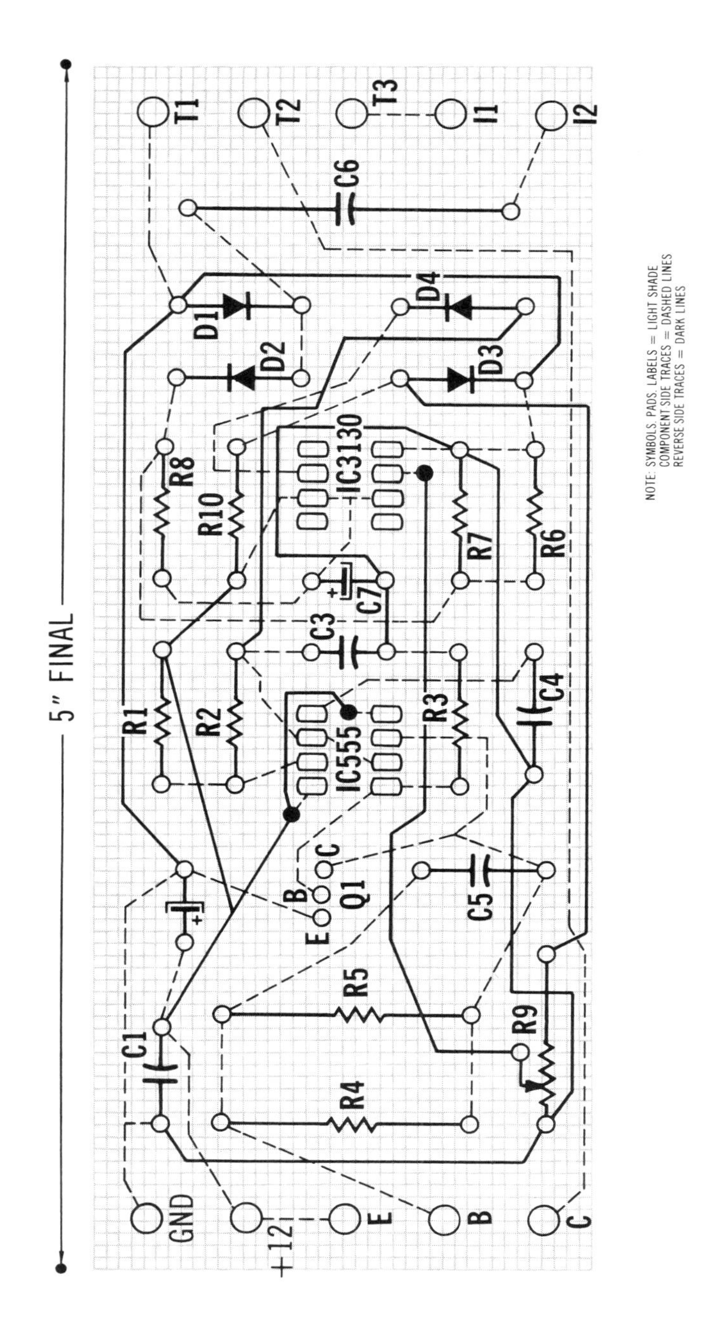

Fig. 3-19. Final analog line layout drawing.

ARTWORK PREPARATION •

Guidelines • The objective of artwork preparation is to create precise scale images of the desired printed circuit patterns to be etched on the surfaces of the board. These images will be photographically reproduced as negative or positive transparencies to be used as masking tools in photofabrication processes such as photo resist and screen printing. Because we are demanding dense, double-sided printed circuit board prototypes, the following guidelines are necessary for artwork generation:

1. Accuracy — The accurate positioning of terminal pads and narrow, closely spaced conductors demands that artwork be prepared on an enlarged scale, and that an accurate background grid be used for reference. As in the layout drawing, a 2× scale should provide sufficient accuracy while minimizing cost and space requirements. Other common scales in use are 1× and 4×, for which standard artwork materials are available. The 1× scale is difficult to work at because of the small size of patterns and tape, and this scale also provides no increase in image accuracy through photographic reduction. (Artwork imperfections are reduced in direct proportion to the amount of image reduction during photography.) The 4× scale is very large and provides unnecessary accuracy for prototype construction. The 2× scale is a convenient compromise, and at this magnification, a simple 0.1 inch (2.54 mm) background grid defines an effective 0.05 inch (1.27 mm) cross-hatch pattern that is ideal for conventional electronic component packaging.

2. Contrast — For good line definition, sharp pad outlines, and clear lettering, commercial transfer patterns and precision circuit tapes should be used to prepare high-contrast black-and-white artwork. Pen-and-ink drawings, either freehand or with templates, are laborious and will not produce the sharp results of standard PCB artwork materials; inked artwork is also difficult to correct or alter. Circuit tapes and 2× scale transfer patterns are relatively inexpensive considering the savings in time and the quality results obtained from their use.

3. Stability — For precise alignment of double-sided PCB artwork, it is imperative that a dimensionally stable substrate (such as polyester film) be used as the artwork base material. Other materials, such as paper or drafting vellum, will contract or stretch with changes in humidity and temperature; this will lead to unpredictable changes between the time artwork is prepared and the time it is photographed.

The Three-Layer Artwork System • The principle of three-layer artwork for double-sided PCBs is presented diagrammatically in Fig. 3-20. A *base figure* contains all patterns common to both sides of the board: outlines, edge connectors, pads, and holes. A *component-side overlay* contains all patterns specific to the component side of the board — interconnection traces and labeling information. When this overlay is superimposed on the base figure, it results in a complete image of the component-side artwork. The third layer of artwork is the *reverse-side overlay,* and it contains all patterns specific to the reverse side of the board. In combination with the base figure, this layer produces the complete reverse-side artwork; however, it is the mirror image of the expected pattern as viewed from the reverse side of the PCB.

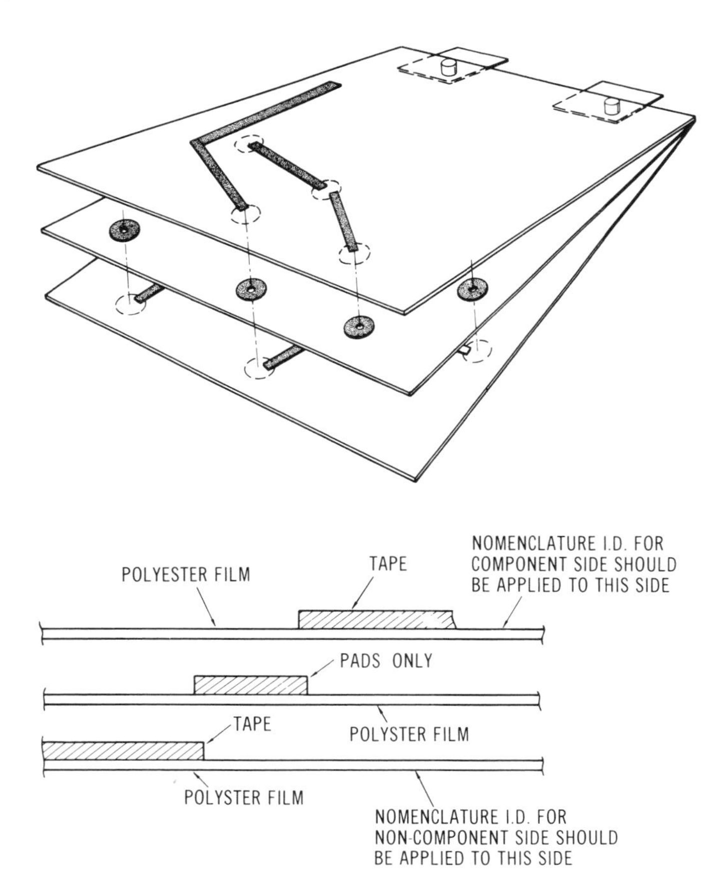

Fig. 3-20. The three-layer artwork principle for double-sided PCBs. *(Courtesy Bishop Graphics, Inc.)*

Fig. 3-21. Layered artwork on a grid background.

An example of finished prototype artwork is shown in Fig. 3-21 corresponding to the line layout of Fig. 3-19. This artwork consists of three clear polyester sheets which have been superimposed in the photograph. The three-layer approach is recommended for double-sided PCBs because it is the simplest method of obtaining exact side-to-side registration of negatives without resorting to special photographic techniques.

During preparation, the layers are taped down on a stable *grid background,* as seen in Fig. 3-21, to aid in precise scale alignment of all patterns. Commercial light tables are available as a background work surface, as shown in Chapter 2, Fig. 2-5, but a desktop is adequate. The layers are aligned with several target patterns on each sheet so that they can always be superimposed accurately after preparation. The three-layer artwork system offers some important advantages over other methods in prototype PCB design as follows:

1. Registration — Because the same pads and holes will be photographed with each overlay, the two resultant negatives *must give* exact superposition of pad holes, if the base figure is not moved in between exposures.

2. Tape-up — Artwork preparation is straightforward because patterns are copied directly from the final line drawing. Red pencil lines are converted to black taped lines on the component-side overlay, and blue pencil traces are taped on the reverse-side overlay. Just as in the line drawing, all artwork images are viewed from the component side of the board — component traces

normal, and reverse traces as their mirror image. The reversed artwork presents no problem to photography, because a mirror image can be obtained by simply reversing the film in the camera. One slight inconvenience of the reverse-side overlay is that any lettering must be accomplished by turning the sheet over and transferring letters from the opposite side to form reverse images as viewed from the front. However, dry-transfer letters are available as their mirror images to eliminate even this small inconvenience!

3. Economy — The most expensive artwork materials are stick-on pads and connector patterns. By using these only once in the artwork base layer, considerable material and time are saved.
4. Photography — Artwork photography is straightforward. Using a process camera, the base layer is mounted on the copy board, and the first overlay is superimposed with targets and photographed. The first overlay is then removed and replaced with the second overlay, which is photographed with film reversed in the camera. This produces two negatives with perfect registration of holes, and with proper orientation of photographic images in the film emulsion. The three-layer artwork system is also ideally suited for *amateur photography* using a 35 mm SLR camera and black-and-white darkroom equipment. As will be seen later in Chapter 4, this procedure calls for three separate photographs of the artwork layers.

Alternate Artwork Systems • Two other systems for preparation of double-sided PCB artwork should be mentioned because they are in common use.

Two-Layer Artwork — The *two-layer* approach starts with two identical sets of base pads on separate sheets of clear film. The component-side traces are then taped on one sheet, and the reverse traces are taped on the second sheet, producing separate, complete artwork representations for each side of the board. This method is not recommended because extra time and materials are required to generate a dual set of pad patterns, and the alignment of these pads is very difficult and tedious. The final side-to-side registration of negatives prepared from this artwork tends to be inferior, leading to alignment problems during PCB fabrication and assembly.

Red and Blue Artwork System — A second common approach is the red and blue taping system that has been extensively developed by Bishop Graphics and Eastman Kodak Company. This system used a single sheet of transparent taping film. Pads and patterns that are common to both sides of the board are first transferred to the film, using an accurate grid underlay; these patterns are black in color. The component-side traces are then added to one side of the sheet using special *red* transparent plastic tape, symbols, and lettering; reverse-side traces are taped on the opposite side of the sheet using *blue* transparent plastic artwork materials. Photographic color separation methods can then convert the composite artwork into its red and blue components to give two photo transparencies in perfect registration, with the desired patterns for the component and reverse sides of a printed circuit board. The red and blue artwork system has the greatest technical potential of all manual approaches for producing high-quality double-sided PCB photo masks; only precision computer-driven artwork plotters can give better results. However, the red and blue system is not widely used for hobby and prototype PCB artwork for the following reasons:

1. Expense — Colored tapes are more expensive than standard black tapes.
2. Flexibility — Transparent colored plastic tapes are not flexible like black crepe tapes, which can be easily contoured to produce smooth curves and bends. Special pre-cut angled *elbows* must, therefore, be used to change direction of lines, or else time-consuming overlapping splices are required to form angled corners.
3. Artwork reversal — Although both red and blue tape can be attached to the same side of the artwork sheet, this leads to a mess when changes must be made to the overlapping traces. For this reason, it is customary to separate red and blue tapes by confining them to opposite sides of the artwork sheet. However, this means that artwork must be *turned over* to complete the reverse-side tape-up; reversing the artwork patterns makes it very difficult to follow the original line layout drawing, which is created from the component-side viewpoint. (If the layout drawing is on semi-transparent drafting vellum, it may be possible to turn it over and view it through a light table during reverse-side artwork preparation.)
4. Photography — Professional photographic color separation work is required to convert red and blue artwork into photo masks. The photographer must have a process camera with a color-corrected lens, special filters, proper light source, and a back-lighted copyboard. Exposure times and film processing using color-sensitive *panchromatic* film become much more critical as compared to simpler blue/green-sensitive *orthochromatic* film.

Artwork Materials •

Sources — Many general-purpose artwork materials are available from your local drafting or office supply house, such as drafting pens, vellum, templates, dry-transfer letters, Mylar film, and masking tape. However, Bishop Graphics, Inc., is recommended specifically for printed circuit drafting materials. Most of the artwork materials discussed in this section are available from Bishop Graphics distributors; the reader is encouraged to obtain their technical manual and catalog, which contain a wealth of diagrams, photographs, techniques, and information for printed circuit designers and draftsmen. Other manufacturers and distributors of special drafting materials designed for printed circuit artwork are the following:

Chartpak Graphic Products
Circuit-Stik, Inc.
Datak Corp.
Feedback, Inc.
GC Electronics
Injectorall Electronics Corp.
Jack Spears, Inc.

Grid Background — An accurate background grid pattern is essential for preparing PCB artwork, because it allows the precise placement of terminal pads and holes to be drilled (see Fig. 3-22). Modern electronic packaging generally conforms to standard lead increments of a 0.1 inch (2.54 mm) grid system. Earlier in this chapter, we discussed the use of gridded drafting vellum for preparing PCB layout drawings. However, for final artwork, a more *dimensionally stable* grid is desirable. *Precision grids* are prepared photographically on stable plastic sheets or glass plates, which can be attached to the work table as a standard background surface. Precision grids are available in many sizes, thicknesses, and grid patterns. The Accufilm grid shown previously in Fig. 3-21 is an 18 by 24 inch polyester film, 0.0075 inch thick, matte finish, with 10 lines per inch, and heavy accent on each 10th line. Transparent artwork sheets are taped over this grid so that it can be used repeatedly; precision grids are too costly to be used as the actual artwork surface. An 18 by 24 inch background grid is adequate for preparing artwork on a 2× scale for PCBs up to 8 by 10 inches in size,

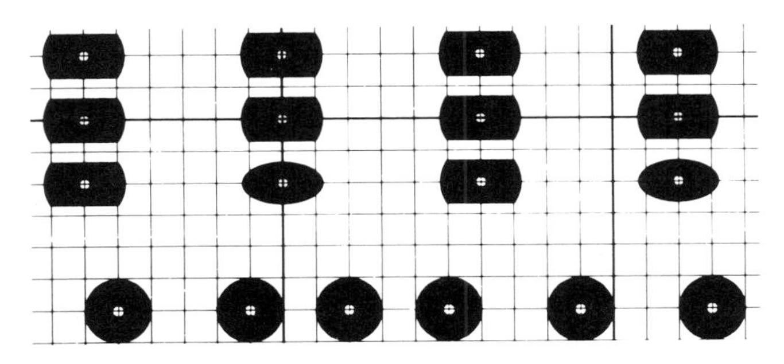

Fig. 3-22. Using the grid system for positioning artwork. *(Courtesy Bishop Graphics, Inc.)*

which is the practical limit for prototypes constructed with techniques given in this book.

Base Material — The base material for artwork preparation is chosen with the following factors in mind:

1. Surface characteristics — Will transfer patterns and printed circuit tape stick to the surface, and will it accept ink?
2. Photography — Can the artwork be easily photographed to provide sharp, clear results?
3. Stability — Will the artwork retain its dimensions with changes in temperature and humidity?
4. Durability — Can the artwork withstand reasonable handling without crumpling or tearing?

Transparent polyester film is generally used for taping printed circuit artwork. The main advantage of a clear base material is that it is convenient for preparing multilayer artwork (see Fig. 3-21), but it can also be *backlighted* during photography to improve contrast and produce better photo masks. Polyester composition provides a temperature and humidity-stable base with excellent surface adhesion properties for transfer letters, pressure-sensitive patterns, and printed circuit tapes. The film is available as sheets or rolls in a variety of sizes and thicknesses. Recommended thickness is 0.005 inch (0.127 mm) for most artwork. Different surface finishes include clear, matte, and antistatic. The antistatic feature is useful for keeping artwork clean of lint, erasures, and other residues that seem to be attracted to plastic surfaces; antistatic also makes it easier to position large transfer patterns, such as IC DIP pads, which may develop their own static charges. A matte finish is usually needed for making inked drawings; this finish also tends to reflect less light during photography. An economical way

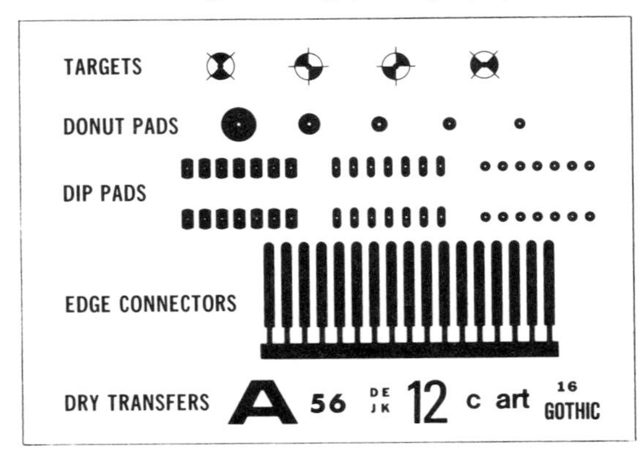

Fig. 3-23. Transfer patterns used in printed circuit artwork.

to purchase taping film is in 36 inch-wide, 20-yard rolls. When ready to use, each sheet is cut to size and taped in place over the background reference grid.

Transfer Patterns — Typical patterns used in printed circuit artwork are illustrated in Fig. 3-23. These patterns are available in standard 1×, 2×, and 4× sizes as pressure-sensitive decals that are removed from their backing paper with a knife edge, and transferred to the taping film with gentle pressure. High-quality decals have a *low-tack* adhesive surface that allows repeated positioning of the patterns until exact placement is made. Following transfer, the patterns appear as sharp, photographically opaque images on a transparent background. The following basic patterns are illustrated in Fig. 3-23:

1. Targets — Several of these patterns are placed on each artwork layer to serve as alignment points, and as precise reduction marks during photography.

2. Donut pads — These are the standard round shapes used for component lead pads and miscellaneous holes. The small center hole is very important for drilling holes. Following etching, it becomes a small dimple in the pad which guides the tip of the drill bit. Without this feature, pad positioning is more difficult during artwork preparation, and final PCB hand-drilling becomes a nightmare. (It should be noted that for commercial PCB artwork with plated-through-holes, photo masks are not permitted to have pads with center holes; this would lead to the introduction of a spot of photo resist or screened ink resist inside predrilled through-holes. Of course, the use of numerically controlled precision drilling machines does not require dimple guides in the laminate surface for accuracy.)

   Most prototype PCB artwork at the 2× scale can be completed with five standard sizes of donut pads: 0.125 inch (3.175 mm) OD (outside diameter) for very small IC pads and small through-holes, 0.156 inch (3.96 mm) OD for small component pads and standard through-holes, 0.187 inch (4.75 mm) OD for standard resistor, capacitor, and diode pads, 0.250 inch (6.35 mm) OD for large component pads, and 0.400 inch (10.16 mm) OD for very large pads, such as I/O terminal points to be soldered with large stranded wire. If room is available, it is desirable to use donut pads as large as possible for prototype PCBs to minimize drilling problems, and to allow extensive soldering without foil delamination.

3. DIP patterns — Dual-In-Line Package IC pads can be laid down in artwork as many separate donut pads, but this approach is not worth the trouble. The DIP patterns provide very fast, accurate

placement of IC terminals. They should be purchased in the least expensive form, which is usually 18 leads per pattern. The patterns are then cut to form smaller DIP patterns, such as the common 8-, 14- and 16-pin packages. Larger IC patterns, such as those for 40-pin microprocessor chips, can be built up out of several smaller strips. Three DIP pad shapes are illustrated in Fig. 3-23: *large terminal area,* best for hand-drilling and soldering, *oval pad* for medium surface area that will allow traces to be routed between pads, and *small round pads* for very dense circuitry.

4. Edge connector patterns — These strip transfer patterns are available for all of the common spacings used in PCB edge receptacle connectors. Some of these patterns are fairly difficult to form by other methods on a 0.1 inch (2.54 mm) grid background, because they use odd spacings such as 0.125 inch (3.175 mm) and 0.156 inch (3.96 mm). The strip shown has a *plating bar* feature that allows the connector finger to be plated with nickel and gold before the board is cut to its final dimensions; the bar provides electrical continuity from the cathode connection to each finger during electroplating.
5. Letters and numbers — Character symbols are used for labels, identification, terminal definitions, and other information to be etched into the board. One typical use for characters in your PCBs is for the copyright message © *John Doe 1982,* which should be included in the artwork if you want to be assured of future copyright protection for your PCB designs. Character symbols are available as precut decals, or as dry transfer "rub-down" patterns. The most common characters used for 2× PCB artwork are 0.125 inch (3.175 mm) in height.

Many other transfer patterns are available for printed circuit artwork, to accommodate almost any lead patterns for any electronic component package. Another important group of artwork designs not pictured in Fig. 3-23 includes component outlines, terminal markings, and schematic/logic symbols used to form screen-printed patterns. An example of the use of these patterns in silk screen artwork is shown in Fig. 3-24. This is the information that will be screen-printed on the component side of a PCB to aid in assembly, component identification, terminal definition, and designation of test points. Generation of screen printing artwork will be discussed later in this chapter.

Tape — *Precision slit tape* is the standard medium used to form interconnection traces in printed circuit artwork. Standard tape is opaque black with a crepe finish. It is flexible, with a pressure-sensitive adhesive that minimizes creeping, lifting, or stretching after attachment to the

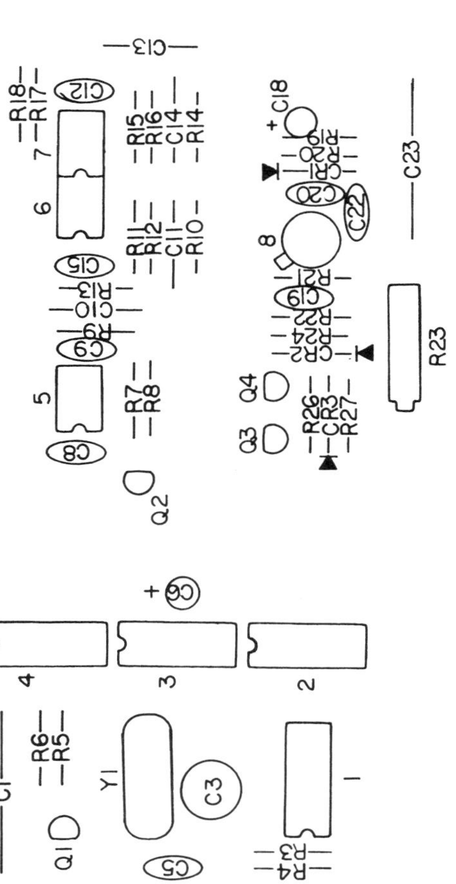

Fig. 3-24. Artwork patterns for screen printing. ***(Courtesy Bishop Graphics, Inc.)***

background film surface. Printed circuit tape is typically available in 20 yard rolls ranging from 0.015 inch (0.381 mm) to 2.000 inches (50.8 mm) in width. For preparing prototype artwork on a 2× scale, the following tape sizes are most useful: 0.031 inch (0.787 mm), 0.040 inch (1.016 mm), 0.050 inch (1.27 mm), 0.080 inch (2.032 mm), 0.100 inch (2.54 mm), 0.160 inch (4.064 mm), 0.200 inch (5.08 mm), and 0.250 inch (6.35 mm). The 0.050 inch (1.27 mm) tape will find the most use as a standard 2× line size in medium density prototype PCBs. The smaller 0.040 inch (1.016 mm) and 0.031 inch (0.787 mm) tapes are needed for dense line patterns, or for running a trace between two closely spaced IC pads. (The 0.031 inch (0.787 mm) tape gives a final line width of about 0.015 inch (0.381 mm), which is the smallest line that should be attempted in prototype work. Most commercial double-sided boards do not use line widths smaller than this.) The 0.080 inch (2.032 mm) tape is occasionally used for broad signal traces. Greater line widths — 0.100 inch (2.54 mm) to 0.250 inch (6.35 mm) — are reserved for power supply traces; the 0.100 inch (2.54 mm) size is standard for digital power and ground lines routed through the center of IC packages. Low-impedance power and ground distribution networks are often provided by routing wide 0.200 inch (5.08 mm) tape around the perimeter of a circuit board. Even wider tape — 0.500 inch (12.7 mm) to 1 inch (25.4 mm) — may be useful for forming ground planes or other wide expanses of unetched copper foil in selected areas of a board.

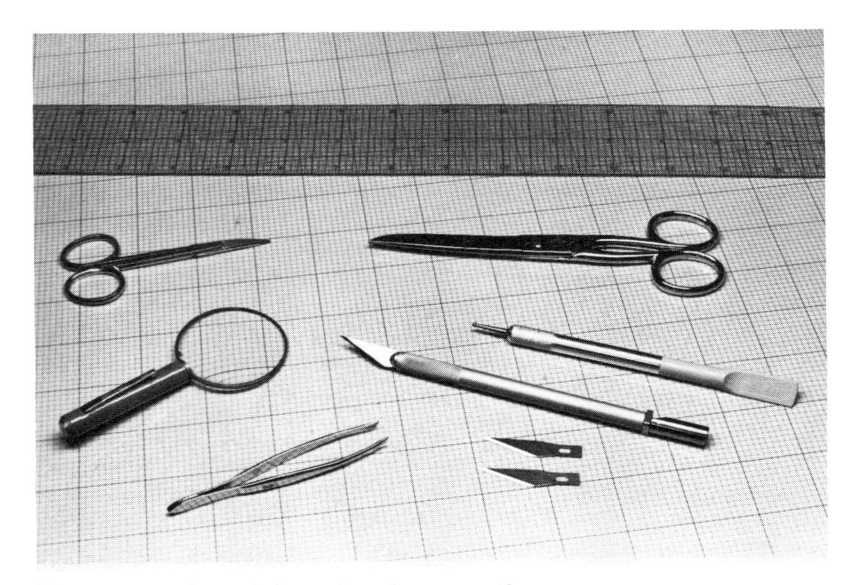

Fig. 3-25. Basic tools for artwork preparation.

Artwork Tools — Basic tools needed for artwork generation are shown in Fig. 3-25. These include X-Acto knives, a burnisher, tweezers, cuticle scissors, medium-size scissors, magnifying glass, and a long straightedge. Another useful tool is a transparent, precision scale for measuring distances. The clear, flexible polyester Accuscales from Bishop Graphics (Fig. 3-26) are ideal for checking dimensions and tolerances in printed circuit artwork, as well as in photographic negatives and masks; they are available in 18 inch, 24 inch, and 36 inch lengths, with divisions down to 0.005 inch for standard scales.

Tape-Up Techniques •

Base Pattern Layer — In the three-layer artwork approach a base pattern, or *pad master* layer, is prepared first and contains all patterns that are common to both sides of the board. These patterns include the PCB outline, terminal pads and holes, alignment targets, and photo reduction marks to guide the camera work. Patterns are applied in the following sequence:

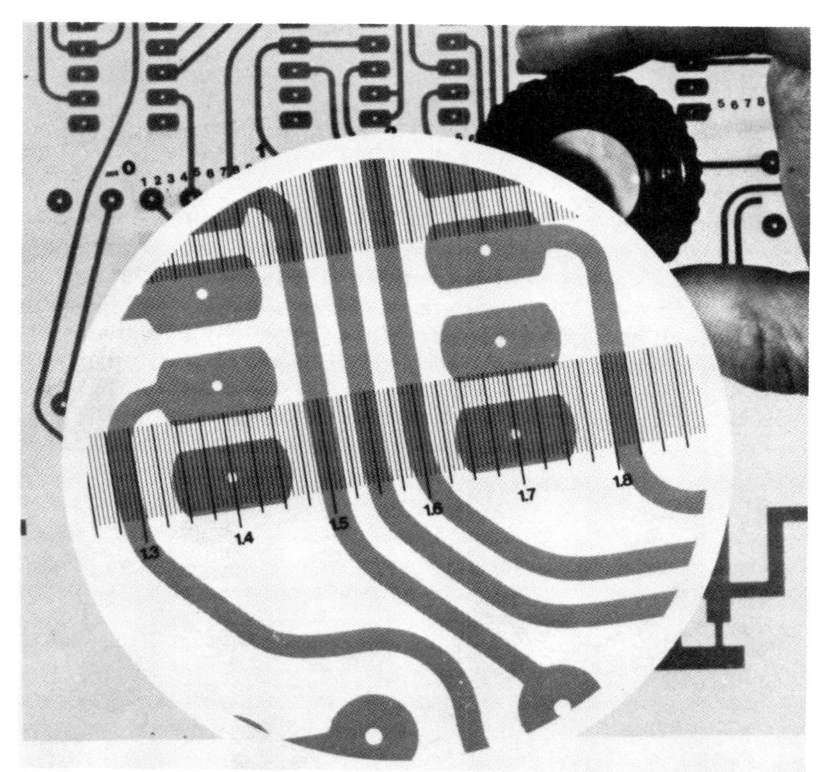

Fig. 3-26. Precision Accuscale for artwork examination. ***(Courtesy Bishop Graphics, Inc.)***

1. Board outline — The board outline should be established first with narrow 0.031 inch (0.787 mm) PC tape applied along the entire outside edges. This technique gives a complete reference perimeter for final board shaping following fabrication; the exact desired shape is obtained through complete removal of perimeter lines by rough shearing or sawing, followed by careful filing.
2. Targets — At least three widely spaced alignment targets should be positioned outside the PCB outline, and duplicated on subsequent artwork layers for purposes of pattern registration. These targets can also serve as photo reduction marks to precisely establish final photo transparency dimensions. For example, if two targets are positioned on grid intersections exactly 10.00 inches (254 mm) apart in the 2× artwork, the photographer can be given instructions to reduce this dimension to 5.00 inches (127 mm) in the final negatives.
3. Connector strips — Edge connector contacts and input/output pads are applied next. Individual donut pads and multiple prespaced patterns can be used to form almost any I/O terminal geometry. If edge connectors will be gold-plated, be sure to connect each contact to a *plating bar* that extends away from the artwork, and around the perimeter of the board. This makes it easy to connect a wire to the plating bar when the edge connector is dipped into a plating solution.
4. IC pads — While following the layout drawing for correct positioning, lay down all IC patterns and other multipad component shapes. The use of stick-on transfer patterns is highly recommended for ICs to save time and give accurate hole spacings. Stick-ons should first be cut to their proper length with scissors while still attached to the protective backing paper. The pattern is

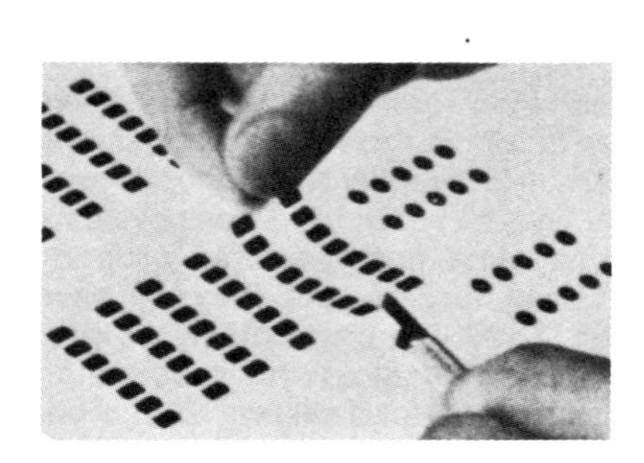

(A) DIP IC pads.

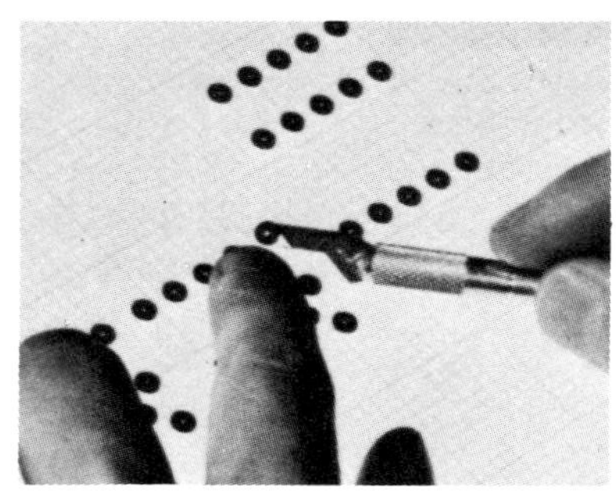

(B) Donut pads.

Fig. 3-27. Transferring artwork stick-on patterns. *(Courtesy Bishop Graphics, Inc.)*

then removed by inserting a knife blade at one end and gently peeling it away from the backing. Position the pattern over the artwork and grid pattern while still stuck to the knife blade, and when correctly lined up, push down one edge and lay down the remainder in a smooth motion while removing the blade (see Fig. 3-27A). Precise alignment is achieved by keeping grid intersections visually centered at each pad center hole. If transfer patterns are laid down incorrectly, it is usually possible to separate them from the clear taping film and reposition them without loss of adhesive quality. However, excess handling will eventually destroy the adhesive characteristic, particularly if fingers are moist or greasy.

Laying down multipad stick-on transfers is sometimes complicated by static charges which attract the patterns to artwork before they are correctly positioned. This problem can usually be avoided by turning the pattern over while it is still stuck to the knife blade, and gently rubbing it against the artwork surface to equalize charges.

5. High-density pads — For prototype work, it is desirable to keep IC terminal pads as large as possible to avoid tolerance and alignment problems during drilling. However, it is sometimes necessary to route interconnection traces *between* 0.1 inch (2.54 mm)-spaced IC pads, requiring the use of small area patterns. A good technique to follow during artwork preparation is to minimize the use of small pads by inserting them only *as needed* into larger IC patterns; this technique was illustrated previously in Fig. 3-12B. Using small cuticle scissors, the complete IC stick-on is carefully cut to leave blank spaces where small-area pads are needed. These blank spaces are then filled in with individual small pads in the base pattern artwork. If small pads are not needed on both sides of the board, a large-area pattern can be inserted in the appropriate artwork layer to *cover up* the small pad in the base layer. With this approach, the final PCB will be etched with a small pad on one side, and a large-area pad on the opposite side of the board; drilling should then start from the side of the small pad to minimize the chances of pad tear-up on drill exit.

6. Individual pads — Pads for discrete component terminals are applied next using individual pre-cut donut pads of the appropriate diameter. A good pad size for most small resistors, capacitors, diodes, and transistor leads is 0.187 inch (4.75 mm) OD for 2×-scale artwork. A donut pad is applied by slipping an X-Acto knife under its edge (see Fig. 3-27B), and lifting it from the backing sheet. The pad is then transferred to the artwork, taking care not to damage its round edges while pushing down firmly.

Terminal pads should have a center hole to aid in artwork alignment, and to serve as a drilling guide. After transferring all component lead pads, any special holes should be added to the artwork using various-size donut pads; these pads include board mounting holes, holes for hardware attachment, and through-holes. A good pad size for 2× through-holes is 0.156 inch (3.96 mm) OD when using No. 22 solid wire to fill holes, and 0.125 inch (3.175 mm) OD pads can be used to conserve even more space. The size of pads for establishing screw holes in a board is not important, since the main function of the pad is to position the initial drill bit; however, it is important to maintain proper clearance *around* these holes during subsequent line tape-up on other artwork layers.

Component-Side Overlay — The first overlay is begun by taping a new sheet of transparent polyester film over the base layer. The orientation of this new layer is established directly by transferring a set of alignment targets exactly superimposed on the base layer targets; when layers are separated, the targets will permit them to be re-positioned exactly as before. The main patterns to be added on this layer of artwork are ground planes, interconnection traces, and lettering. The component-side overlay forms a normal image, so all patterns are created as if they were being viewed from the top side of the PCB.

1. Ground planes — *Ground planes* are wide areas of unetched copper foil, usually surrounding pads on the component side of the board, with the purpose of improving high-frequency circuit performance by establishing a grounded, reflective metal surface on a plane directly beneath components. Many PCBs utilizing ground planes are essentially single-sided boards with all line

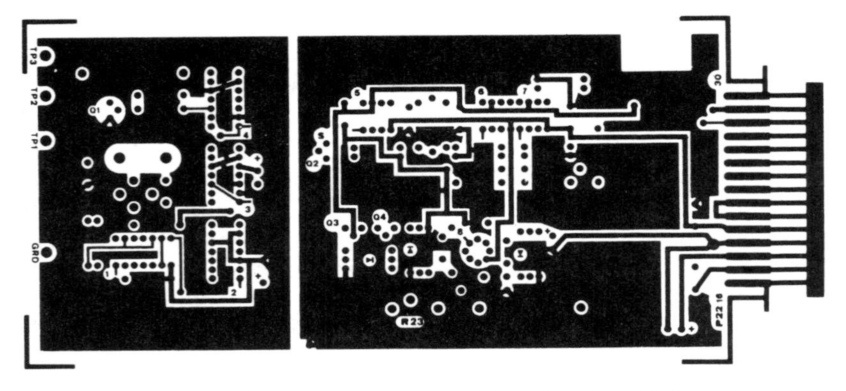

Fig. 3-28. Ground plane on component side. *(Courtesy Bishop Graphics, Inc.)*

traces on the reverse side. The component side is then used strictly to establish the ground plane, with only an occasional etched trace. A typical ground plane pattern is illustrated in Fig. 3-28. This artwork is generated with the help of special transfer patterns and wide black tape. Very large ground plane areas may be formed easier with the use of Rubylith knife-cut masking film.

2. Interconnections — Precision slit tape is used to lay down interconnecting lines of predetermined width in the artwork. These trace patterns are copied directly from the layout drawing, which contains red penciled lines to indicate component-side traces. The background grid pattern is helpful in laying down neat, evenly spaced taped lines. Four tape sizes are often sufficient to complete line artwork: 0.031 inch (0.787 mm) and 0.050 inch (1.27 mm) for signal traces, and 0.100 inch (2.54 mm) and 0.200 inch (5.08 mm) for power/ground lines. Good general practices in taping artwork patterns are illustrated in Fig. 3-29.

The use of slit tape is illustrated in Fig. 3-30. It is good practice to keep hands particularly clean while using slit tape, because grease on the film surface or tape will prevent good adherence. A few inches of tape are first unrolled and the end pressed down over the pad with which it connects. The tape is then routed to the next connecting pad, while running a finger over the trace to press it down permanently (Fig. 3-30A). (Tape should not be stretched, which causes creeping.) The final end is cut over its underlying pad by carefully snipping it to length with cuticle scissors, or by pressing it down and pulling the free roll of tape back across a firmly held knife edge (Fig. 3-30B). In any case, it is important that a sufficient amount of tape overlap each pad (Fig. 3-30C) to avoid problems when slight misalignment of artwork layers occurs during photography. Avoid the use of too much overlap also, which could cover up the center holes needed for drilling.

Changing directions with PC tape is a frequent necessity during routing, and this is accomplished with three basic methods: *smooth curves, angled cornering,* or the use of individual overlapping *angled elbow patterns.* This last approach is generally unnecessary for prototype artwork, but it is seen often in high-quality commercial artwork.

Slit tape made from black crepe material is very flexible and can be bent into smooth curves with little problem. Gentle tension is exerted on the tape with one hand, while forming the curve with the index finger of the other hand (see Fig. 3-30D). If tape is not stretched excessively, or if its bends are not too sharp,

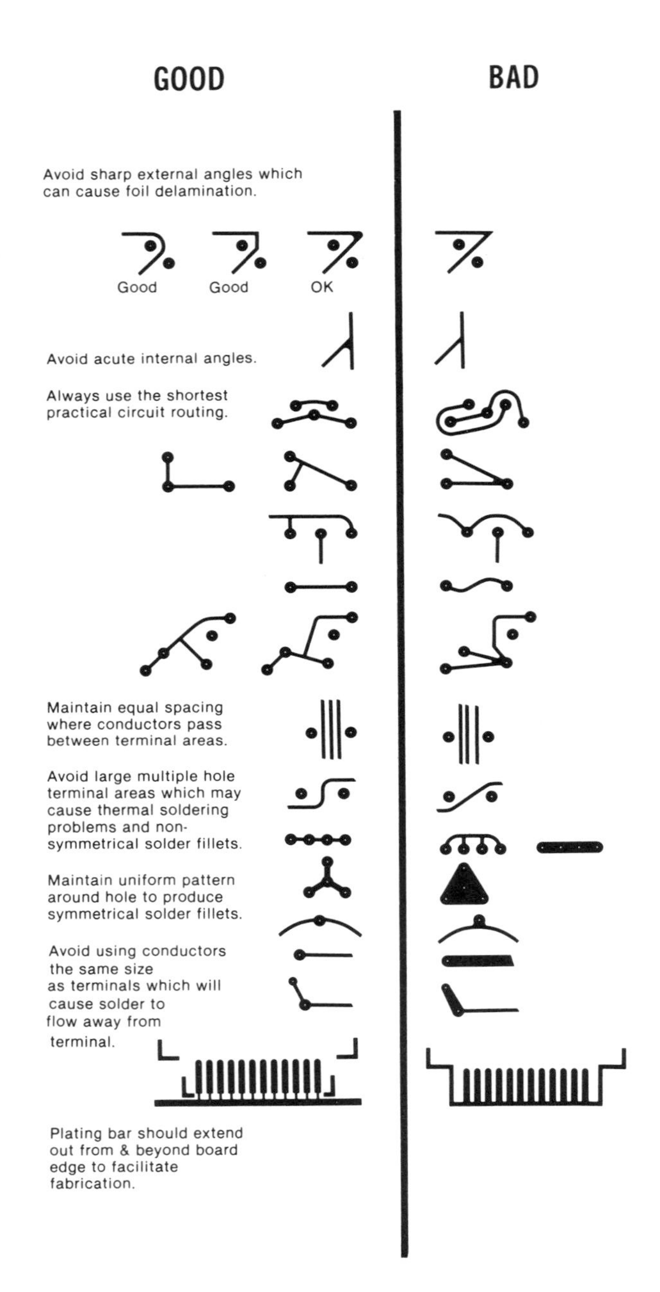

Fig. 3-29. General practices in taping artwork. ***(Courtesy Bishop Graphics, Inc.)***

(A) Routing a trace.

(B) Cutting tape end.

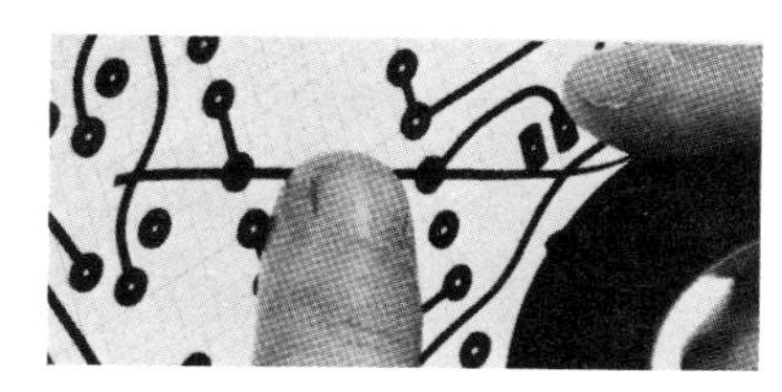

(C) Tape should overlap pad before cutting.

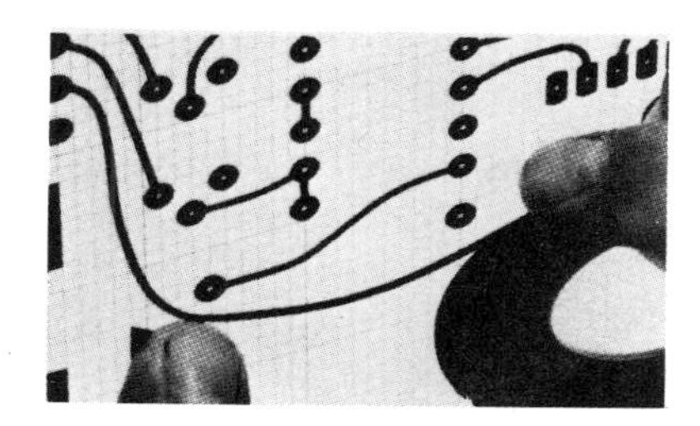

(D) Forming smooth curves.

Fig. 3-30. Using PC slit tape. ***(Courtesy Bishop Graphics, Inc.)***

tape has good adhesive characteristics and will hold its position without creeping.

The angled corner method for forming bends is more time-consuming, but it is necessary when sharp angles are needed; corners of any angle may be formed by overlapping tape and trimming the overhanging ends with a knife. However, angles greater than 90° should be avoided because this creates uneven tension in the final etched traces, which could lead to foil separation (delamination).

3. Lettering — After all other patterns have been completed in the artwork, letters and numbers can be added as space is available. The main requirement of etched lettering is that it be visible following PCB assembly, and that it not interfere with conductive patterns by *bridging* signals. Letters 0.125 inch (3.175 mm) in height are about the smallest symbols that can be accurately photographed and etched from 2× artwork. A variety of valuable information can be etched into the board using transfer characters: pin numbers, component polarity marks, test points, component labels, board identification, and copyright message. These symbols are usually found on the component side of the circuit board as a guide during assembly and troubleshooting; however, some information for multiterminal components and con-

nectors is also useful on the reverse side, such as location of IC pin No. 1. If enough information can be included in the etched patterns, there may be no need for screen printing similar symbols as a separate operation.

Reverse-Side Overlay — Following completion of the component-side overlay, it is removed from the artwork table and replaced with a new sheet of clear taping film for the reverse-side overlay. Reverse-side patterns are formed in the same manner as discussed for the component-side; the appropriate patterns are found in blue pencil in the line layout drawing. If any symbols are to be added to the reverse-side overlay, remember that patterns are being viewed as their *mirror images* during tape-up. Therefore, the overlay sheet must be turned over before applying dry transfer characters. To aid in their positioning, it is helpful to temporarily *underline* text locations with slit tape before turning the artwork over; alternatively, the base pattern can be turned over and positioned under the second overlay for a lettering guide.

Proofing • The unforgiving rigidity of the PCB system demands a thorough final inspection of the master artwork, even though some of this checking is also done following the completion of line layout.

Signal Connections — After all three layers of the artwork have been completed, the accuracy of the final line connections should be verified. This is done by overlaying all three artwork sheets and systematically comparing each connection in the schematic with signal traces in the artwork. A complete list of electronic components is helpful for this proofing operation, which proceeds sequentially through every pin on every component. A second list should also be prepared of *common signals* that occur at more than three separate terminals, because these connections tend to remain incomplete.

Geometry — The geometry of component lead patterns should also be verified following artwork preparation. It is important that packages can be inserted easily in their PCB holes, that hardware has correctly spaced mounting holes, that all parts have adequate clearance, and that lead patterns do not become confused. A common problem that occurs with multiple-lead components, such as transistors and ICs, is that the pad pattern is laid down correctly, but signal lines are taped as the *mirror image* of what is really desired.

Assembly Considerations — If leads cannot be soldered on the component side because of lack of access, the artwork should be checked to make sure that connecting traces originate from the reverse side of the board. This is an important principle in designing high-density prototype PCBs that lack plated-through-holes.

Artwork for Screen Printing • Screen-printed patterns and symbols in white ink are often found on the component side of PCBs to aid in assembly, component location, pin assignments, and general troubleshooting. Screen-printed information is handy, especially when other people will be involved with the assembly, testing, and troubleshooting of a new circuit board design. An example of silk-screen artwork was shown previously in Fig. 3-24. The ink that is used is typically solvent-resistant, nonconductive white epoxy, and Fig. 2-19 in Chapter 2 shows how it appears on a finished PCB.

The main consideration in preparing artwork for screen printing is to avoid printing over solder pads, and to ensure that symbols will be visible following PCB assembly. Just as in the case of master artwork, a variety of drafting aids are available to help with this preparation; popular transfer patterns include component outlines and electrical symbols. However, screen patterns are not functional, but rather informational in nature. If the final ink is readable, then the artwork serves its purpose. For this reason, templates, drafting pens, and lettering sets are satisfactory for preparing this artwork. The actual drawing is accomplished by laying a matte-finish polyester sheet over the base pattern artwork, which serves as a positioning guide for symbols and part outlines; the matte finish accepts ink well.

Solder Mask Artwork • Another optional item in prototype PCB artwork preparation is the *solder mask.* This pattern is usually screen-printed on the reverse side of high-density boards as an aid to prevent

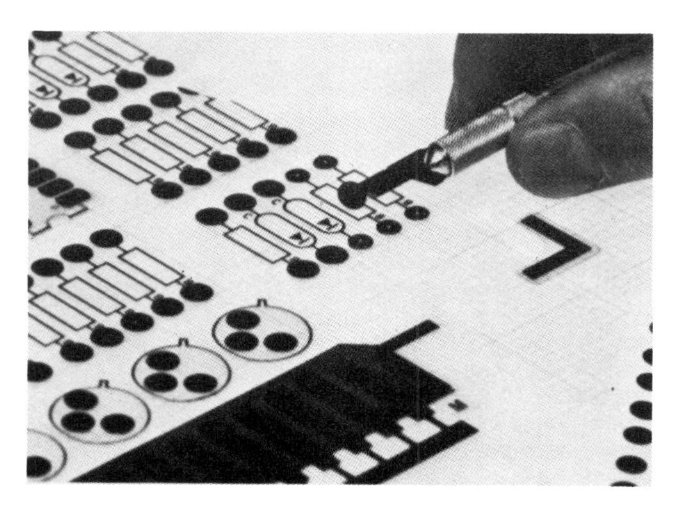

Fig. 3-31. Preparation of solder mask artwork. ***(Courtesy Bishop Graphics, Inc.)***

bridging of conductors and excess solder build-up during wave soldering. Prototype boards that are hand-soldered do not require a solder mask, but it is readily available from the three-layer artwork system. Since the solder mask pattern is usually equivalent to the base pattern image of *common pads,* the base pattern can be photographed separately as a solder mask for screen printing high-temperature solder-repellant epoxy ink. Some touch-up of the resultant photo mask may be necessary, such as removing all pad center holes to prevent a spot of ink in the center of each terminal area. Also, it may be desirable to protect some board areas that will not receive ink, such as edge connector patterns. In professional PCB design, the solder mask artwork is prepared separately with mask pad areas *slightly larger* than component pads, as shown in Fig. 3-31, to prevent even the slightest ink contamination of pad areas that are to receive solder.

CHAPTER 4
Photography

INTRODUCTION • The methods to be discussed in this chapter are a part of what is generally termed *graphic arts photography,* which involves the generation of *high-contrast* film images, as opposed to *common photography* which produces *continuous* tone images. Graphic arts film is used to obtain dense, opaque/transparent reproductions of original artwork, which could be a map, lettering for a sign, an engineering drawing, or a silk-screen pattern. The film is also known as *line film,* and the original artwork is also known as *line copy.* The end use for most line film is as a positive or negative transparency; if it is to be used in photomechanical reproduction processes, the transparency is also known as a *photo mask.* For more detailed treatment of this subject, the reader is referred to other books[12] and trade publications[13].

To produce an image transfer tool for photo resist or screen printing processes, the master artwork must be converted into a suitable transparency using photographic methods. Three approaches to printed circuit photography are feasible:

1. Make a simple *contact negative* from the transparent artwork. The film is placed underneath and in contact with the artwork, and is exposed to light and developed. Since the resultant negative image is exactly the same size as the original artwork patterns, no size reduction is possible and the original artwork must, therefore, be prepared on an exact 1:1 scale. Printed circuit artwork materials are available on a 1× scale, but it is difficult to work at this level. For best results, particularly with double-sided PCBs, enlarged artwork at two to four times the final size is recommended; this makes artwork easier to prepare, gives greater accuracy, and provides image sharpening through photographic reduction. Contact printing is mentioned because some printed circuit designers are willing to painstakingly prepare prototype artwork at the 1× scale, avoiding the need for camera work.
2. Leave the photographic work to a local graphic arts printer. The photographer will use a large *process camera* that is designed to

make precise photo reductions of high contrast artwork. If you understand the methods used, and can specify exactly what you need, good results are guaranteed because the commercial shop has the experience, equipment, and process controls needed to produce quality transparencies under standardized conditions of camera setup, lighting, exposure, and film processing.

3. Learn to photograph your own artwork with a *35 mm SLR camera,* and prepare transparencies using simple black-and-white darkroom techniques. Photo masks suitable for dense prototype PCB fabrication can be made with this approach, up to 8 by 10 inches in size. The hobbyist and experimenter with a limited budget will find the necessary graphic arts photographic techniques are easily learned and require little equipment outlay. This approach may also appeal to the experienced prototyper because it offers him or her a dependable route for *fast* conversion of artwork into working photo masks; outside professional service may take a few days.

OBJECTIVES • The objectives of this chapter are as follows:

- Discuss the operation of commercial process cameras, showing the variables involved and what results can be obtained.
- Outline a complete approach to artwork photo reduction using 35 mm photography and simple darkroom processing.
- Detail the exposure and development procedures for high-contrast graphic arts film.
- Detail the use of an enlarger for printing accurate, properly registered transparencies for fabrication of double-sided PCBs.
- Discuss contact printing techniques for combining images and for the interconversion of positive and negative transparencies.

PROFESSIONAL GRAPHIC ARTS PHOTOGRAPHY • Large commercial cameras are available to produce transparencies directly from expanded scale artwork. The simplest version of this equipment is known as a *copy camera,* which has limited features — manual adjustments, small film size, and inexpensive lens. More refined equipment is referred to as a *process camera* that allows larger film size, greater reduction capability, quality optical system, and extremely precise adjustments for focusing and reduction. When printed circuit artwork is taken to an outside photographer for preparation of transparencies, the photographer will probably use a large process camera to get accurate size control and uniform, fast results. The exposures taken with some expensive commercial cameras have tolerances within ±0.001 inch (0.025 mm) across the entire film sheet, have excellent line resolution, and show extreme contrast with very few imperfections, such as pinholes in dark areas. The photographer will probably also use an automatic film processor to minimize variations in development, washing, and drying. Architects, engineering organizations, and color printers are the most likely groups to have process cameras suitable for printed circuit work. The following section describes the type of equipment that will be used when artwork is taken to a professional graphic arts photographer; this knowledge should help you specify the results you want, and it should help you deal with photographers who might be unfamiliar with the particulars of printed circuit graphics.

The Process Camera • A sophisticated process camera was shown previously in Chapter 2, Fig. 2-6. Typically, such a camera can make reductions down to one-eighth of the original size, or enlarge up to 2×. The camera back can accept film sheets up to 20 inches (508 mm) by 24 inches (609.6 mm) in size. The copyboard will hold artwork up to 40 inches (1.016 cm) by 48 inches (1.219 cm) in size, and includes a vacuum frame to give positive copy hold-down. The copyboard moves on a sturdy track to adjust the distance from lens to copy, which determines reduction ratio for a particular optical system; the lens is mounted on a large bellows assembly for focusing and aperture adjustment (see Fig. 4-1). The camera elements are precisely aligned so that the copyboard and film are parallel, and both perpendicular to the axis of the lens.

Fig. 4-1. Lens and bellows assembly on a process camera. ***(Courtesy David Leonard Associates PC, Kingsport, TN)***

The entire camera construction must be sturdy and vibration-free throughout to permit long timed exposures. Movable parts such as the copyboard and the lens bellows assembly must have secure locking devices. Other support equipment, such as vacuum pumps and air conditioners, are located at a distance from the camera and may require special vibration-free mounting supports.

Most large process cameras are installed through a wall so that the lens, copyboard, and lights are located in one room, and the camera back with a full set of controls is located in the next room. This setup allows safelight conditions to be maintained in the back room, where film is handled and mounted in the camera; the back room can thus be used as a darkroom, while the front room with the copyboard is available continuously lighted.

The Camera Lens • The lens system requires that no distortion or loss of focus occurs at any point on the film plate. Color correction must also be designed into the lens if it will be used for color photography, since different wavelengths of light have different refractive indices in the lens material. The technical terms for these problems that affect lens quality are *spherical* and *chromatic aberrations.* In most process lenses, the effects of *field curvature* or *astigmatism* can be minimized by closing down the diaphragm aperture, but diffraction effects become more pronounced at very small aperture settings, which will eventually decrease image sharpness.

Most process lenses do have built-in compromise corrections for color work, since one of the main applications of graphic arts cameras is in color separation photography for color printing. Similar color separation capability is needed for double-sided printed circuit artwork prepared by the red and blue tape method discussed previously in Chapter 3. The photographer uses red and blue filters to separate the combined artwork patterns into their red and blue components, which represent opposite sides of a PC board. To check a lens for color correction, the following simple test may be performed:

Place different-colored transparent patterns (such as yellow, red, blue, and green photographic filters) on the copyboard, and run strips of narrow black printed circuit tape across their faces. Backlight this test pattern and observe it on the ground glass focusing surface of the camera. If the black lines are straight, then the lens is suitable for color separation work, but if the lines are wavy or discontinuous, color correction is not built into the lens system.

Lighting • Proper lighting is extremely important to obtain sharp images and uniform exposures. *Front* lighting usually consists of two or four lamps attached to the copyboard so that a specific illumination geometry is maintained as the copy is moved relative to the camera. *Rear* lighting is also standard with transparent copy, because this provides the best contrast between clear and opaque areas of the artwork. It is estimated that backlighting can increase contrast by a factor of 20 as compared to front lighting alone. A typical copyboard with vacuum hold-down and backlighting capability is shown in Fig. 4-2.

The four important factors in lighting are uniformity of illumination, intensity, lack of reflected glare, and color characteristics:

1. Uniformity — Lighting becomes more even as the distance between light sources and copy increases. Uniform lighting is needed to maintain the same exposure index across the entire copy face.

2. Intensity — To avoid long exposure times, high-intensity light sources are used such as pulsed-xenon, white-flame arc, quartz-iodine, or photoflood lamps. Banks of fluorescent tubes are also common. Graphic arts film is low-speed and excessive exposure times tend to decrease image sharpness through vibration effects.

3. Reflection — Reflective glare is avoided by keeping the angle between front lights and copyboard at not more than 45°. Lamps should also be mounted with proper reflectors to give wide, glare-free lighting.

4. Color — Color separation work requires white light. Since the light source directly affects filters and filter factors, some commercial lighting equipment has color-adjustable output. However, for

Fig. 4-2. Vacuum copyboard with fluorescent lighting. ***(Courtesy David Leonard Associates PC, Kingsport, TN)***

black-and-white photography, optimum results are obtained with a *single color* light source; this minimizes chromatic aberration in the lens, and simplifies focusing. Green light has the best overall properties because it occurs in the middle of the visible spectrum and makes best use of the compromises built into the lens system. Most of the photographic film to be used for exposure is *orthochromatic,* or sensitive to blue-green light. Green light is less scattered than blue light in the photographic emulsion containing silver halide particles, so green is preferred and leads to sharper images. Another characteristic of green light is that the human eye is most sensitive to this wavelength; during focusing, a green image by visual perception is most likely to be in true photographic focus. Therefore, copyboards for orthochromatic photography are commonly illuminated with many rows of green fluorescent tubes; an alternative is the use of a green filter material between the lamp sources and the copy, front and back.

Reduction, Focusing, and Exposure • Process cameras are equipped with a ground glass plate on the camera back, which allows the photographer to observe the final image on the film plane (see Fig. 4-3). Most cameras also have calibrated devices to determine the proper positioning of lens and copyboard for a particular reduction ratio. However, final precise adjustments for focus and image size are usually made by checking the ground glass plate, often with the aid of a magnifier and a precision scale. Focusing should be done with the

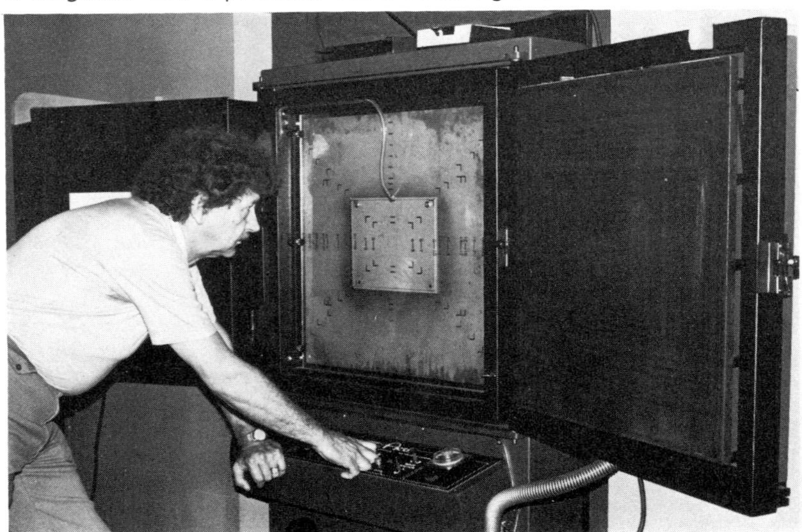

Fig. 4-3. Examining the ground glass focusing plate. ***(Courtesy M. J. Osborne and David Leonard Associates PC, Kingsport, TN)***

lens diaphragm pre-adjusted to the desired aperture size for proper film exposure. Test pattern charts are available to be mounted next to the copy as an aid in focusing. If reduction targets are provided in the original artwork, this makes the photographer's adjustments much easier. The ground glass allows complete observation of composition, uniformity of lighting, focus, and dimensional accuracy. Once these variables are satisfactory, film is inserted in the camera back and an exposure is made.

Exposure level is controlled by adjusting the shutter speed and lens aperture under given lighting conditions. Although there is a great deal of exposure latitude for graphic arts film, the current trend is towards the use of automated film processors with standardized time-temperature variables; this approach demands that exposure level also be standardized. An experienced photographer using a good process camera will use calibrated controls with his considerable background knowledge to obtain correct film exposure at almost any lighting and camera reduction conditions. Control scales, also known as photographic step tablets, can be mounted next to the artwork, and test exposures using these patterns can be used to standardize exposure level and processing conditions. Results are evaluated with the objectives of obtaining dense black areas, clear transparent areas, sharp line edges, correct size and shape of patterns, and minimal amount of pinholes in dark areas.

Film • High contrast grahic arts film is used to make transparencies for photofabrication. Film is usually of the conventional silver-halide type, and must be capable of reproducing very fine detail. Both positive and negative-acting films are available. Another requirement of film is that it must be on a stable base material that does not distort, contract, or expand excessively with environmental changes; polyester is therefore the most commonly used substrate material. Glass plates are also occasionally used because of their high stability, but their fragility and lack of flexibility make them generally unsuitable for photofabrication of PCBs.

Commonly used films for printed circuit work are the Kodak 0.004 inch (0.102 mm) Precision Line films LP4 (negative-acting) and LD4 (positive-acting), which are also available in 0.007 inch (0.178 mm) base material as the LP7 and LD7 versions. These films are on Kodak's Estar base. Material may be purchased in rolls or in individual sheets, up to 30 by 40 inches (0.76 by 1.02 meters) in size. The thicker base material is preferred as a final photo mask because of its durability. The Kodak LP-series films can be exposed directly to the emulsion, or indirectly through the base plastic layer to obtain the desired image-to-film geometry. Focusing corrections may be required depending on which way the film is turned in the camera. Precision Line film is also suitable for making contact reproductions. Mechanized processing is recommended by the manufacturer.

Another type of photographic film coming into common graphic arts use for printed circuits is the molecular photopolymer-based diazo imaging film, which is developed in a dry process using ammonia vapors. The final transparencies are roughly equivalent in properties to silver halide film, but the diazo film is available with a distinct difference: opaque areas can be a transparent dark red in color, which allows much easier alignment of photo masks on printed circuit laminates during image transfer. Both positive and negative-acting diazo films are available.

Film Image Geometry • When submitting printed circuit artwork to a photographer for generation of transparencies, you must be very careful to clearly specify final image orientation in the emulsion. The correct orientation is particularly important when using thick 0.007 inch (0.178 mm) masks, whose emulsion side should always face downwards during contact printing onto photo resist. The accepted terminology for emulsion geometry is as follows:

Images are compared to the original artwork while holding up the transparency (negative or positive) with its emulsion (dull side) facing *away* from you; in other words, you are looking through the base of the film material. If the image you see is identical to the artwork as it was photographed, then the transparency is said to be *right-reading*; if the image is reversed (mirror image) then the transparency is said to be *wrong-reading*. It is recommended that these terms be discussed with the photographer before work is begun so that both parties will understand exactly what is being specified. Negatives for double-sided PCBs should be right-reading for the component side, and wrong-reading for the reverse side; this assumes that the original artwork was all prepared as viewed from the component side of the board.

Contact Printing • Contact printing is used to make direct 1:1 reproductions of transparencies. If negative-acting film is used for the exposure, then contact printing will generate a positive from a negative, and vice-versa; if positive-acting film is used, the reproduction will match the original with respect to opaque and transparent areas. The general purposes for contact printing are as follows:

1. Interconvert positives and negatives. For example, a positive transparency for screen-printing may be prepared using the negative obtained from the process camera. A positive paper print may also be prepared for project documentation.
2. Generate duplicate transparencies. The master transparency is contact-printed with positive-acting film for this purpose.
3. Change image orientation. By exposing the contact film directly

to its emulsion or indirectly through its base, a wrong-reading or right-reading copy can be obtained.

4. Convert original multilayer transparent artwork into photofabrication transparencies. This operation is limited to 1:1 scale artwork because no reduction or enlargement is possible with contact printing.

Professional equipment for contact printing consists of a vacuum frame to hold materials in tight contact, and a point source of light for exposure. A typical system is illustrated in Fig. 4-4. The light should have bright, variable intensity with an automatic timer. It is desirable to keep the light source and copy holder as far apart as possible to give even illumination over the entire area to be exposed.

Processing • Mechanized equipment is used to automate and control processing variables for large sheets of graphic arts film. An example is the compact processor shown in Fig. 4-5. Although film can be developed in large open trays, the usual variables of time, temperature, developer strength/activity, and agitation are more carefully maintained with an automated processor. Warm, forced air and squeegee rollers eliminate the problems of streaks, uneven drying, and film distortion. During processing, filtered solutions are continuously circulated across the face of the film emulsion to sweep away dust particles that

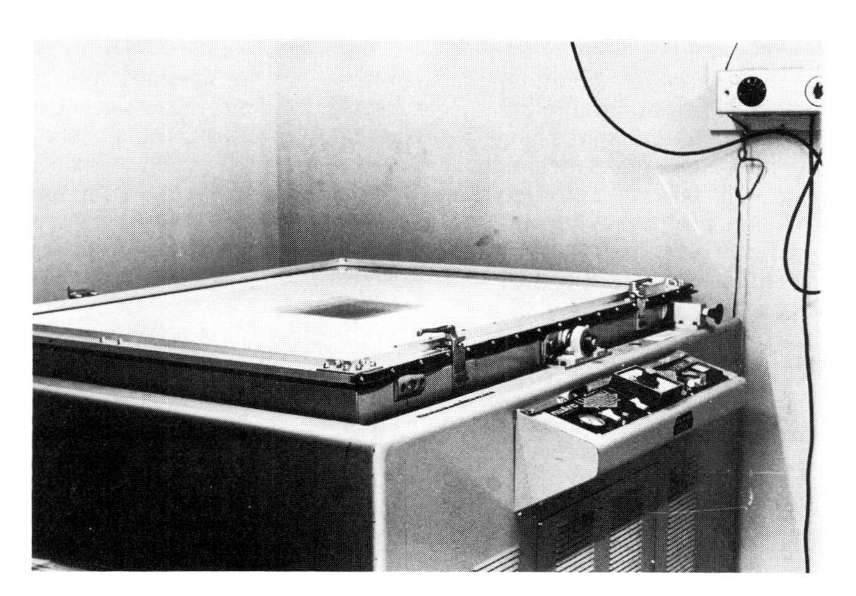

Fig. 4-4. Professional contact printing equipment. *(Courtesy David Leonard Associates PC, Kingsport, TN)*

Fig. 4-5. Automated film processor. ***(Courtesy M. J.Osborne and David Leonard Associates PC, Kingsport, TN)***

can lead to pinholes and scratches. When they occur, these imperfections require extensive touch-up; fine patterns may be difficult to correct by hand brushing opaque on final transparencies.

It should be mentioned that airborne dust is another major cause of pinhole imperfections in transparencies, and professional photographic installations require special efforts to avoid dust which can settle on artwork, in the camera, and on unexposed film. The air conditioning system filters incoming air and establishes a slight positive pressure to keep dust out. Floors are smooth-surface to allow complete cleaning and the entire photographic area is vacuumed frequently. Dust gratings and dust-attractive mats are located at each entrance to catch dust from foot traffic.

35 mm PHOTOGRAPHIC REDUCTION PROCESS • The following steps summarize a complete approach to obtaining negative or positive transparencies from black-and-white artwork, using 35 mm photography and amateur darkroom equipment. The method is fundamentally limited by the camera frame size, which is 35 mm by 24 mm (approximately 1 inch by 1.5 inches), and by the film's resolution. As larger and larger artwork is compressed into the 35 mm frame, line edges begin to lose sharpness, and small images, such as the center dots in pads, may be blurred or lost. As was stated earlier, the largest PCB that can be satisfactorily etched with negatives prepared from 35 mm film is about 8 inches by 10 inches in size. The 35 mm camera itself may also introduce process limitations if it is not focused accurately, or if the lens causes image distortion.

The basic 35 mm procedure for double-sided PCBs is as follows:

1. Photograph each of the three artwork layers separately against a white background, or backlighted if possible.
2. Develop film and verify sharpness and contrast of all three negatives.
3. Enlarge and print the base pattern onto sheet film at proper final dimensions.
4. Enlarge the component-side pattern so that it superimposes exactly on the base pattern transparency, and print it on a separate piece of sheet film.
5. Enlarge and print the reverse-side pattern in a similar manner.
6. Combine the base and component-side patterns, and make a contact negative transparency.
7. Combine the base and reverse-side patterns, and make a contact negative transparency.

This approach offers considerable control over the dimensions and quality of final negatives because of the following factors:

1. Careful dimensional measurements are made under the enlarger before the base pattern is printed.
2. Touch-up corrections can be made at the intermediate positive steps.
3. Exact side-to-side registration is maintained because the same base pattern is contact-printed for both final negatives.

35 mm PHOTOGRAPHY •

Taking Pictures • The three basic requirements for taking pictures are a suitable camera, a method for mounting artwork, and a source of even light.

Camera — A 35 mm camera with the single lens reflex (SLR) focusing system is necessary to obtain good photographs. The author uses a Pentax K-1000 camera with a 100 mm lens and all manual features. The important characteristics of the camera include a distortion-free lens, the ability to focus precisely through the lens, an aperture that stops down to f/16 or f/22, and the ability to make timed exposures. The camera should have standard base mounting threads so that it can be fixed to a tripod or copy stand. Some useful accessories are a locking cable release, which prevents camera movement during timed exposures, and closeup lenses that permit close focusing. The simplest closeup lenses screw over the front of the master lens like filters; they are needed for photographing small-size printed circuit artwork, such as the patterns routinely published in books and magazines.

Mounting Artwork — A good frontlighted system for mounting artwork is a copy stand, such as the one shown in Fig. 4-6. This apparatus allows copy to lie flat on its base while the camera is attached to a movable arm, permitting easy adjustment of vertical distance, and accurate parallel alignment of lens and artwork. Some stands come equipped with flood lamps for even illumination of artwork. When choosing a copy stand from your photographic distributor, be sure that the base is wide enough (typically 18 inches by 24 inches or 4.57 by 6.10 cm) to accommodate the largest size artwork you expect to generate. The vertical shaft should also be long enough to keep the entire image in the viewfinder of the camera. A useful accessory is a table clamp that allows the stand to be mounted on any table top; however, it is usually most convenient to place the copy stand on the floor with its own base, which makes the camera accessible for adjustments.

An alternative frontlighted mounting system is to tape the artwork to a white background against a wall and mount the camera on a large, sturdy tripod; this setup is illustrated in Fig. 4-7. Although it is somewhat more difficult to adjust the camera lens parallel with the artwork, this arrangement allows photography of very large copy, such as

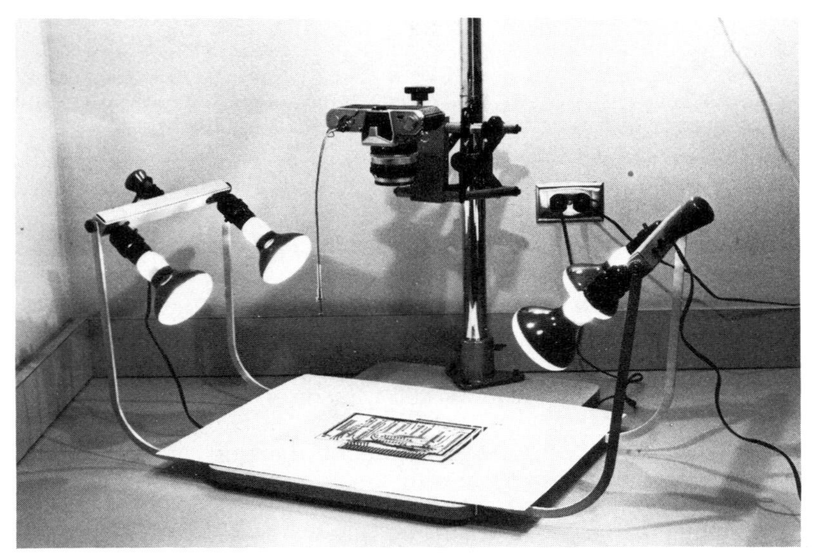

Fig. 4-6. A copy stand for 35 mm graphic arts photography.

printed circuit artwork prepared on a 4× scale. Vertical wall-mounting of artwork may require some transparent tape at strategic spots across the back of the sheet to keep copy flat against the wall without sagging or bowing.

For photography of backlighted artwork, some sort of transparent frame is necessary to sandwich the copy between two pieces of glass. The light source is then positioned behind the frame. To obtain even, diffuse lighting across the entire image area, a sheet of semitransparent paper can be taped to the back glass plate, or white glass can be used.

Whatever mounting system is chosen, be sure to photograph each sheet of multilayer artwork in the *same position* relative to the camera viewfinder. Do not move the camera once photograhy begins. Some light pencil marks on the background sheet can serve as reference marks for mounting artwork layers in identical location during each exposure; this practice will ensure good side-to-side registration of double-sided PCB patterns, even when the camera lens introduces some overall image distortion, because the distortion will be equivalent for all artwork layers.

Lighting — It is important that the artwork be evenly illuminated during photography, so that exposure is uniform across the entire negative surface. Bright light is also desirable because the high-resolution graphic arts film is very low speed (typically ASA 8) and exposure times can become very long. Two 500-watt reflector-type photolamps at 4–5 feet from the center of the artwork should be adequate for most work.

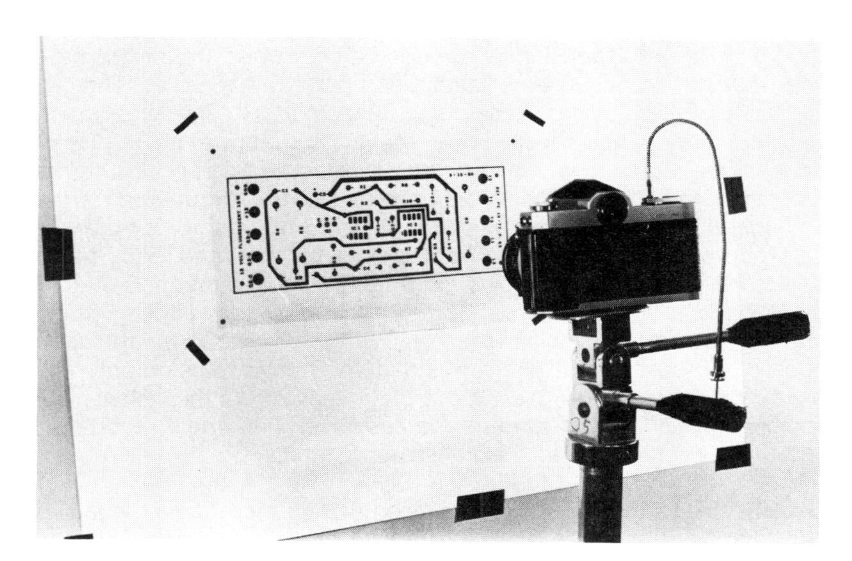

Fig. 4-7. Vertical wall-mounted artwork for tripod photography.

The author has used four standard 100-watt tungsten bulbs successfully with the wall and tripod mounting system, positioning the bulbs on two lamp stands located on each side of the artwork at a 45° angle, 6 feet from the wall. However, the exposure times were typically 90 to 120 seconds at f/22. Time is no problem if the tripod is sturdy and no camera vibration occurs during exposure.

Another approach to obtaining uniform lighting is to use a *single* source of light, but move it around during the timed exposure so that the net effect is an even illumination of the artwork. The author has used a single 100-watt table lamp to achieve proper exposure in 40 seconds; the bulb was held 2 feet away from the artwork for 10 seconds at each corner. During long exposures, the camera lens may be blocked for an instant while moving the light; however, the lens should never see the bare light bulb directly, only reflected light.

Glare is a frequent lighting problem which must be completely avoided to prevent dark blotches in negatives. The black tape and transfer patterns used as printed circuit drafting aids seldom produce glare, but the clear polyester backing sheet may be a source of problems. For best photographic results, use clear taping film with a *translucent*, *matte* finish to avoid glare. If shiny Mylar film is used, pay special attention to glare patterns and reflections while focusing the camera and positioning lights. Make adjustments to completely eliminate these effects. Artwork should be smooth and flat against its backing surface to avoid ripples that cause glare, as well as background shadows, image distortion, and uneven focusing.

Lighting is usually simplest with a copy stand that comes equipped with four clamp-on lamps that can be tilted to various angles, as was shown in Fig. 4-6. Table lamps can also be positioned around the base of a copy stand until an even illumination pattern is achieved. The fact that artwork lies flat on a copy stand helps matters considerably.

Backlighting is as effective in 35 mm photography as it is in professional process camera work. Increasing contrast by a factor of 20 times can eliminate fuzzy edges and loss of resolution that occur when larger and finer PCB patterns are crowded into the 35 mm frame. For the very best photographic results, artwork should be mounted on a transparent background that allows extra lights to be positioned directly behind and at some distance from the copy. A free-standing frame made from wood and glass can be used in combination with tripod photography. In the case of a copy stand, the stand can be mounted on a table or workbench with a cut-out top, over which a thick glass plate has been placed. Lights can then be positioned underneath the table, and the artwork taped to the transparent surface.

Graphic Arts Film • The film recommended for 35 mm negatives and final transparencies is *Kodalith ortho film, type 3* from Eastman Kodak Company. The 35 mm *6556* film is on a stable 0.0053 inch

(0.135 mm) acetate base, and has the dense, high contrast characteristics needed to prepare photofabrication masks having entirely transparent/opaque patterns. Graphic arts film is available in many sizes; for our purposes, a 100-foot roll of 35 mm film and a box of 8 by 10 inch (203 by 254 mm) sheets are needed. The recommended sheet film is *2556,* a stable 0.004 inch (0.102 mm) Kodalith ortho film on Estar base that can be exposed from either side. This film is low-speed (ASA 8) and orthochromatic, which means that it responds to light in the blue-green end of the visible spectrum. It is fairly insensitive to red light. This also means that the film can be easily handled and processed under red or yellow safelight illumination. A small, 10-watt red light bulb from your local hardware store will serve as adequate lighting for loading reusable 35 mm film cartridges in a darkened room; cut off the length of film needed, tape it to the reel, wind it, and seal the cartridge. Equipment for winding film under daylight conditions is also available.

Exposure • The camera is focused and the aperture stopped down to its smallest opening (typically f/16) to obtain a sharp image. Unless conditions have been previously well-established, a wide range of timed exposures should be made with the 35 mm camera. Even when you have a good feel for proper exposure time, each shot should be *bracketed* to allow for variations in artwork size and processing variables. For example, if optimum exposure time is estimated to be 20 seconds, take five shots at 10, 15, 20, 25, and 30 seconds. This guarantees that at least one negative will result with the proper contrast. Following film development, examine the negatives on a light table with a magnifying glass, and look for image detail, sharp line edges, and density; proper density should make all negative areas either clearly transparent, or completely opaque. Some examples of good and bad exposures are shown in Fig. 4-8.

35 mm Film Processing • A booklet[14] listed in Appendix A is recommended as a guide to developing, printing, and general darkroom work with black-and-white photography. Following exposure, film should be processed according to directions in the original package. Processing of Kodalith film is basically the same as for most black-and-white films, except that a special developer is used. The following equipment and supplies are needed to process 35 mm film:

- Darkroom with red or yellow safelight, sink with running water
- Developing tank and reel
- Timer
- Thermometer
- 600 mL graduated plastic beakers
- Chemicals: Kodalith developer parts A & B, Kodak indicator stop bath, Kodak fixer, and Kodak Photo-Flo solution

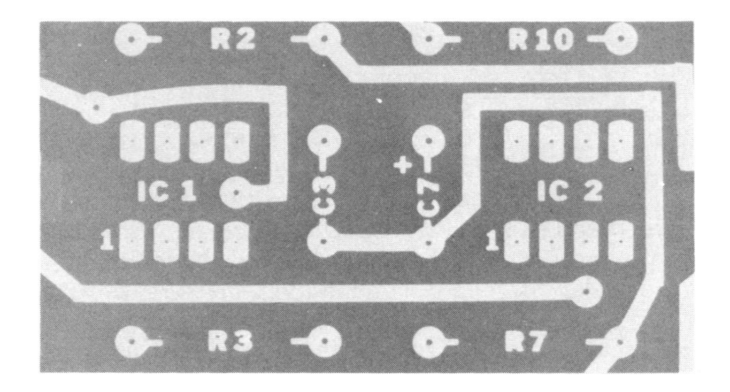

(A) Underexposed 35 mm negative on Kodalith film.

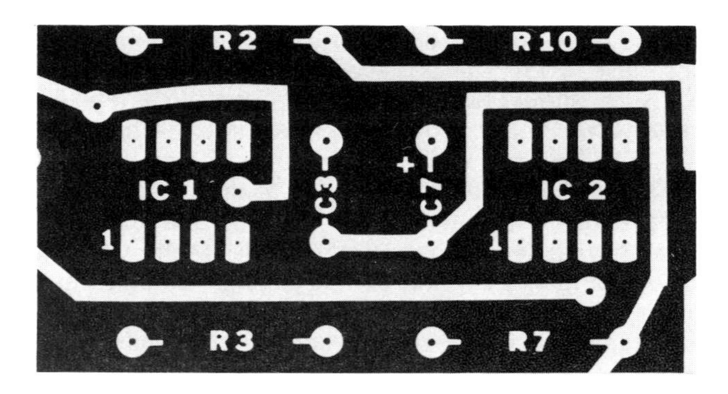

(B) Properly exposed 35 mm negative on Kodalith film.

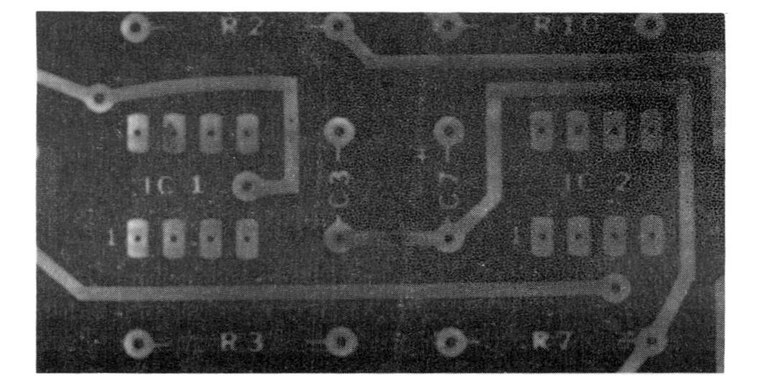

(C) Overexposed 35 mm negative on Kodalith film.
Fig. 4-8. Examples of good and bad exposures.

A roll of 35 mm film can be processed and dried in about one hour. The developing operation is straightforward because Kodalith film does not have to be handled in total darkness; only a few minutes of safe lighting are actually required while transferring the film onto the reel and into the lightproof developing tank (see Fig. 4-9). This operation can be done in a closet with a red light, or it can be done inside a black plastic glove bag available from photographic dealers. Additional items that you may need, such as sponges, scissors, masking tape, waste bucket, etc., are not listed here because they are normal household items.

Photographic chemicals should be premixed according to their package instructions. This normally requires a hotplate, large metal mixing can, thermometer, funnel, and plastic gallon jugs for solution storage. The Kodalith developer is prepared as A and B solutions that are combined just prior to use, since the mixture is unstable and is deteriorated after standing six hours. Other solutions are stable and can be reused many times before discarding. About 200—400 mL of each solution is needed for processing one 35 mm reel of film.

Although no more hazardous than many household items, photographic chemicals should be handled with care. Plastic gloves are recommended, because the mixtures do contain acids, bases, and other materials that require precautions. The most common handling problem is skin irritation or a rash that occurs after contact, so wash your hands frequently and clean up any spills that occur. To prevent

Fig 4-9. Reel and tank for developing 35 mm film.

contamination of food, a laundry room is more desirable than the kitchen as an area to develop film.

After film has been processed and air-dried, it should be cut up and stored in protective plastic envelopes to prevent damage due to scratching of the thin emulsion layer. A recommended source of plastic envelopes for safe storage of film, prints, and transparencies is 20th Century Plastics, Inc. (see Appendix B).

PREPARATION OF TRANSPARENCIES • Transparencies are prepared using standard enlarger printing techniques, but Kodalith film in sheets is used in place of photographic paper.

Darkroom Setup • A darkroom area must be established to make prints with an enlarger. Because of the need for running water and a sink, most amateur darkrooms are located near a bathroom, laundry, or kitchen area. Permanent installations are ideal, but you can keep your equipment in a closet and set it up in an appropriate sink area when needed. A typical darkroom plan is shown in Fig. 4-10. A layout such as this one helps the operation because wet and dry areas are separated, and every item has its place so that things can be readily located in dim light. A safelight should be installed above the workplace for situations that require controlled lighting. If the room is not lightproof, it may be necessary to do your processing at night. However, Kodalith film is not very light sensitive; adequate protection may be obtained by blocking major light leaks with paper, masking tape, and towels. The following equipment is needed for a basic black-and-white darkroom:

1. Sink with running water
2. Black-and-white enlarger with timer
3. Safelight — a 15-watt lamp fixture with a red or amber filter; for example, a Kodak safelight filter No. 1A, light red
4. Four 11 by 14 inch (279 by 356 mm) plastic trays for processing
5. One large tray for washing
6. 1-quart measuring container
7. 11 by 14 inch (279 by 356 mm) glass plate for making contact prints, or a contact frame
8. Light-tight film and paper storage box (optional)
9. Paper cutter
10. Miscellaneous: thermometer, plastic tongs, clips for hanging film to dry, soft brush for cleaning negatives, ruler, cloth, scissors, paper, etc.

Enlargement • The enlarger resembles a slide projector mounted on a vertical stand; by moving the unit up and down, the size of the projected image is adjusted at the base. The black-and-white enlarger is the most expensive piece of darkroom equipment needed, a typical model costing about $120. It consists of a flat base, an adjustable stand which allows the lens assembly to move up and down, a light source, a lens with a diaphragm aperture and focusing adjustment, a film carrier

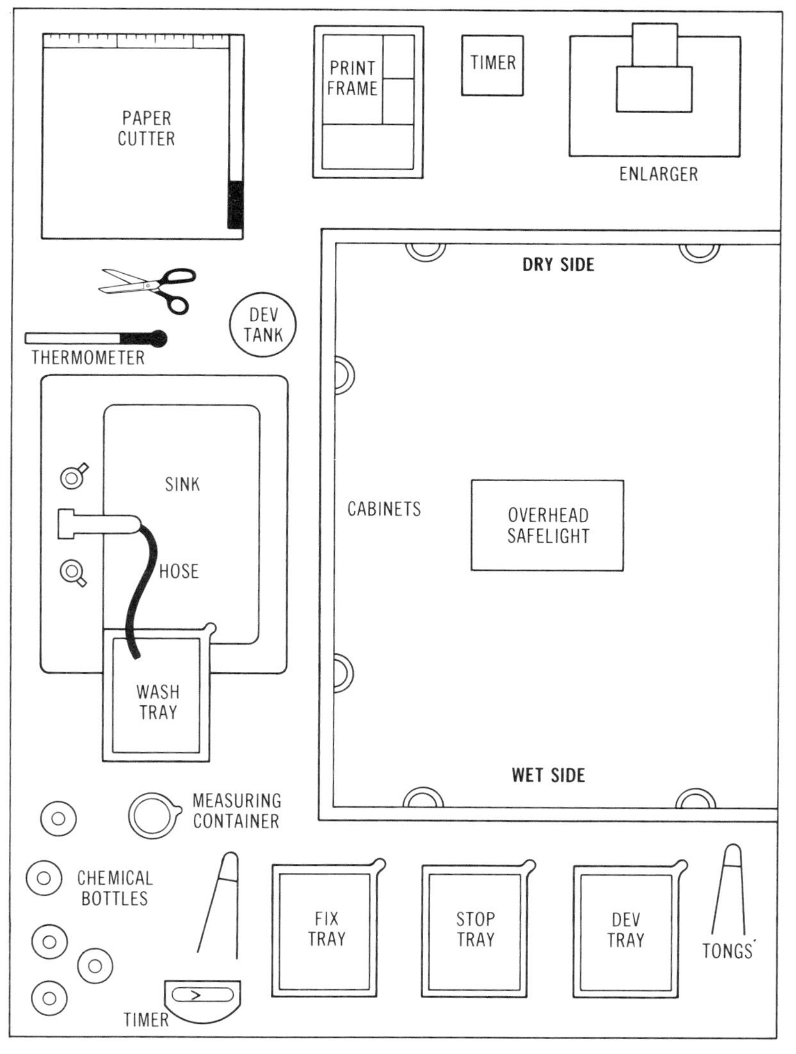

Fig. 4-10. A typical darkroom plan.

for holding the negative, and a timer for controlling the light exposure (see Fig. 4-11). The following steps are necessary to make a positive enlargement from 35 mm film negatives:

1. Handling the negative only by its edges, dust it with a soft brush and place it in the film carrier with emulsion side (dull side) down.
2. Turn on the enlarger light, open the diaphragm to its wide-open position (low f/stop number), and focus the image on a white piece of paper.
3. Adjust the enlarger height and focus until the image forms the desired final dimensions. Printed circuit artwork should have some widely spaced reduction targets whose final spacing is known exactly; these targets serve as guides to the photographer during 35 mm enlargement, or during reduction with a process camera. Image adjustment is illustrated in Fig. 4-12 with the use of an Accuscale precision scale from Bishop Graphics, Inc.

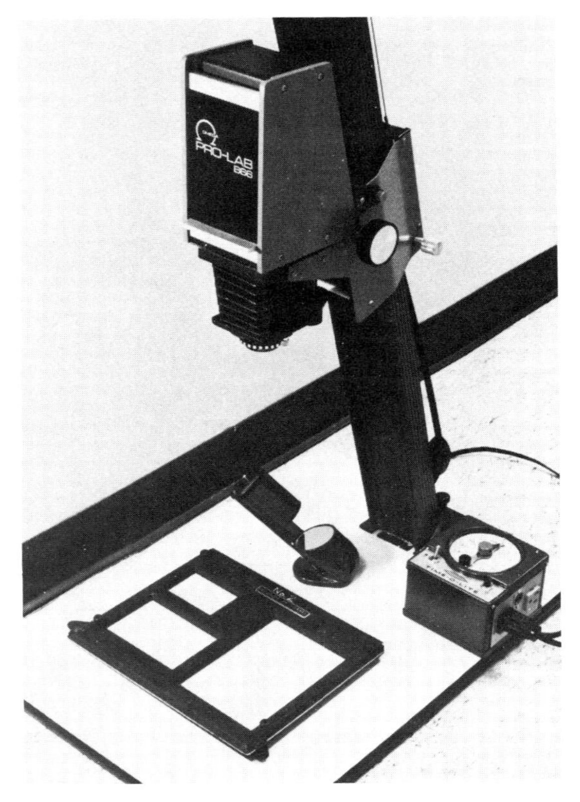

Fig. 4-11. A black-and-white enlarger.

4. Mark the final image outlines on the background paper and adjust the enlarger diaphragm and timer to their proper film exposure settings. A proper exposure will give full image development after film has been in the developing bath for about one minute; this typically requires 10 seconds of enlarger exposure at aperture f/8. The light sensitivity for Kodalith film is similar to most black-and-white papers.
5. With room lights off and safelight on, place a sheet of Kodalith film under the enlarger and expose it with the timer. It is not important that the film emulsion be facing up or down, since in this case the resultant positive transparency is not the final mask for photofabrication. However, note that the emulsion side of unexposed Kodalith film is a dull pink, while the base side is a glossy dark red. Under a yellow safelight these colors appear as light tan and dark brown.

When printing transparencies from the three-layer artwork system recommended in Chapter 3, it is best to prepare the base layer positive first so that it can serve as a background image for the other two layers. This makes it easier to print the component- and reverse-side positives, because the enlarger is simply adjusted to align their targets with those of the underlying base positive.

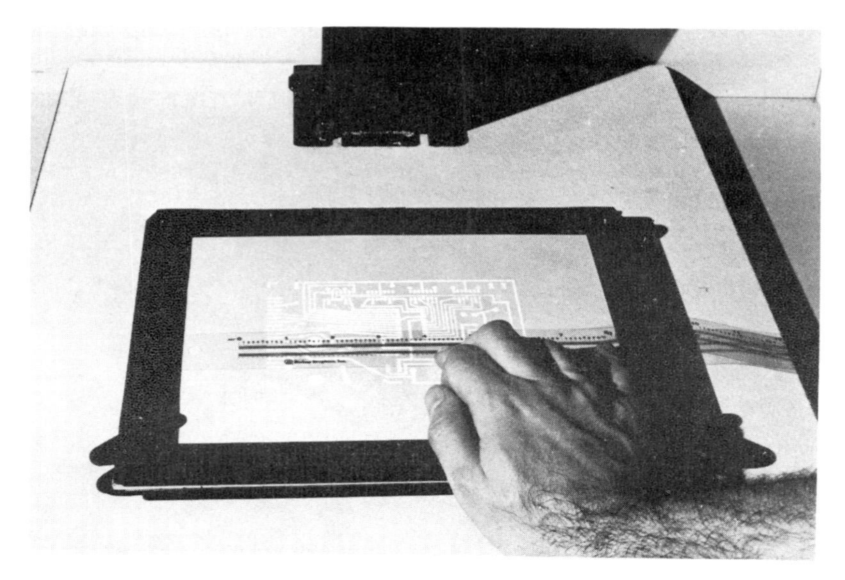

Fig. 4-12. Use of an Accuscale for image size adjustment during enlargement.

Processing Sheet Film • Processing Kodalith film in sheets proceeds in the same manner as with 35 mm film, except that trays are used instead of a tank. Trays should be filled with chemical solutions ahead of time and arranged from left to right in the order of processing: developer, stop bath, and fixer. After fixing, the film is thoroughly washed and dipped momentarily in Kodak Photo-Flo solution to minimize drying marks during air-drying; hang the sheet from a clip or clothespin to dry.

The developing bath is the most critical phase of processing Kodalith film. As stated before, an optimum exposure requires about one minute in the developer; however, this time can be shortened considerably, or it can be extended to almost three minutes. Luckily, the low light sensitivity of the film permits fairly bright safelight illumination, and this allows the developing image to be readily observed in a white processing tray (see Fig. 4-13). The developing solution should be agitated constantly as soon as film is immersed, and the film is transferred to the stop bath as soon as proper image density has been formed. Although 15 seconds development time may give proper results, a longer time is recommended because this allows more even development and less chance for over-development. Good and bad results for positive transparencies are illustrated in Fig. 4-14; as seen in the figure, underdevelopment or underexposure cause poorly defined images of insufficient density, and overdevelopment and or overexposure result in images spreading out from their intended boundaries. With a little experience, the proper results will be readily apparent in the developing tray, under widely varying conditions of light exposure,

Fig. 4-13. Tray processing a sheet of Kodalith film.

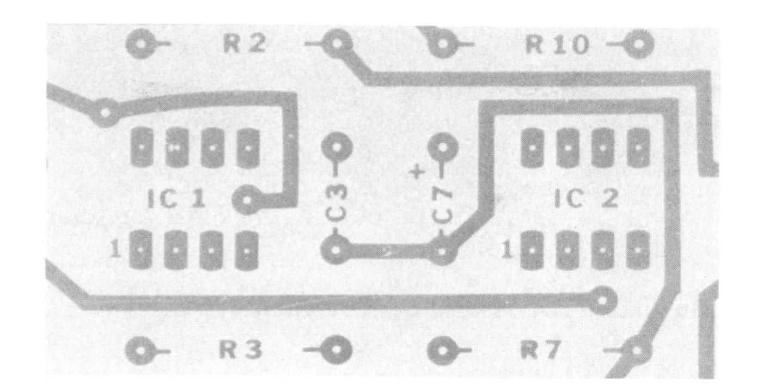

(A) An underdeveloped positive transparency.

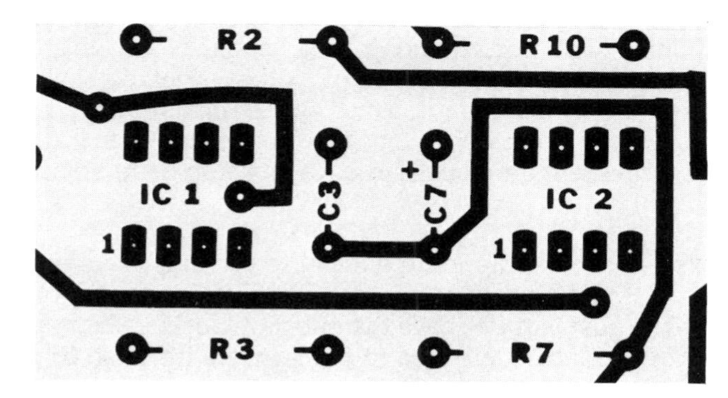

(B) Proper development for a positive transparency.

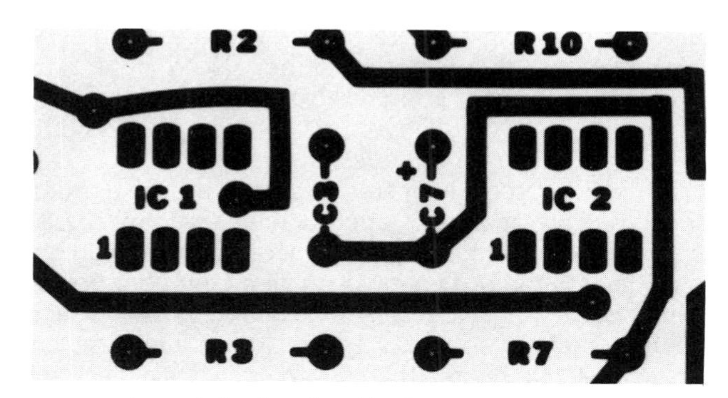

(C) An overexposed or overdeveloped positive transparency.

Fig. 4-14. Good and bad results for a positive transparency.

developer strength, developer temperature, solution agitation, and solution decomposition. The author has successfully processed Kodalith film after mixing A and B developer solutions that have stood for over a year, long past the recommended shelf life, but this is not recommended.

A complete set of positive transparencies prepared from 35 mm negatives of three-layer PCB artwork is shown in Fig. 4-15. These transparencies are exactly equivalent to the original artwork layers, but at the proper scale for image transfer to a circuit board. They will be used to make *contact negatives* to be used as the final photofabrication masks for photo resist exposure.

Problems with Processing of Kodalith Film • Two common problems are seen when processing graphic arts film: *fog* in clear areas and *pinholes* in dark areas.

Fog — Fogged areas occur from extraneous light exposure, or from chemical processing problems. The typical cause is an improper safelight condition that occurs during film handling and enlargement. White light leaks must be avoided, although considerable red light can be tolerated in orthochromatic films. Always minimize the length of time that film stays outside its light-proof package. Chemical fogging can result from spent developer, absence of a stop bath, or insufficient fixing.

Pinholes — Pinholes are more difficult to avoid, and are usually the result of dust accumulating on the film surface during or before exposure. The dust particles have the effect of blocking light during exposure, thus causing a clear area to occur where it should be opaque. Some photographers use a can of compressed air to clean film just before enlarger exposure. Improper handling can also cause *scratches* in the film emulsion that result in light areas, so handle sheet film carefully and avoid dragging the emulsion layer across other surfaces. During development, the film should be placed in the tray with emulsion side up to prevent scratches from occurring as dust particles in the solution rub between film and tray. When pinholes do occur in transparencies, they can be opaqued over (see Fig. 4-16) with a fine artist's brush and water-soluble red pigment available from photographic dealers. This opaque is applied with the film emulsion side *down* on a light table surface, so that (if necessary) the pigment can be swabbed off later without damaging the film emulsion. Scratches and pinholes in *positive* transparencies should always be corrected with opaque before making contact *negative* transparencies; it is much easier to paint over a light area than it is to eradicate a black area in the emulsion. Black dots in a contact negative are the direct result of clear pinholes in the positive original.

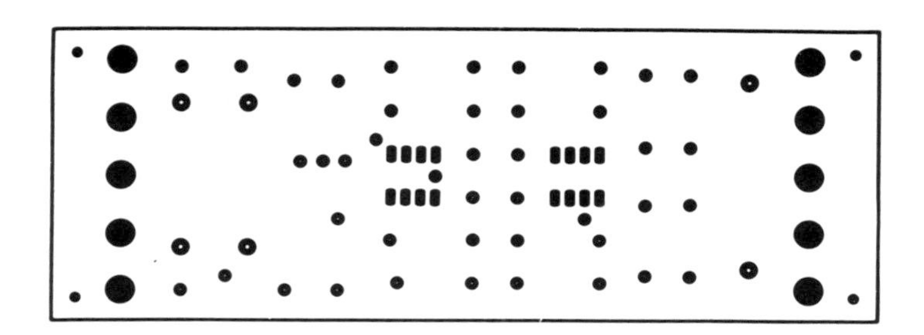

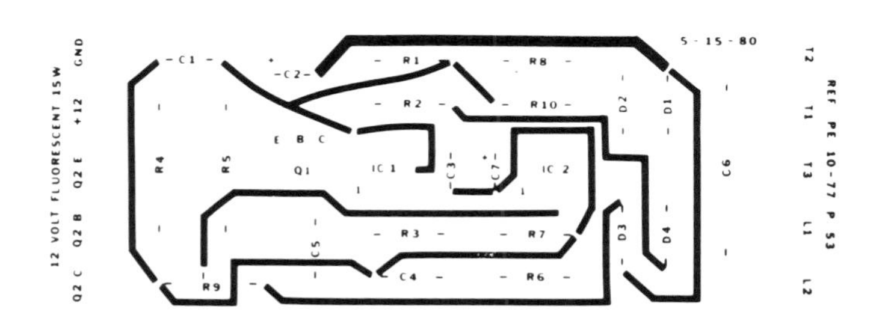

Fig. 4-15. Complete positive transparencies from three-layer artwork.

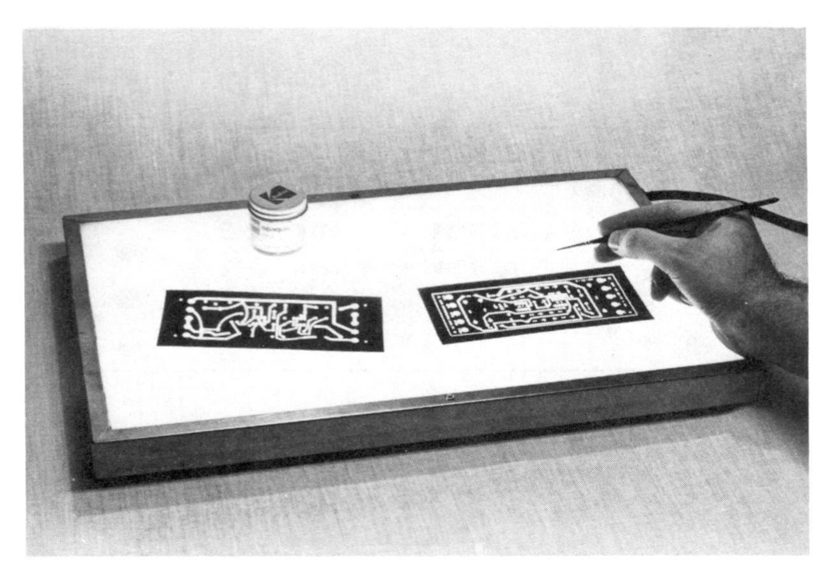

Fig. 4-16. Examination and touch-up of final negative transparencies using a light table.

Contact Prints • *Contact printing* is necessary to convert positive transparencies into negatives, and vice-versa. The PCB fabrication approach described in this book requires *negative* photo masks, but other processes require *positive* photo masks. For example, screen printing requires the use of a positive image for preparing photo stencils, and some types of photo resist coatings are positive-acting. In any case, it is frequently necessary to prepare one or both types of transparencies from the original artwork, and contact printing is the photographic operation that makes these masks interconvertible.

General Procedure — To prepare a contact print, the transparent image to be duplicated is simply placed on top of a fresh sheet of film, which is then exposed to light and developed. Two variables must be controlled to make the operation successful: the two material layers must be in *intimate contact,* and the light must be an even *point source.* If the sheets are not in close contact and light is diffuse (such as sunlight), the contact image tends to spread and lose good definition. What we are after is a perfect shadow; therefore, the best approach is to place a piece of flat glass over the original and the film to hold them tightly together during exposure. A slightly refined version of this can be purchased from a photographic dealer and is called a *contact printing frame* (see Fig. 4-17). Exposure should use the even, collimated light from an enlarger.

Fig. 4-17. A simple contact printing frame.

Emulsion Orientation — Another consideration in making contact prints is the orientation of the final film emulsion layer. Kodalith film in sheets may be exposed from either side, so the film is turned in the direction to give desired image/emulsion geometry: right-reading, or wrong-reading. To understand the importance of emulsion orientation, consider the photo resist image transfer operation.

A negative transparency will be used to transfer the printed circuit pattern onto a copper-clad laminate surface, which has a light-sensitive coating known as *photo resist.* The photo resist operation is in itself basically a form of contact printing. A negative mask is held tightly against the coated copper surface and exposed to ultraviolet light. However, in this case the light source is rarely a collimated point source; it is more likely to be a couple of UV lamp tubes that emit diffuse light. Therefore, the film/surface contact geometry becomes increasingly important to prevent light-scattering, which leads to poor image resolution. The important point is to keep the film emulsion layer in direct contact with the copper surface, on the *bottom* side of the negative, so that incident light does not have a chance to scatter through the plastic base of the film before reaching the resist coating. If film emulsion is on the *top* surface during exposure, this has the same effect as loose contact. Therefore, final transparencies intended for use in photofabrication processes should have proper emulsion orientation.

Negatives from Three-Layer Positives — In preparing negative transparencies for fabrication of double-sided PCBs, the last step in the 35 mm photographic approach is to make contact prints from three positive transparencies. The *component-side* negative is obtained by contact printing the *component trace image* superimposed on the *base pattern image.* In this case, the underlying sheet film should have its emulsion layer facing *down* during contact exposure, which results in a right-reading negative. All three layers of film are held down securely with a glass plate during exposure, as seen in Fig. 4-17, and the use of a collimated light beam from the enlarger lens minimizes scattering effects.

The *reverse-side* negative is obtained in similar fashion by contact printing the *reverse trace image* superimposed on the *base pattern image.* The desired result in this case is a wrong-reading negative, so either the film emulsion must be turned up during exposure, or the positives must be turned over as the *mirror image* of the original artwork.

Following exposure, contact negatives are developed and processed in the normal fashion for Kodalith film. Proper exposure should give complete, dense, resolute images after approximately one minute of development time as discussed before. Following drying, the negatives should be examined carefully on a light table (see Fig. 4-16) for high density (opaque background), and dimensional accuracy of images. The stability of polyester Estar film base should not pose any dimensional tolerance problems for prototype PCB construction as temperature and humidity changes occur in the masks. Side-to-side registration is the main concern, and since both masks are contact-printed from the same base positive, they should maintain alignment if stored together. It is likely that some pinholes will exist in the final negatives and these should be corrected with opaque before using the masks for photofabrication of PCBs.

SPECIAL ARTWORK CONSIDERATIONS •

Direct Use of Transparent Artwork • It should be obvious after the previous discussion of contact printing why some prototype designers are willing to painstakingly prepare PCB artwork on a 1:1 scale with tiny transfer patterns and very narrow slit tape. Contact negatives can be prepared directly from this transparent artwork without the need for camera work or enlargement; the only photographic equipment requirements are a glass plate, a lamp, and several film processing trays. Even the need for contact negatives can be avoided if *positive* image transfer processes are used for PCB fabrication, such as screen printing or positive photo resist. The use of 1:1 artwork is particularly attractive for producing single-sided PCBs; initial taped artwork can be neatly prepared on a single transparent plastic sheet, and is seldom very complicated or crowded. However, for the most consistent and satisfactory results with double-sided PCBs, 2× artwork and negative-acting photo resist are recommended.

Published Artwork • An important benefit from learning 35 mm photographic techniques is the ability to make PCB transparencies from any scale artwork published in electronic books, magazines, and manuals. Most of these patterns are designed for single-sided circuit boards, so a simple sequence of photography, enlargement, and contact negative will give quick results. Some considerations for photographing published PCB artwork are as follows:

1. Remove the printed page from its binding so that it can be mounted completely flat on the copyboard.
2. Most published artwork has flaws in printing. Touch up the patterns carefully with pencil, ink, or white opaque before taking the photograph. Corrections are easier on paper than on the transparency.
3. Fill the camera viewfinder completely with the image to maintain resolution. Most patterns are published at a 1:1 scale or smaller, so a closeup lens may be necessary to focus the camera.
4. Many published patterns will not have dimensional targets. If the scale is known, draw in your own targets for reference during enlargement to a positive. If the scale is unknown, use a precision Accuscale during enlargement to key in on known dimensions, such as DIP IC lead spacings.
5. Observe copyright laws. Do not photograph published artwork for commercial use without the permission of the author or publisher. Printed circuit patterns may carry the same legal protection as printed text. However, the "fair-use doctrine" does permit certain exceptions to the copyright laws; you should see your lawyer for advice.

CHAPTER 5
Photo Resist

INTRODUCTION • Photo resists are light-sensitive coating materials used in *photofabrication* techniques to reproduce photographic images on other surfaces, primarily metals. The sensitized surface is exposed to light through a transparent mask, and then developed with a solvent to remove unwanted areas of the coating. The resultant resist pattern duplicates the original photo mask, and is resistant to chemical action; it forms a stencil of openings and chemically resistant barriers that direct the subsequent surface treatment such as etching, electroplating, dyeing, or chemical diffusion. Photo resists are the basis for many fabrication processes, including ones for the following products:

- Printed Circuits
- Integrated Circuits
- Chemically Etched Metal Parts
- Nameplates and Panels
- Decorative Designs
- Lithographic Plates

The outstanding characteristic of photo resists is their ability to reproduce extremely fine detail and sharp lines; in fact, photofabrication is often the only reliable method for producing intricate parts, such as microelectronic devices.

The most popular negative-acting liquid resists are Kodak photo resist (KPR) from Eastman Kodak Company, and DCR from Dynachem Corporation. We will direct our attention to the use of Kodak photo resist, type 3 (KPR 3), which was designed for dip coating, and is a general-purpose resist for etching and electroplating metal surfaces. When using Kodak photo resists, several associated publications [16-19] listed in Appendix A are recommended reading. The following source and distributor (see Appendix B) are recommended for PCB laminates, photo resist, processing chemicals, and related materials for the process described in this chapter:

Kepro Circuit Systems, Inc.
Jack Spears, Inc.

OBJECTIVES ● The objectives of this chapter are as follows:

- Explain the principles of photofabrication.
- Detail the properties and use of KPR 3, a solvent-based negative-acting liquid photo resist.
- Show the importance of properly preparing a laminate surface for photo resist coatings.
- Detail a simple procedure for in-house coating of KPR 3 photo resist on blank laminate, providing greater freedom in PCB fabrication.
- Detail image transfer procedures for exposure, development, and post-treatment of the resist pattern.
- Discuss some alternative forms of photo resist, designed for positive photo action and aqueous processing.

PHOTO RESIST PROCESSES ● Photo resist is an excellent basis for low-volume PCB fabrication, since it is economical in short runs and requires little equipment or setup procedures. Photofabrication techniques will produce high image resolution, whether a simple manual process or elaborate high-volume equipment is used. In general, the use of better accessories makes the job easier, but the beginner can obtain good results with principal tools being care and patience. There are three basic variations in photo resist products to consider.

Liquid Solution or Dry Film • Photo resist has been traditionally used in liquid form to coat blank circuit boards by dipping, spraying, or roller coating. However, the latest technique in photo resist PCB applications is to purchase the material in thin, dry sheets or rolls which are then *laminated* to the copper surface. A typical low-volume dry film laminating machine is illustrated in Fig. 5-1. The main advantage of dry films is that they allow a much more *even* coating to be obtained, and thicker coatings with superior chemical resistance are also possible. The typical thickness of photo resist coated in liquid form is 0.0001 inch (2.54^{-3} mm), while that of dry film resist is closer to 0.001 inch (2.54^{-2} mm). Because a laminating machine is recommended for applying dry film sheets, and can cost close to $2000, the use of dry film resist will not be emphasized in this book. However, if the prototyping shop can afford such equipment, it certainly makes the image transfer process smoother; otherwise, liquid resist must be used because it can be coated with simple manual methods. (With care, dry film photo resist can be successfully coated on copper boards using a hot-air gun and a hard rubber roller, but machine application is preferred.) One advantage of liquid resist is that it is more stable and less light-sensitive in *solution* as opposed to the *dry* form.

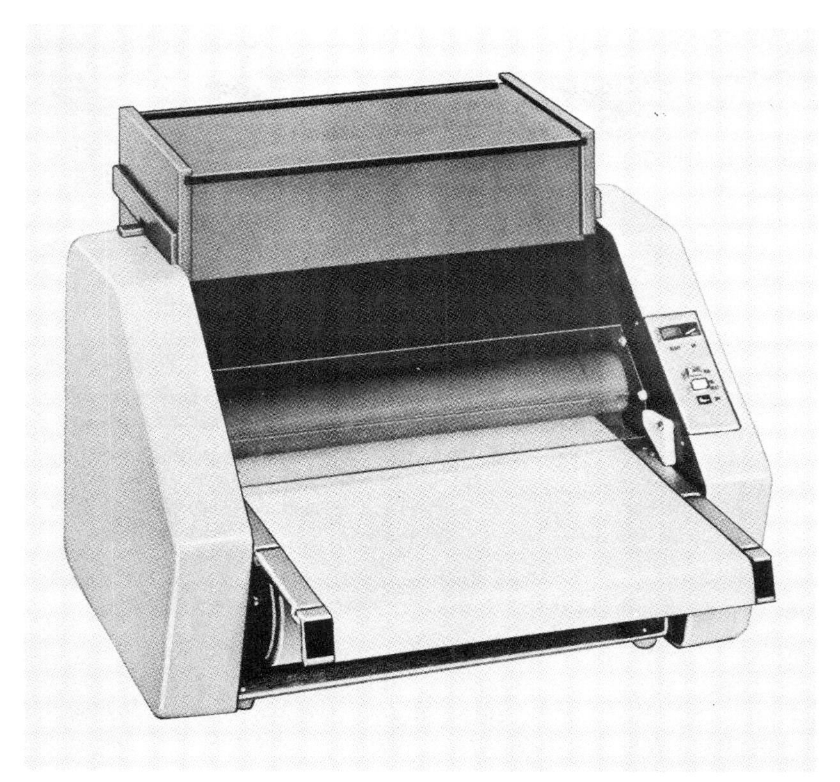

Fig. 5-1. Hot roll laminating machine for dry film photo resist. ***(Courtesy Kepro Circuit Systems, Inc.)***

Positive- or Negative-Acting • Just as with photographic films and papers, photo resist products are formulated in both positive- and negative-acting forms. The working difference between them is illustrated in Fig. 5-2. In general, the negative-acting varieties are preferred because of greater stability and chemical resistance, because more information and experience is available concerning their properties and applications, and because their processing steps are less critical. A negative photo mask is required for exposing negative-acting resist, but photography can generate either form of mask equally well as was seen in Chapter 4. There is only one situation where the use of positive-acting photo resist may give an advantage in PCB prototype construction; this is where transparent artwork is prepared on a 1:1 scale and used directly as the fabrication mask, thereby circumventing the need for photographic work. Dense circuit patterns for double-sided boards are difficult to prepare on a 1:1 scale, but as was pointed out in Chapter 1, wire-wrap and printed circuit techniques may be suc-

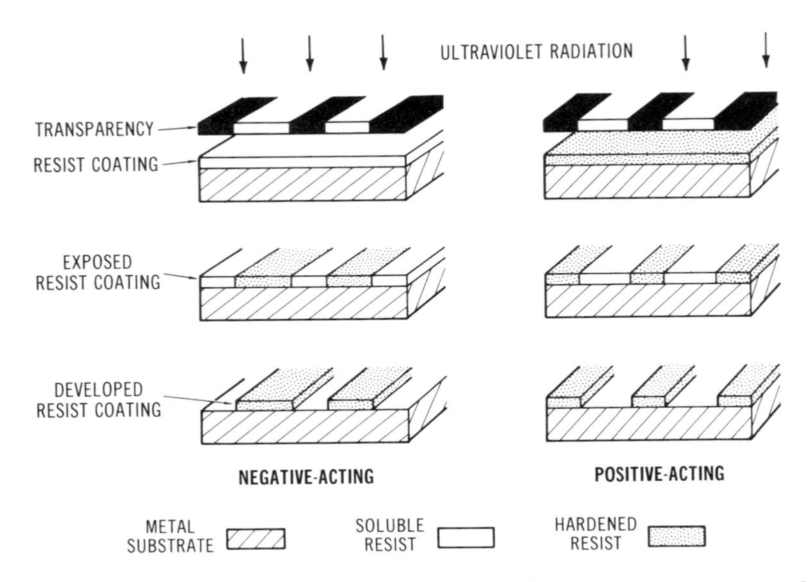

Fig. 5-2. Effects of exposure and development on negative-acting and positive-acting photo resist. ***(Courtesy Eastman Kodak Co.)***

cessfully combined by designing a single-sided board containing only component solder pads and power supply traces, with the bulk of interconnections left to wire-wrap methods. The use of a positive-acting photo resist will, therefore, be mentioned at the end of this chapter.

Solvent or Aqueous Processes • Conventional photo resist materials require organic solvents for coating, development, and stripping of the images. These solvents (such as xylene, trichloroethylene, methylene chloride, and acetone) create flammability, handling, and disposal problems. They are best handled in a well-ventilated area, such as a laboratory fume hood (see Fig. 5-3). To avoid these problems, photo resists are available that use aqueous processing solutions, such as water, alcohols, alkali, and bleach. Such resist coatings are often based on azide polymers or protein chemistry, which may not be overall as satisfactory compared to the negative-acting solvent-based systems. However, aqueous processing is attractive to the hobbyist and will be discussed at the end of this chapter along with positive-acting photo resist.

GENERAL PROPERTIES OF KPR 3 PHOTO RESIST • KPR 3 photo resist is initially received as a dilute monomer solution in dichlorobenzene, and is relatively inactive in this form. However, for safe extended storage the solution should be kept in a tightly closed, amber bottle to exclude light and prevent gradual monomer decomposition. The evaporation of solvent should be avoided because this changes the concentration and thus the viscosity.

After the liquid resist has been coated on a blank board and dried, it becomes sensitive to light and *polymerizes* with exposure to light of the proper wavelength. The polymer forms a tough, cross-linked, insoluble coating that protects the underlying metal surface during chemical etching. The unexposed resist can be dissolved in *developing solvents* that wash away the monomer, leaving the desired resist pattern coated on the laminate surface. If copper surfaces are properly cleaned and prepared, polymerized resist is strongly adherent and requires sanding or *stripping solvents* for removal. Adherence and toughness can be improved further by *postbaking* the coated image, which is typically 0.0001 inch (2.54^{-3} mm) thick. The final coating is a light yellow transparent color, but it can be easily dyed to allow visual inspection and touch-up.

Several factors combine to make KPR 3 an ideal photo resist for manual or low-volume PCB fabrication:

1. Stability — The resist is sensitive to a narrow range of light, and

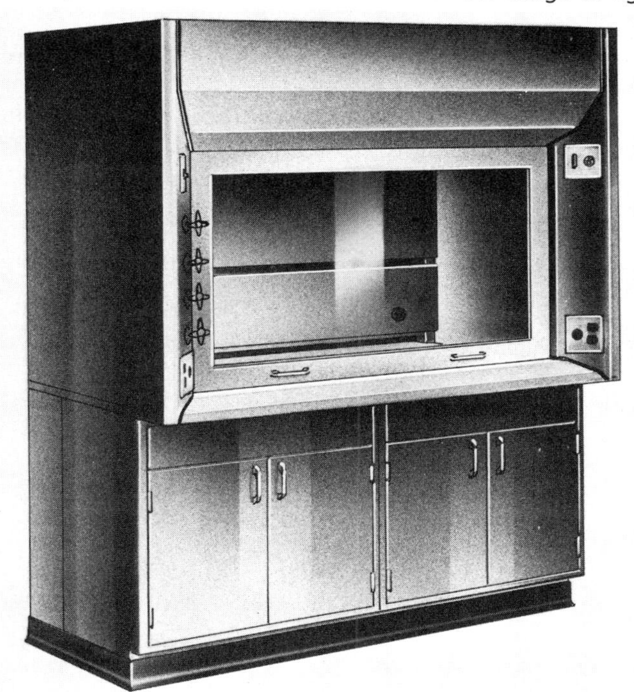

Fig. 5-3. A commercial laboratory fume hood. *(Courtesty Kewaunee Scientific Equipment Corp.)*

then only in the dry form. Other conditions, such as oxygen exposure and moderate heat (less than 120°C or 248°F), do not affect the coating. Boards coated with KPR 3 photo resist can be stored under light-free conditions for *years* before using, and they can be handled under subdued room lighting for several minutes without problems.

2. Toughness — Properly prepared coatings are not easily scratched, and they are strongly resistant to most chemical etching and plating solutions.
3. Ease of Use — Coating and processing are straightforward using manual techniques.
4. Visibility — After dyeing, resist patterns appear sharp with good contrast, allowing easy examination of coating quality before the irreversible etching process begins.

Spectral Sensitivity • A spectral sensitivity curve for KPR 3 photo resist is found in Fig. 5-4. This curve shows that the resist is sensitive to a particular area of the electromagnetic energy spectrum in the near-ultraviolet region. Light sources that emit radiation at these wavelengths include black lights, carbon arc lamps, photoflood lamps, and mercury vapor lamps. Sunlight includes a substantial ultraviolet component, and white fluorescent tubes contain a small component. Eye protection is required for ultraviolet lamps, which should never be observed directly.

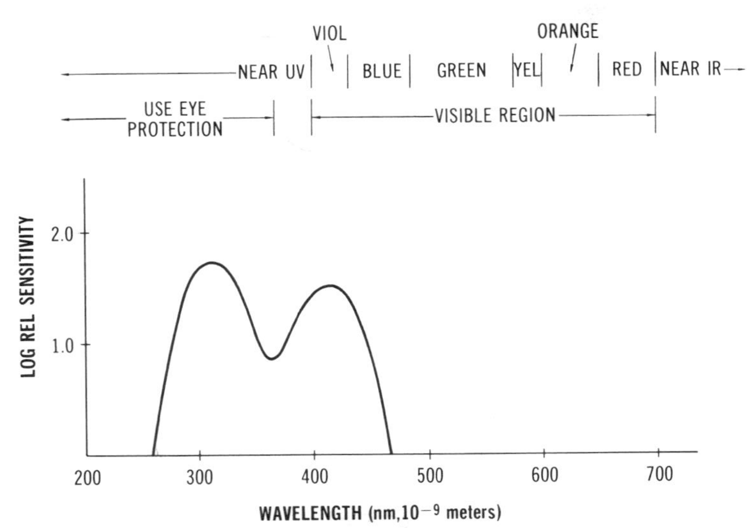

Fig. 5-4. Spectral sensitivity of KPR 3. *(Courtesy Eastman Kodak Co.)*

An important consequence of the specific sensitivity of KPR 3 photo resist is that it can be safely handled under any light that excludes radiation below about 460 nanometers, which is in the blue visible region. For extended safe exposure, this photo resist may be used under yellow incandescent lighting, or under white fluorescent tubes wrapped in yellow or red plastic sheets. Brief exposure to normal incandescent room lighting will not affect most KPR 3 resist coatings; fluorescent lights are more likely to cause image *fogging* (slight polymerization of resist).

Viscosity • At room temperature, KPR 3 resist has a viscosity of 32–36 centistokes, which is comparable to that of light weight oil. The solids content is 9–10% by weight. Viscosity is important in controlling coating thickness, particularly if boards are dip coated; the viscosity of KPR 3 resist is about right for this operation if the stock solution is diluted 1:1 with solvent, such as Kodak ortho resist (KOR) thinner. Other coating methods, particularly spraying, may require further dilution. A simple method for standardizing viscosity is to observe the rate at which a measured amount of resist spreads out when poured onto a clean, flat copper laminate surface.

Flammability • KPR 3 resist, KOR thinner, and KOR developer are all flammable liquids with flash points of 30 °C, 46.5 °C, and 40 °C, respectively. The flash point is the temperature at which vapors can ignite in air when exposed to a flame, electrical spark, or other ignition source. Although these liquids may be handled safely in small amounts with adequate ventilation, resist coatings should be dried in an oven which does not provide an ignition source for combustion to occur. The KPR dyes are dissolved in a solvent similar to the KOR developer.

Coating Thickness • Coating thickness affects exposure, development, and final resolution of photo resist images; it also affects the durability of the protective layer during chemical etching or plating. If resist is too thin, it tends to break down during etching and exposes the metal that should remain protected. However, a coating that is too thick is difficult to expose and develop without loss of resolution and image distortion, which result from light scattering and resist swelling. Thick coatings also tend to remain incompletely polymerized, and may streak and blotch during development. For these reasons it is important to standardize your approach to coating, and maintain a resist solution of constant viscosity. Solvent evaporation is the main problem to avoid.

The flow coating technique to be described in this chapter will produce KPR 3 resist layers of 0.0001 to 0.0002 inch (2.54^{-3} to 5.08^{-3} mm) thickness which is ideal for most printed circuit etching. These dimensions provide photographic resolution of lines and spaces down to 0.002 to 0.005 inch (0.051 to 0.127 mm) wide, smaller than any lines needed for PCB traces. Dyeing of resist coatings provides a key estimate of thickness; if any sign of a shiny underlying metal surface can be seen after dyeing, the coating is probably too thin. Blue or black dye should make the coating entirely opaque. On the other hand, if images show distortion and pad center holes begin to become obliterated, then the coating is probably too thick. Note that these latter symptoms can occur due to other reasons: poor contact of the negative during exposure, insufficient exposure time, or excessive development time.

LAMINATE PREPARATION •

Stock Laminate • The most economical method for purchasing laminate is in bulk packages, copper clad on both sides. This same material can then be used for both double- and single-sided circuit boards, with one side requiring complete etching in the latter case. Recommended material is one-sixteenth inch, 1-ounce, FR-4 or G-10 glass epoxy composition. Bulk packaged copper clads from Kepro are furnished as 12 by 24 inch (305 by 610 mm) panels which must be cut or sheared to desired size. A small shear is illustrated in Chapter 9, Fig. 9-1. As a second choice, laminate can be cut with a hacksaw and edges filed smooth. Precut circuit board blanks are also available down to 3 by 6 inches (76 by 152 mm) in size, in small quantities but at a substantially higher unit cost.

Kepro also supplies copper clad laminate precoated with KPR 4 photo resist. This coating is essentially identical to KPR 3 resist, but it is roller coated from a more viscous solution. The cost of photosensitized laminate is 30 to 50% higher than blank stock; however, a factory coating provides the user with an even, reproducible coating thickness, and saves time and trouble during image transfer. Pre-sensitized boards are particularly desirable for low-volume *single-sided* applications. However, the prototyping process described in this book for *double-sided* PCBs requires that at least one side of the board be user-coated; this is a direct consequence of separating the etching operation into two separate steps. Another important reason why users should have the ability to coat laminates in-house is that it allows more freedom in correcting errors during image transfer, without wasting stock material. An unsatisfactory coating is merely stripped and reapplied with fresh photo resist.

Surface Preparation • The most important step in photo resist fabrication is to properly clean and prepare metal surfaces, so that coatings will be strongly adherent. For copper clad laminate, this means four distinct considerations:

Surface Roughening — A glossy, polished surface is undesirable because it doesn't allow the coating to "grab" the metal. The first step in laminate preparation is therefore to lightly roughen the surface with fine steel wool, or a gentle abrasive cleanser. The desired result is a smooth surface with very fine scratches. (Deep scratches should be avoided.)

Cleaning — The copper surface must be thoroughly cleaned to prevent resist failure. This means complete removal of oil, grease, fingerprints, and metal oxides, which form readily on exposed copper. Industrial cleaning processes often include three or more separate steps to ensure complete removal of surface residues. For our purposes, copper foil can be adequately cleaned by scouring with a stiff bristle brush and a wet slurry of fine pumice, dishwashing detergent, and water. This operation is illustrated in Fig. 5-5. Note that many household abrasive cleansing powders contain additives (such as bleach) that will interfere with surface adhesion of resist, so these materials should be avoided. Steel wool should also be avoided during the final cleansing step because it often contains a light film of oil. The use of pumice eliminates the need for surface pre-roughening, since it serves as an abrasive scouring agent.

Thoroughly wash the laminate surface with running water after scouring. Water will provide a time-honored test for cleanliness of metal surfaces — the *water break* test. After rinsing the metal, hold it level and observe the manner in which water wets the surface. If any breaks occur, or if water "beads up" in any spots, this is evidence for unsatisfactory cleaning; the photo resist will probably behave in the same manner as the water during coating.

Acid Dip — Organic coatings adhere best to a metal surface that has been acidified. Acid also helps dissolve oxides and other inorganic contaminants. Following cleaning, the laminate should be briefly treated for 30 seconds with a 5–10% solution of hydrochloric or sulfuric acid. The acid is then removed by rinsing the board under running water.

Drying — Moisture must be completely removed before attempting to apply photo resist. The best drying method is to blow water off the metal surface with clean compressed air, followed by a 10–15 minute bake at 40–50 °C (104–122 °F). The metal can also be wiped with a soft cloth before heating. In any case, do not use excessive temperatures or time during drying, which can cause re-oxidation of the copper surface. Once the laminate is dry, proceed immediately to photo resist coating.

FLOW COATING • A wide variety of methods has been recommended for coating liquid photo resist on blank circuit boards. Most of these methods reduce down to two basic techniques in low-volume applications — dipping or spraying. More novel approaches are sometimes encountered, such as spin coating, where the board is mounted on a rotary disk and photo resist poured in the middle is slung outwards to give a supposedly even coating. The primary and most difficult objective in coating is always to achieve a uniform surface thickness.

If only one side of the board will be coated at a time, *flow coating* is the method of choice. This approach requires no special equipment, little skill, and few variables to control; it also conserves photo resist material. Simply outlined, the method involves pouring a measured amount of liquid resist across the metal, smoothing it out with a wide, flat, soft brush, and laying the board horizontally on a level surface to allow the coating to spread evenly before drying. It is remarkable that such a simple procedure gives such good results, considering the number of more difficult methods that are widely recommended. The flow coating procedure is described in more detail as follows:

1. Cleaning — A clean surface is absolutely essential as in all approaches (Fig. 5-5). If the surface passes the final water break test, it will also allow liquid resist to flow smoothly and evenly without creating voids. Following drying, the clean board should be allowed to come to room temperature before applying resist. (A hot surface will cause solvent to start evaporating before the coating has been evenly spread.)

Fig. 5-5. Cleaning blank copper clad laminate.

2. Level surface — The blank board must not be warped, as observed on a smooth, flat table top. If the board is badly warped, it should be allowed to flatten out in an oven at 150 °C (302 °F) before cleaning. Minor distortions can often be corrected by gently bending FR-4 or G-10 laminate, which is very flexible.
3. Photo resist — The only critical variables in flow coating are the viscosity and amount of resist. For best results, KPR 3 resist should be diluted in half to about 5% solids content. The stock liquid is mixed with an equal volume of solvent (KOR thinner or toluene) in a small bottle just before use. The dilute KPR 3 resist solution can be used openly under normal room lighting for several hours before it begins to become cloudy through polymerization and precipitation of solids. Provided the stock solution is kept tightly closed, its viscosity should not change appreciably due to evaporation, and this variable will not be significant. (If evaporation should occur, the result will be gradually thicker coatings.) Viscosity can be readjusted based on experience, trial-and-error, or an accurate percent solids measurement obtained by evaporating and weighing a measured amount of stock solution.
4. Coating — The recommended volume of 5% KPR 3 resist solution is 1 milliliter for every 10 square inches of board space. The liquid can be measured in a small beaker or other graduated container, or an eyedropper can be approximately calibrated. Laying the board flat on a sheet of absorbent paper, deposit the necessary amount of liquid in small puddles across the copper surface. The liquid is spread out evenly (see Fig. 5-6) by gently stroking the board with a wide, flat camel hair brush, such as the one shown in the photo. (This brush is commonly available in stationery and artist supply stores.) The board is then allowed to stand on a level surface for 30 seconds to let the liquid flow into a smooth, even coating. The coating operation can be carried out under incandescent room lighting, since photo resist is fairly insensitive to light while in solution. Once the coating appears to be level, hold the board up to the light and observe reflection patterns across the smooth surface. There should be no void areas or air bubbles which lead to holes, or "frog eyes" in the coating. Frog eyes are usually the result of improper cleaning, which prevents photo resist from "wetting" certain areas of the surface. If this condition occurs, it can sometimes be corrected by dabbing some more resist on the affected area with a brush.
5. Drying — After coating, the panel should be dried in an oven to remove all solvents and leave a thin, light-sensitive coating. Light should be excluded from the drying oven because photo resist becomes sensitive as soon as the solvents evaporate. A typical tray

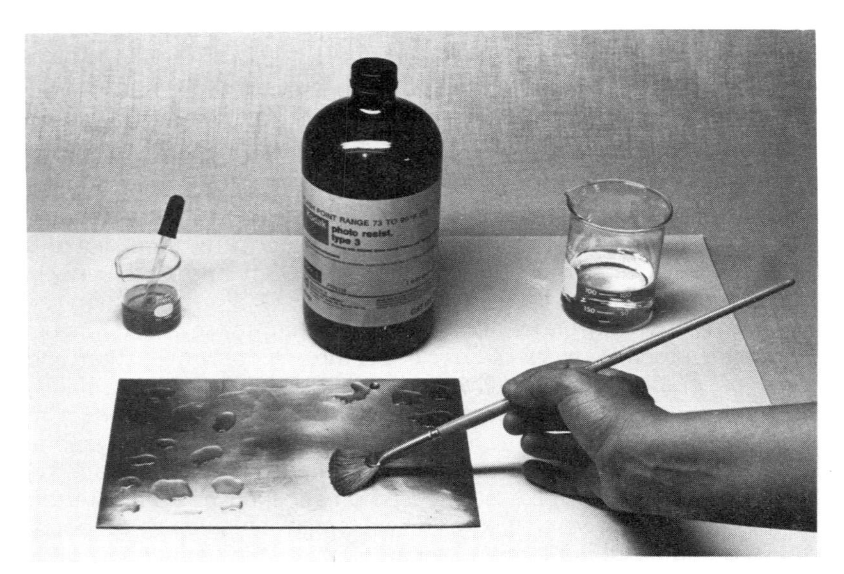

Fig. 5-6. Flow coating liquid photo resist with a brush.

oven is shown in Fig. 5-7. Recommended baking conditions are 10–15 minutes at 40–50°C (104–122°F); the board is placed face-up on an oven tray to give a level coating. Some precautions should be noted with regards to having flammable vapors present in an electrically heated oven. The amount of solvent is small enough that forced-air ovens are safe, since no vapor buildup occurs. However, static ovens with a thermostat and exposed resistance coils could generate a spark or hot wire temperatures sufficient to ignite organic vapors. Therefore, it is prudent to turn off the oven briefly when the solvent-wet board is first placed inside. A few minutes later, open the door and fan away vapors before turning on the oven to complete baking.

6. Multiple coats — Some users like to apply two thin resist coatings instead of a single layer to achieve more uniform results. The resist may be thinned out for this purpose. Each successive coat should be applied after the previous layer has been completely dried.

7. Cleanup — Items that require cleaning after contacting photo resist, such as brushes and beakers, should never be allowed to dry before cleaning. The dry resist will leave a film coating that is virtually impossible to remove from brushes; brushes should therefore be soaked in toluene or KOR thinner when not in use, and rinsed in clean solvent before drying. Dried, polymerized KPR 3 resist can be removed from smooth surfaces by softening in a

Fig. 5-7. Drying oven for baking resist coatings.

solvent such as acetone, methylene chloride, or commercial paint strippers, and scouring with a brush or steel wool.

IMAGE TRANSFER ● Image transfer for KPR 3 photo resist involves exposing the sensitized surface through a negative photo mask to an appropriate light source, and developing the image to leave a sharp, polymerized pattern on the copper surface.

Light Source • Any high-energy light source that is rich in the near-ultraviolet region of the spectrum may be used. A practical light source is an unfiltered long-wave ultraviolet fluorescent tube, type BL, mounted in a desk lamp that clamps to the edge of a table and swings out over the work area. This arrangement is as illustrated in Fig. 5-8 with a fixture containing two 15-watt bulbs from General Electric, type F15T8-BLB Blacklights. A No. 2 photoflood lamp also makes a good light source. Any UV-emitting source is potentially dangerous to the eyes, and these lamps are best operated unattended with a timer. High-energy UV radiation cannot be detected visually and permanent eye damage may occur before any physical symptoms, such as a burning sensation in the eyes, become apparent.

Exposure • The negative mask must be in tight contact with the sensitized surface of the circuit board during exposure to prevent light leakage, which results in poorly defined, low-resolution images. A simple exposure approach is shown in Fig. 5-8. With the laminate face-

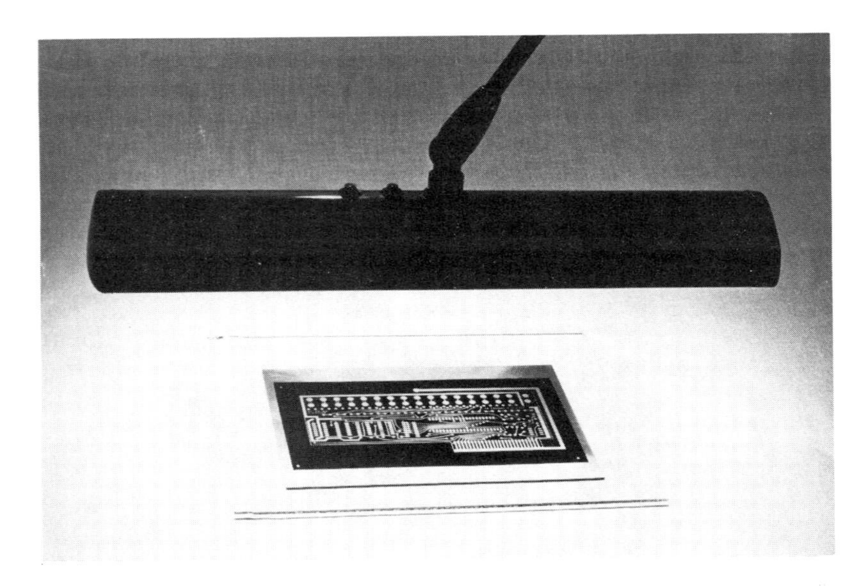

Fig. 5-8. A practical UV-light source and exposure method for photo resist.

up on a table top, the negative is positioned on the copper surface and a sheet of heavy plate glass is placed over the negative to hold it down securely. This method works well if the laminate is not warped. Otherwise, it may be necessary to add weights along the edges of the glass plate, or use a commercially available *contact frame* that clamps the sandwich securely together. A professional UV-exposure frame designed for prototype PCBs is shown in Fig. 5-9, complete with light source. If you use the plate glass approach, be sure to obtain glass that has not been treated to absorb UV radiation. The author once wasted considerable time when a new piece of glass was used for exposures, and only very faint resist images could be developed; every other photo resist processing variable was examined before the glass plate was finally suspected!

Oven-dried sensitized laminate should be allowed to come to room temperature before exposing, because a hot plate will distort and curl the film negative. The entire image transfer process can be accomplished under normal room lighting conditions if you work fast and take a few precautions. The laminate is first removed from the oven and placed face-down on a clean cloth to cool. The board/film/glass plate sandwich is then prepared and a UV exposure is made. When processing double-sided PCBs, the first exposure is easier because the film mask is simply laid on top of the board and approximately centered. However, when exposing the second side of the board, this negative must be very carefully positioned over 3–4 reference holes that are predrilled in various pads following the first etching step. After alignment, tape the negative down or very carefully position the glass plate without disturbing the negative. Closely examine the alignment before turning on a UV lamp source. If the negative can be satisfactorily positioned within a minute's time, room lighting will not have any undesirable effect. (If you find this process taking much longer, yellow or red safelights should be used.)

Exposure time is critical. Without sufficient radiation, the coating will not develop properly, and the board will be streaked and smeared with partially polymerized photo resist that swells up in the solvent but does not really dissolve. You will probably need to make some test exposures and reapply a few coatings before the optimum exposure time is established for your particular conditions of coating thickness, light source, and light-to-board distance. In general, 5–10 minutes exposure time is required using a photoflood lamp or two 15-watt blacklights positioned 6–10 inches from the mask surface. This time will increase with thicker coatings and gradual deterioration of the light

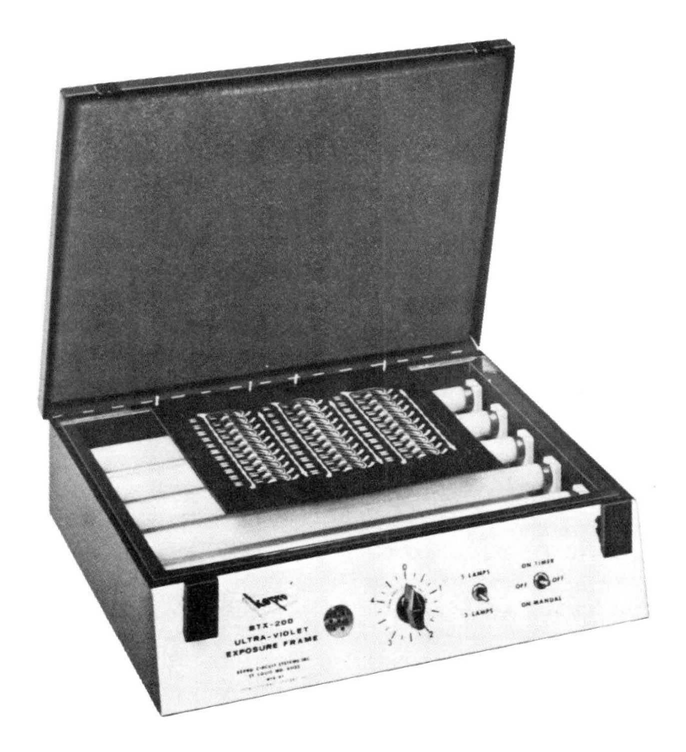

Fig. 5-9. A commercial UV-exposure frame for prototype PCBs. ***(Courtesy Kepro Circuit Systems, Inc.)***

bulb intensity. One of the real advantages of processing double-sided PCBs one side at a time is that little time or materials are wasted when a problem does occur during image transfer. The poor image is easily removed with steel wool and acetone, and the board can be cleaned, recoated with fresh resist, and ready for a new exposure in about 30 minutes. Overexposure is less of a problem than underexposure, particularly if the negative is high in contrast with very dark opaque areas. The only evidence for overexposure in these circumstances is a slight loss of resolution where excess light begins leaking through at the edge of lines. However, if the negative is not sufficiently dense, overexposure may lead to "fog," which is a thin film of photo resist polymerized evenly across the entire board surface.

Development • Following exposure, the unpolymerized areas of resist that did not receive light are dissolved away in a solvent to leave unprotected copper foil surfaces, which will be chemically etched in the next step. This solvent action is termed *development.* The KOR developer sold for KPR 3 photo resist is primarily xylene with a few additives. This solvent is not extremely volatile, but it is flammable, so use the developer in a well-ventilated area. Methylene chloride is also a good developing solvent, and is not flammable; however, it is very volatile and will evaporate rapidly when used in an open tray. Other chlorinated solvents, such as the traditional trichloroethylene used in vapor degreasers, are now considered too toxic and hazardous for use except in very well-controlled conditions.

The exposed PCB laminate is placed face-up in a glass or aluminum tray filled with developing solvent. The tray is gently rocked back and forth to develop the image completely in about 2 minutes. (Thicker coatings may take a little longer to develop.) The KOR developer has a high capacity for resist and can be used over and over for many boards. However, it is wise to rinse each board in fresh solvent to avoid a thin film of resist that could remain on the surface after drying. Such a rinse is carried out conveniently with clean solvent in a polyethylene squirt bottle. The developed board is raised from the tray, and lightly sprayed across its surface to rinse the image (see Fig. 5-10). The excess solvent that runs back into the tray will also serve to compensate for gradual losses due to evaporation. Pour the developer back into its bottle following each use to minimize this loss. Developer will eventually become too loaded with resist to remain effective, and it should be disposed of through incineration. Remember to wear rubber gloves while handling solvents that can be toxic if absorbed through the skin.

After developing the image, the board is briefly air-dried before inspection. The polymerized resist becomes soft in the developing solvent and should be handled carefully while wet. Rather than shake or blow-dry the board, it is safer to simply stand it on one edge for a minute under conditions of dim light. The image can then be visually inspected to

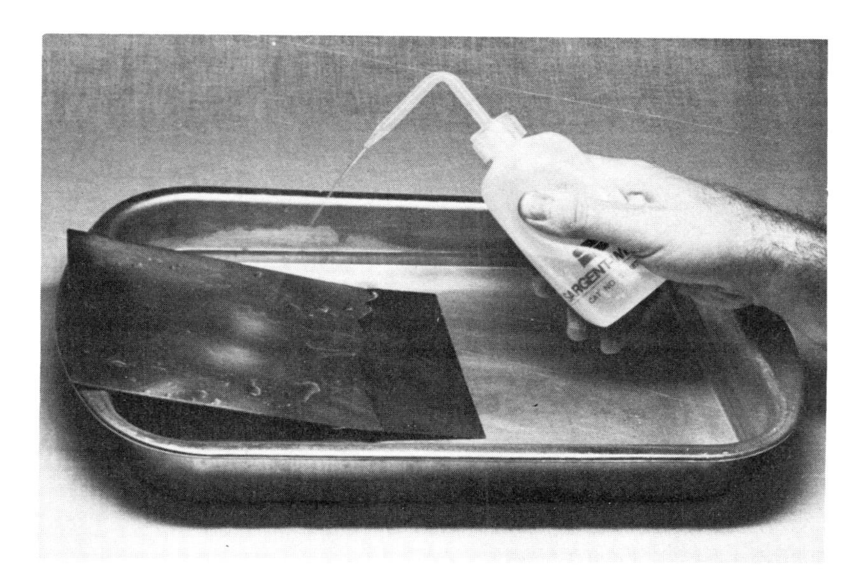

Fig. 5-10. Developing photo resist images in a tray.

ascertain if patterns are sharp and well-defined, or if further development may be necessary. A frequent problem is that small blobs of incompletely dissolved resist will stick to the copper surface, but these can be washed away with the squirt bottle. Once development is complete, dry the board for a few minutes at 40–50°C (104–122°F) in preparation for dyeing.

An improvement over simple tray development is *spray development* in an enclosed tank, as illustrated in Fig. 5-11 with a prototype PCB spray developer from Kepro. It is fairly difficult to develop both sides of a PCB simultaneously in a tray because the soft coating is easily disturbed. However, the equipment shown in the figure suspends the board in a light, even spray of solvent coming from both directions, which provides high agitation at the material surface, and continuous rinsing of the resist from developed areas. This method is particularly useful for thicker photo resist coatings.

PREPARATION FOR ETCHING • Four steps are recommended in preparation for etching PCBs coated with KPR 3 photo resist: dyeing of the image, inspection/touch-up, protection of the opposite side, and postbaking.

Image Dyeing • Examination of KPR 3 resist coatings is greatly enhanced by dyeing the polymer with a coloring agent. Kodak supplies both a black and a blue dye solution for this purpose. Dyeing is most effective if the coating has not been baked, so dyeing should be done

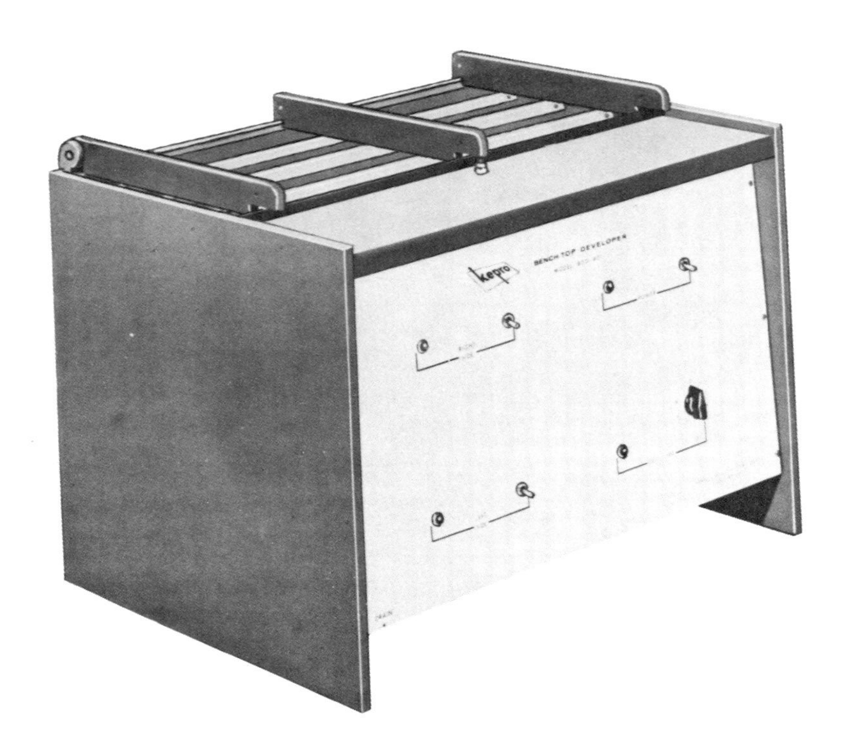

Fig. 5-11. Spray development equipment for prototype PCBs. *(Courtesy Kepro Circuit Systems, Inc.)*

as soon as the developed image is reasonably dry. Simply immerse the board in the xylene dye solution for 30 seconds, and wash off excess dye under cold running water. If it is impractical to maintain a tank of dye solution, which can be messy and tends to evaporate, an alternative approach is to contain this operation in a large aluminum pan of the type used for roasting turkeys. The dye is squirted on the surface of the board with an eye dropper and rolled around until it completely covers the resist pattern (see Fig. 5-12). After a few seconds, the excess dye is drained into the pan and allowed to evaporate, while the board is washed in a sink.

A slight scum of dye often tends to remain across the copper surface following the water wash. This is easily removed by dipping the board in isopropyl alcohol and again washing it under running water. The board is then ready for drying and inspection; wipe off the excess water with a soft cloth and dry at 40–50°C (104–122°F) for 10 minutes. The final coating should be completely opaque against a bright copper background, as shown in Fig. 5-13.

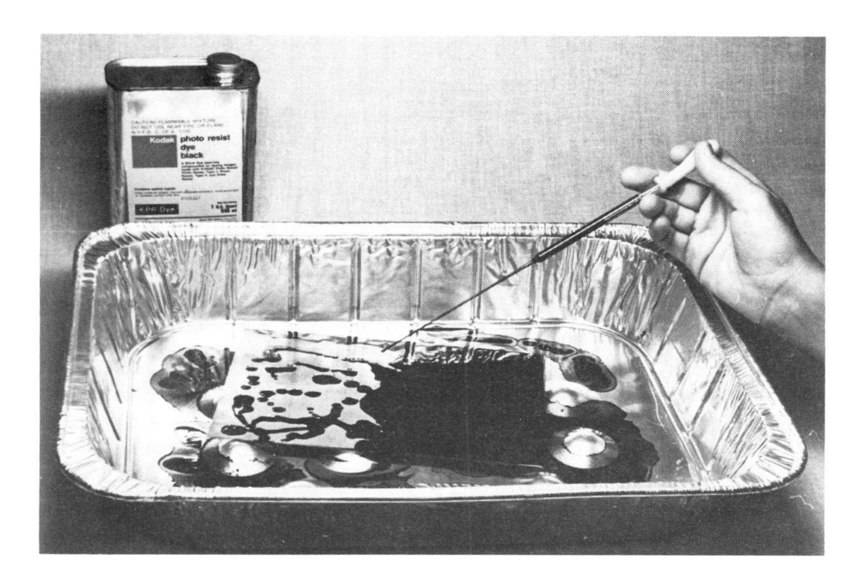

Fig. 5-12. Dyeing the photo resist image for clarity.

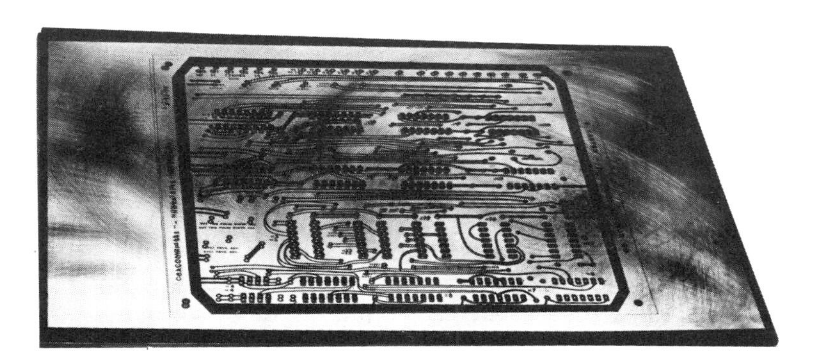

Fig. 5-13. Final dyed photo resist coating.

Inspection and Touch-Up • The printed circuit image is now ready for careful examination. Two types of problems may be apparent at this stage: minor ones that can be corrected, and major ones that will require the coating to be stripped, and the image transfer process repeated. Major problems and their causes are presented in Table 5-1. Remember that etching is an irreversible process, and you do not want to subject the board to an etching bath until the resist coating is completely acceptable. The following problems are minor and can be corrected:

Table 5-1. Typical Processing Problems with KPR 3

Problem	Cause
Coating has "orange peel" texture.	Too rapid drying, or excessive resist applied to surface.
Image washes off during development.	Poor surface preparation, insufficient exposure, or resist too thick.
Image does not develop, even after long soak.	Resist is fogged from light exposure, or transparency has low density.
Developed image is distorted, with poor resolution and ragged line edges.	Coating too thick, development time too long, insufficient exposure, or poor contact with transparency.
Developed image does not accept dye, or dyes darker in some areas than others.	Coating too thin, or coating uneven due to poor surface preparation and poor leveling of the liquid resist.
Resist is evident in nonimage areas.	Fogged coating, insufficient time in developer, or spent developer.
Resist image peels or breaks down during etching step.	Insufficient coating thickness, poor adhesion due to unsatisfactory surface preparation, or lack of postbake.

Pinholes in the Image — Voids in the resist pattern are easily detected after dyeing, since the bright copper surface shows through distinctly. Pinholes are caused by a poor coating, and by dust and dirt in the processing steps; they are usually present to some extent dependent upon the quality of housekeeping. Correct pinholes by dabbing on black resist ink or black enamel paint with a very fine artist's brush, such as a 000 size. Corrections are made easier with the use of a *linen tester,* which is a small magnifying glass on a stand (see Fig. 5-14). The linen tester allows precise examination of resist patterns, and allows the paint brush tip to be carefully observed through the magnifying glass during touch-up.

Scratches in the Image — These may be very fine and are more difficult to detect; hold the board up to a strong light and tilt it at various angles to catch the glint of copper showing through the dark coating. Scratches are the result of physical damage to resist during processing, and are a more serious problem than pinholes. A long scratch can cause complete electrical discontinuity in several etched lines. Cover up scratches and damaged traces with paint, a small brush, and a magnifying glass as before. Errors due to imperfections in the original negative can also be corrected in this manner.

Fig. 5-14. Inspection and touch-up with a linen tester.

Surface Scum — This includes dirt, water marks, and dye scum in areas of the board that should be clean, unprotected copper surfaces. Such problems are readily apparent when etching begins; for this reason, the board should always be examined a few seconds after placing it in the etching bath. Exposed copper will quickly change from a bright sheen to a dull, red-brown color in cupric chloride or ferric chloride etchant. However, surface scum will protect the copper and show up as shiny patterns; scum can usually be removed by lightly rubbing the board with a soft cloth soaked in water or isopropyl alcohol. If the problem is actually due to a thin coating of fogged resist or resist that was not washed away during development, this material may be more difficult to remove. Affected areas can often be corrected by rubbing with a pencil eraser, or by scratching the metal surface with a razor knife.

Opposite Side Protection • When etching double-sided circuit boards one side at a time, the opposite side must be physically protected. Spray paint is a simple coating that can be quickly applied and removed for this purpose; wide adhesive tape is a good alternative. A spray enamel that is quick-drying and is easily stripped with acetone or paint thinner is recommended, such as Krylon enamel. The circuit board is placed face-down on a newspaper background and sprayed with two successive coats of the enamel, 2 or 3 minutes apart. Examine the paint coating and retouch any pinholes or voids that occur due to surface dirt. To remove the paint later after etching, soak the board briefly in acetone or paint thinner and wipe off the softened material; repeat this process until the surface is clean. Never soak printed circuit laminates in any organic solvent for extended periods of time, because some solvents will gradually attack bonding resins and cause delamination of the copper foil.

Postbaking • A postbake of the resist image will substantially increase its durability during etching or plating processes. Postbaking will improve the chemical resistance of photo resist, as well as its ability to adhere to metal surfaces without flaking. In the next chapter we will recommend a cupric chloride etching system, which requires 20–30 minutes chemical exposure at room temperature. Baking of the resist is particularly desirable to maintain coating resistance over such an extended time in an etchant that contains hydrochloric acid.

Postbaking at 40–50°C (104–122°F) for several hours should be sufficient for KPR 3 resist coatings; the same temperature oven can then be used for all coating and baking steps. An alternative for fast preparation of resist is 120°C (248°F) for 10 minutes. Properly cured resist is tested by firmly scratching with a fingernail; it should be difficult to damage the coating in this manner.

ALTERNATIVE PHOTO RESIST PROCESSES •

Positive-Acting Photo Resist • As previously noted, the use of positive-acting photo resist may be desirable if a positive transparency is to be used directly for image transfer. The recommended material for this process is Microposit 340 photo resist manufactured by Shipley Company and formerly known as AZ-340 photo resist. The Microposit resists are soluble in organic solvents, from which they are coated using similar techniques as for KPR 3 resist. The material is often available packaged in the form of spray cans from electronic distributors. Unfortunately for hobbyists, Microposit 340 resist is not as stable as negative-acting KPR 3 resist; the shelf life is only about 1 year from the time of manufacture, so users may be disappointed with results from this resist unless they obtain it directly from the manufacturer.

Microposit resist is sensitive to ultraviolet radiation like KPR 3 resist, although its sensitivity extends further into the visible region. However, once the image is exposed to light, it becomes *soluble* rather than *insoluble* in the developing solvent, an alkali-water solution; this is the basis for the positive process. After developing the image, the resist is baked at 65°C (149°F) and the board can be etched in all of the common etching solutions.

An inconvenience in using Microposit resist is that the coating is still light-sensitive after processing, and should be handled under yellow safelight conditions. This makes examination and touch-up of the final image difficult. If the board is exposed to white light after development or during etching, the coating changes form and becomes more soluble in etching solutions, an unfortunate consequence of the positive photofabrication process. While it may be possible to etch exposed boards in acidic solutions successfully, alkaline etchants require that white light be completely excluded following image development.

Because inspection and touch-up are so important in low-volume PCB fabrication, the author recommends that positive masks be converted into negatives so that a less-critical negative photo resist process can be used. The long-term storage instability of Microposit resist also makes it less desirable for low-volume use.

Aqueous Processing • The use of organic solvents presents problems in flammability, vapor exposure, and waste disposal. For this reason, there is constant incentive to formulate photo resist processes that are entirely based on aqueous chemistry. Two such recent processes will be briefly mentioned for the benefit of readers who do not have a fume hood or other well-ventilated work area; fume hoods can be fairly expensive since they include the enclosure itself as well as a fan and ductwork that are chemically resistant and explosion-proof.

Water-Soluble Dry Film — A good negative process that uses alkaline developer and an aqueous stripping solution is marketed by Kepro. Copper clad panels are available with the dry film prelaminated, but double-sided PCB fabrication is more versatile if a hot roll dry film laminator is used in-house. No organic solvents are needed in coating, developing, or in any other step using this system.

Plastic-Based Photo Resist — Alcohol-miscible, negative-acting liquid resist is available consisting of a vinyl acetate suspension and an inorganic activator. A recommended process that the author has recently tested is based on Re-Solv photo resist from Coval Industries. The green emulsion is brushed into a clean blank board and air-dried at room temperature. When exposed to light, the activated material is polymerized into a rubbery-like consistency that adheres to the copper surface and can be postbaked to give a good etch-resistant coating. Unexposed areas are developed by simply brushing away unreacted material with running water. The simple, nonhazardous nature of this process makes it very attractive to electronic experimenters, but be aware of the following drawbacks relative to a time-proven process using KPR 3 photo resist:

1. Re-Solv resist coatings tend to incorporate small air bubbles due to the nature of the thick suspension. These bubbles will result in voids or pinholes in the final image, which may require extensive touch-up work.
2. The resist formulation is not stable with time and the activated suspension must be used and processed as soon as it is mixed. Because the material is sensitive to both light and heat, it cannot be baked prior to exposure, and must be air-dried.
3. Exposure times tend to be long. More than 20 minutes may be necessary to complete the chemical hardening reaction with common light sources.
4. The resist does not produce high-resolution images, but they will be adequate for most prototyping work. Lettering and pad center holes tend to lose definition. This is a consequence of a rather thick resist coating and a long exposure time. Thinner coatings are difficult to apply evenly and are not adequately etch-resistant.

CHAPTER 6
Etching

INTRODUCTION • Printed circuit *etching* is the process of exposing a circuit board to a chemical solution that attacks and dissolves copper metal in unprotected areas, leaving behind the desired trace patterns intact. The process is dependent upon the use of an *etch resist,* as detailed in Chapter 5, which is applied as an adhesive, chemically resistant coating during image transfer. Individual etching steps include resist inspection, exposure of board to etchant, resist removal, and final surface cleaning.

Many chemical formulations and mechanical methods are used to achieve the general objectives of printed circuit etching. In low-volume prototype work, the most important factors in developing a suitable process are as follows:

1. Simple equipment
2. Low cost
3. High yield
4. Control of etch rate
5. Ease of process control
6. Minimal handling and waste disposal problems
7. Compatibility of etchant with resist
8. Acceptable line resolution and pattern definition

Etching equipment and chemistry must be chosen with these factors in mind. Despite all of the possibilities, hobbyists and experimenters will usually end up with about the same system: *ferric chloride* or *cupric chloride* used in a tray, or in an enclosed container that provides some degree of agitation and possibly heating. The metal chloride formulations offer the most stable, simple etchant chemistry while minimizing chemical exposure hazards and process control problems.

As work volume increases, better equipment can be purchased to improve etching efficiency, and different chemical etchants may be chosen to complement the characteristics of a particular resist. For example, a prototype *spray etcher* system from Kepro Circuit Systems is shown in Fig. 6-1. This equipment allows a double-sided PCB up to 12 by 24 inches (305 by 610 mm) in size to be etched on both sides in just 2–3 minutes, as opposed to the 15–30 minutes per side required for most open baths. Spray etching will also produce superior results: even etching, finer line definition, less undercut, and less resist breakdown due to shorter chemical contact time. Many chemical solutions can be used in the spray etcher, provided that they are not corrosive to the construction materials of the system. For example, sodium persulfate is a good low-volume etchant used with solder plating; solder is not compatible as a resist with the metal chloride etchant formulations. More information on solder plating used in conjunction with professional plated-through-hole PCBs is found in Chapter 2.

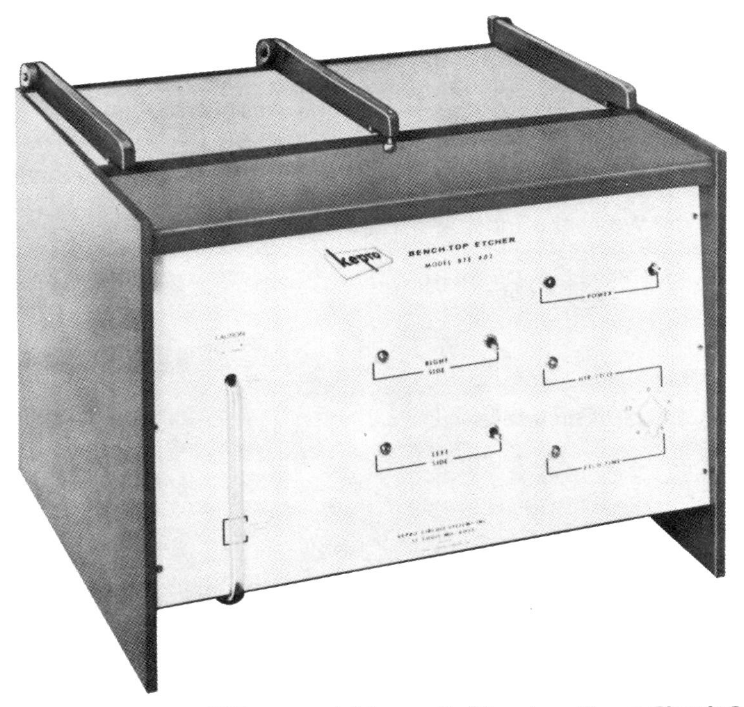

Fig. 6-1. A prototype PCB spray etching unit. *(Courtesy Kepro Circuit Systems, Inc.)*

OBJECTIVES • The objectives of this chapter are as follows:

- Recommend commercial sources for equipment, supplies, and chemicals needed in small-scale printed circuit etching.
- Explain the use of common ferric chloride solutions for tray etching.

- Detail the characteristics of a cupric chloride bath, which allows chemical regeneration and minimizes waste disposal problems.
- Explain *bubble etching* techniques and associated equipment.
- Give simple procedures for the chemical maintenance of a cupric chloride bath.
- Discuss the important post-treatment steps in etching, such as resist stripping and surface cleaning.
- Discuss the common problems that occur in PCB etching, and their typical causes.

SOURCE OF MATERIALS • Materials needed for PCB etching can be roughly divided into four groups: equipment, chemicals, glassware, and laboratory supplies. Depending on the complexity of a particular operation, the reader may obtain most materials locally, or buy equipment and supplies from a vendor who deals specifically with the printed circuit trade. Two commercial sources will be recommended for each of the four groups of materials, and two other sources will be given for PCB supplies in general; company addresses are found in Appendix B.

Prototype Etching Equipment •
Kepro Circuit Systems, Inc.
Feedback, Inc.

Chemicals •
Eastman Kodak Company
MCB Reagents

Glassware •
Lab Glass, Inc.
Ace Glass, Inc.

Laboratory Supplies •
American Scientific Products
Cole Parmer Instrument Co.

General PCB Etching Supplies •
Jack Spears, Inc.
GC Electronics

FERRIC CHLORIDE ETCHANT • Ferric chloride ($FeCl_3$) solutions are widely used by hobbyists to etch copper-clad PCBs because this etchant is low in cost, has a high tolerance for dissolved copper, presents few chemical exposure hazards, and is completely stable. Ferric chloride is compatible with most types of photo resists and screen-printed ink resists. The typical etchant composition is 35% $FeCl_3$ by weight in water, although commercial formulations may include additives to promote even etching (surfactants), to prevent sludge formation (hydrochloric acid), to reduce odor and foaming (antifoam), and to extend etchant life (oxidizers). The raw material is available as inexpensive lump $FeCl_3 \cdot 6\ H_2O$, but the reader is advised to spend a little more for concentrates or pre-mixed solutions designed expressly for printed circuit applications. In their catalog, GC Electronics states that 6 ounces of ferric chloride solution will etch about 180 square inches of PC board, so this material is ideal as an expendable etchant for occasional use.

Ferric Chloride Chemistry • Many users do not obtain full performance from etching solutions because the chemistry is not well known. Therefore, a brief discussion of ferric chloride chemical etch action will be included to help distinguish between the various problems that may occur; type of equipment, changes in solution composition, and conditions of use can all affect etchant performance.

The initial etching reaction is the oxidation of copper metal to cuprous chloride by ferric ion:

$$FeCl_3 + Cu \rightarrow FeCl_2 + CuCl$$

However, as etching action continues, the reaction changes. Cuprous chloride is converted to cupric chloride, which can also oxidize copper metal into solution:

$$FeCl_3 + CuCl \rightarrow FeCl_2 + CuCl_2$$
$$CuCl_2 + Cu \rightarrow 2\ CuCl$$

Since these reactions are all oxidations, the best etch action is obtained by promoting conditions for oxidation: high air content, and elevated temperature (up to 49°C or 120°F). This is one reason why spray, splash, and bubble etching methods are all so much more effective than simple tray immersion; these methods encourage the continuous introduction of air into the system. Another positive factor is improved agitation, which allows spent solution to be continuously replaced by fresh material at the copper surface. The hobbyist may obtain satisfactory results by placing a circuit board in a glass tray filled with etchant and gently rocking it back and forth for 20 minutes, but a spray etcher operating at 38°C (100°F) can complete the same job in 2–3 minutes!

Unfortunately, the use of hot etchant can cause problems that may make room temperature operation more practical for the occasional experimenter. In addition to producing foam, fumes, and odors, hot solutions will lose hydrochloric acid (HCl), which is a desirable component for promoting oxidation and for dissolving sludge. Therefore, HCl is often added to ferric chloride solutions (up to 5%) for this reason, and the etchant life is greatly extended by combined air/acid oxidation of ferrous and cuprous ions to their higher state:

$$2\ Cu_2Cl_2 + 4\ HCl + O_2 \rightarrow 4\ CuCl_2 + 2\ H_2O$$
$$4\ FeCl_2 + 4\ HCl + O_2 \rightarrow 4\ FeCl_3 + 2\ H_2O$$

HCl also discourages the formation of ferric hydroxide sludge:

$$Fe(OH)_3 + 3\ HCl \rightarrow FeCl_3 + H_2O$$

However, excess HCl must be avoided because it can attack and degrade resist coatings.

Ferric chloride solutions tend to form dark sludge on or around the work as it is being etched, due to precipitation of metal oxides and hydroxide salts. These solid residues are more than just an inconvenience, because they can build up and interfere with etch action. Therefore, some form of agitation is always desirable.

Practical Use of Ferric Chloride • Ferric chloride is highly recommended for occasional PCB etching because it is readily available in a form that requires little maintenance. The solution is simply used until its ability to dissolve copper diminishes significantly, or until the sludge build-up becomes intolerable. The problem of waste disposal then arises, but this is not serious for small quantities of etchant.

The simplest etching equipment is an open glass tray (see Fig. 6-2) that can be heated with a hot air blow dryer; a warmer solution will increase etch speed. After immersing the PCB in solution, steady agitation is necessary to obtain complete, even etching within 15–30 minutes time. If uneven etch patterns become apparent, some change in the method of agitation should be made to correct this problem, because it will lead to excessive etch times, excessive image undercut, and resist breakdown. Agitation may consist of gentle rocking, stroking

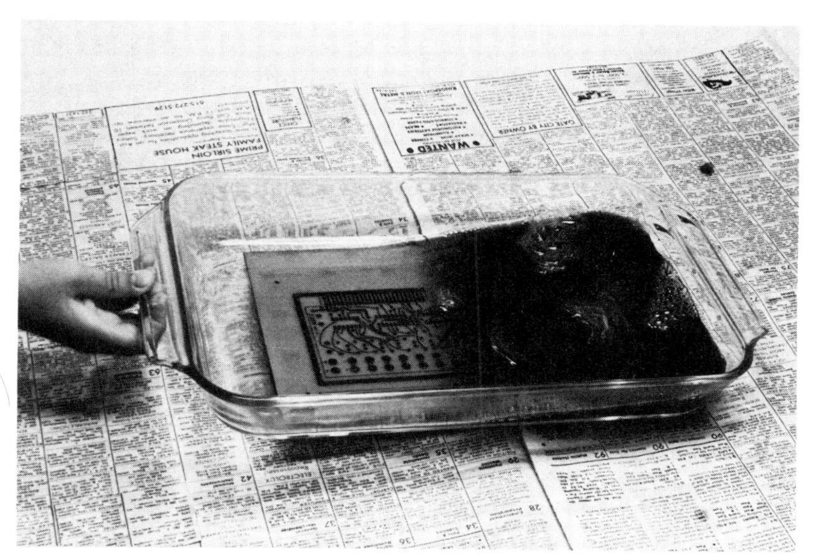

Fig. 6-2. Tray etching with ferric chloride.

with a plastic brush, or splashing solution on the surface of the board with a small cup. In any event, the board should not be exposed to etchant any longer than necessary. Gloves should always be worn when handling harsh etchant solutions.

The next improvement in $FeCl_3$ etching technique is the use of a bath agitated with a steady stream of *air bubbles,* which enhance oxidation and improve solution flow across the surface of the board. The use of air bubble equipment will be discussed in more detail later in this chapter in conjunction with cupric chloride etchant.

Etchant Disposal • Spent ferric chloride etchant is not easily regenerated chemically because of the buildup of sludge, and because the copper and iron ions are difficult to separate. However, disposal is a difficult problem, and accounts for the growing unpopularity of this solution in industrial applications. Metal drain systems will be quickly attacked by $FeCl_3$ and its fumes, the copper content will create environmental problems, and sewer lines may become clogged with sludge when ferric chloride combines with alkaline materials, such as caustic drain cleaner. These problems explain why ferric chloride etchant is only recommended for occasional use. On a small scale, exhausted etchant may be properly disposed of by neutralizing with caustic or sodium carbonate, and mixing with a suitable amount of concrete or mortar mix. After the mass hardens, it can be safely thrown in a landfill dump with the etchant converted to an immobile form.

Stripping and Post-Cleaning • A final cleaning treatment is important for circuit boards etched in $FeCl_3$, because the solution leaves a variety of salts and metal oxides deposited on the laminate surfaces. These contaminant materials will affect the insulation resistance of the circuit, and can cause gradual deterioration of the copper traces. Salt contamination is particularly troublesome with open tray etching because the long contact times tend to promote lateral etching, or *undercutting* of resist patterns; the resultant overhang (see Chapter 2, Fig. 2-16) is a good hiding spot for residual etchant. A frequent symptom of this problem is circuit trace edges that begin turning green a few weeks after the PCB is fabricated.

For effective removal of $FeCl_3$ etchant contaminants, the board should be stripped of resist using a suitable solvent (such as acetone or methylene chloride) that softens the coating and allows it to be brushed away. The board should then be rinsed in 5% HCl or oxalic acid to dissolve the salts. Scrubbing with steel wool (see Fig. 6-3) or abrasive pumice helps this cleaning operation, and protective rubber gloves should be worn to prevent acid skin contact. A final rinse in flowing water should then complete the cleaning.

CUPRIC CHLORIDE ETCHANT • Cupric chloride solutions are very similar to ferric chloride as etchants, but the cupric system offers some distinct advantages in medium volume PCB prototype applications as follows:

1. Simple regeneration of spent solution
2. No waste disposal problems
3. Lower cost
4. Simple process control
5. No sludging

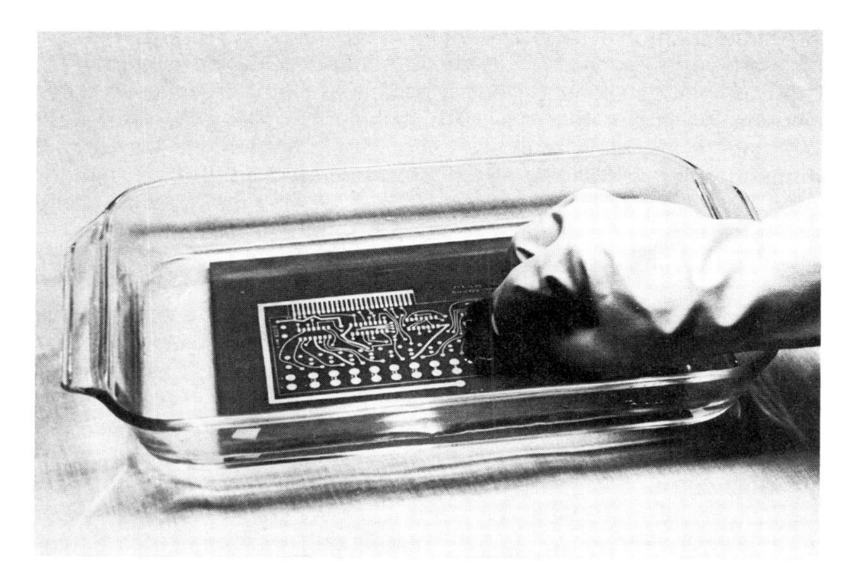

Fig. 6-3. Cleaning the etched board with 5% HCl and steel wool.

At the same time, the desirable properties of ferric chloride are retained: high capacity for dissolved copper, and fast etch rates with suitable techniques, such as bubble, splash, or spray etching. One-ounce copper boards can be completely etched in less than one minute using cupric chloride at 49–54°C (120–130°F) in spray etching equipment. At these high temperatures the problems of HCl fumes and photo resist attack become more likely, but the metal chloride etchants can be used at room temperature as well if a fume hood or other exhaust system is not available; typical processing time for 1-ounce copper boards in a cupric chloride bubble etching tank at 25°C (77°F) is 20 minutes.

Formulation and Chemistry • The following formulation is recommended for a cupric chloride etching bath:

Cupric chloride, solid ($CuCl_2 \bullet 2\ H_2O$)	200 g
Hydrochloric acid, conc. (HCl, 37.5%)	100 g
Water, to make	1000 mL

Unless this bath becomes contaminated with undesirable materials (such as organic solvents, which can soften resist coatings), it will never require waste disposal. The reasons will be apparent later during the discussion of chemical control and regeneration.

Cupric chloride etches copper metal through the following chemical reaction:

$$CuCl_2 + Cu \rightarrow Cu_2Cl_2$$

The initial solution color is a bright emerald green, and it turns to murky olive-brown as cuprous chloride is generated during etching. The cuprous chloride product is not as soluble as cupric chloride, so excess chloride ion is added to the bath to help complex and dissolve the cuprous form, preventing sludging. Sodium chloride (NaCl), ammonium chloride (NH_4Cl), and hydrochloric acid (HCl) are commonly used as the complexing agents. HCl is recommended because it also serves to reverse the etching reaction through air oxidation:

$$2\ Cu_2Cl_2 + 4\ HCl + 0_2 \rightarrow 4\ CuCl_2 + 2\ H_20$$

This air oxidation reaction is slow because oxygen is fairly insoluble in the etchant bath, but an air regeneration approach is ideally suited for bubble etching in a small laboratory operation. Since the bath is already equipped with an air inlet to generate bubbles, it is only necessary after etching to continue passing air through the solution until its color changes back to the characteristic bright green of cupric chloride.

Spray etching is a very effective commercial technique with cupric chloride because the etchant is partially rejuvenated when it mixes with air as a fine mist, activating the solution and decreasing etch time. It should be noted that in commercial high-volume applications regeneration of cupric chloride by air oxidation is too slow to be completely practical, so faster methods are used to reverse the reaction, such as direct chlorination, electrolytic action, or intense oxidation with hydrogen peroxide or sodium hypochlorite.

Air Bubble Etching • Etching with air bubbles is an ideal approach using cupric chloride, with the air providing several improvements over simple immersion in a bath. The action of swarms of tiny air bubbles surrounding a circuit board results in faster, even etching because agitation is increased, oxidation is enhanced, and the etchant is continuously converted back to its active chemical form. A practical air bubble tank can be easily constructed with the air source being a small laboratory pump, a large fish tank aerator, or even a vacuum cleaner with the hose reversed. A recommended approach is to generate an even flow of bubbles rising from the bottom of a tank, and to float the circuit board on top of the solution, allowing air bubbles to support and surround it. This approach is consistent with the general principle maintained throughout this book that prototype PCBs are most conveniently etched *one side at a time.* The mechanical action of air bubbles continuously rising to meet the copper surface is in effect very similar to spray etching, the optimum commercial approach.

A practical cupric chloride etching tank can be made from a medium-size plastic ice chest, as shown in Fig. 6-4. The plastic material is compatible with the etchant solution, the container prevents evaporation, the drain plug is useful, and the lid can be lowered during etching to contain mist and acid fumes. If an immersion heater is added to bring the solution to 38°C (100°F) for fast etching, the heater and thermostat must be constructed of chemically resistant material, such as quartz or titanium. Stainless steel does not stand up to cupric chloride. As shown in Fig. 6-4, the author's etching tank is used inside an old chest freezer that provides even greater containment for the corrosive etchant, and this allows the bath to be used in a workshop area

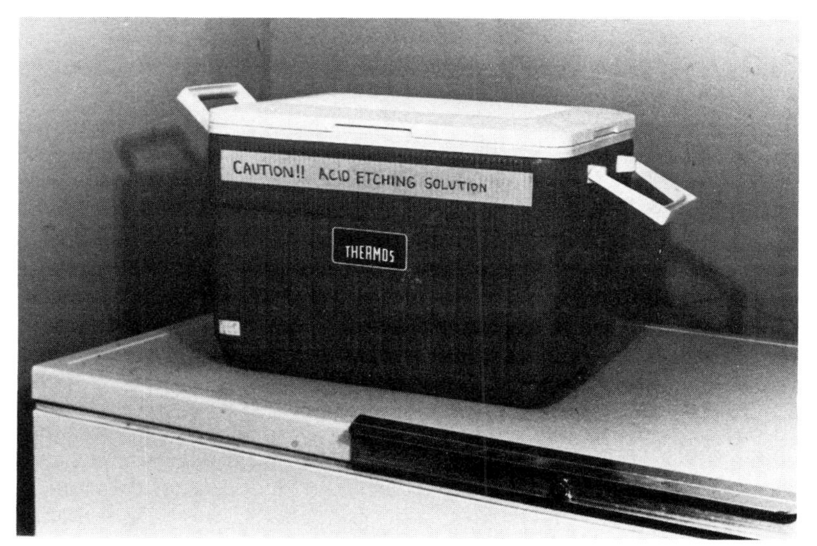

Fig. 6-4. A plastic ice chest makes an ideal etching tank.

without the benefit of a fume hood. The plastic walls of most freezers are acid resistant, and a large chest freezer makes an ideal sink for containing rinse trays, acid dip tanks, dirty gloves, and other potentially corrosive items.

An internal view of the air bubble tank is shown in Fig. 6-5. In order to generate an even flow of small air bubbles, coarse fritted glass air bubblers (gas dispersion tubes) are spaced evenly on the bottom of the

Fig. 6-5. Fritted glass air bubblers in the bottom of an etch tank.

tank and connected with Tygon tubing and plastic "T" connectors. The tubing network is brought out of the tank in one spot as an air inlet, where a low pressure (2–5 lbs or 0.9–2.3 kg) air supply is connected during actual etching. The full tank is shown in operation in Fig. 6-6, illustrating the pattern of air bubbles swarming to the surface. Large circuit boards are easily floated in this tank, but smaller boards may tend to sink and must be supported on some plastic rails, or attached to larger pieces of scrap board. An inexpensive aeration system can be assembled from materials available at pet stores for use in fish tanks, such as sintered glass tubes and a small air compressor.

An alternate approach to generating air bubbles in an etching tank is to line the bottom with coils of plastic tubing connected to an air inlet, and punch tiny holes in the tubing with a miniature drill or with a red-hot needle. The overall objective is to get an even flow of small bubbles throughout the tank so that all areas of a floated circuit board will receive about the same etch action. Some laboratories use a deep, narrow tank instead of a wide, shallow container; in this case, the work is suspended vertically in the tank and air bubbles travel upwards across both faces of the board. Vertical air bubble etching tends to produce less even etching because the bubbles do not encounter the work uniformly. It is usually necessary to rotate circuit boards periodically on all four edges to produce evenly etched results in such a tank.

Cupric Chloride Bath Maintenance • The cupric chloride bath is initially prepared by mixing ingredients in the proportions previously given. The resulting solution has three important parameters that can be adjusted to maintain the bath at its optimum operating conditions: color, density, and HCl content. The proper control sequence consists of the following steps.

1. Air-oxidize etchant to a bright green color
2. Measure and adjust density to 1.17 g/mL
3. Measure acid content by titration
4. Adjust acid level with concentrated HCl

Fig. 6-6. Flotation etching with air bubbles.

Color (Oxidation State) — Bright emerald green is characteristic of acidic cupric chloride solutions. If any cuprous chloride is present, the color will become a darker olive-brown, but this copper form will always revert to its active cupric form with sufficient exposure to air bubbles under acidic conditions.

Density (Copper Content) — The initial operating density (specific gravity) is 1.17 grams/ml. Density will gradually increase as copper dissolves into solution during etching, and as water evaporates. Etch rate will slow down and sludge may form with increasing density, so density must be monitored and periodically adjusted by adding water to the tank. A convenient device for measuring solution density is the *hydrometer* shown in Fig. 6-7. This device is available from scientific supply houses and operates on the same flotation principle used in common sulfuric acid battery fluid checkers. The hydrometer shown in the figure has a narrow, expanded density range from 1.15 to 1.25 g/ml, and is only 6 inches (152 mm) long; it can be floated in the etching bath, and

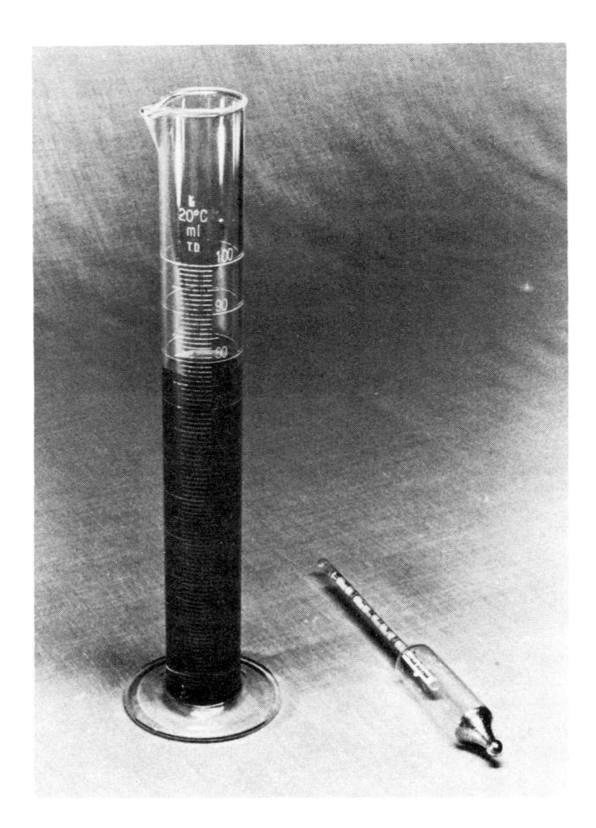

Fig. 6-7. A hydrometer and glass cylinder for measuring solution density.

the flotation level indicates density directly. If the bath is too shallow, a long narrow container must be used to hold a solution sample while the hydrometer is floated.

When density adjustments are made to cupric chloride etchant, the following equation can be used to add water back to the bath for dilution purposes, with the standard density objective being a specific gravity of 1.17g/mL:

$$Y = \frac{X(d_x - 1.17)}{0.17}$$

where,

X is volume of present solution at (measured) density d_x,
Y is volume of water to be added.

This dilution approach requires that the bath volume be easily measured, using level marks or a calibrated dip stick.

Acid Concentration (HCl Content) — The initial hydrochloric acid concentration should be in the range of 3.5 to 4.0%, or 1.1 to 1.3 Molar. If the acid concentration gets much higher, problems such as resist degradation and excessive fuming will be observed. A lower acid level is not intolerable, but it will result in slower etch rates and a more difficult regeneration of cupric chloride from the cuprous form; sludging may also occur, because HCl helps dissolve etch by-products through complexation.

Therefore, an analytical method must be available to periodically monitor HCl concentration. The simplest approach is by *titration,* a technique easily accomplished manually with the aid of a few pieces of volumetric glassware. Note that acid concentration should not be measured *before* the etchant is completely converted to its oxidized bright green form through air bubble agitation, because acid is consumed in this regeneration process.

An acid-base titration requires that a measured amount of acid sample be mixed with a suitable indicator that changes color near the neutral pH region. A base solution of known composition is then slowly added to the acid with mixing until the indicator just changes color, showing that the acid has been neutralized. The initial acid concentration can be directly calculated from the volume of base consumed. Since our HCl analysis does not require extreme accuracy ($\pm$10% is fine), the titration procedure should not take more than 5–10 minutes and does not require extreme care. The following equipment and materials are necessary, and are pictured in Fig. 6-8:

1. 1.0 mL pipet — This calibrated glass tube has a mark to show when exactly 1.0 mL of liquid has been drawn in by suction.
2. Rubber suction bulb — Never use your mouth to fill a pipet with chemical solutions; the risk of pulling material into your mouth is not worth taking.
3. 50 mL beaker.
4. Magnetic stirring apparatus — The solution can also be stirred manually during titration, but a laboratory stirrer using a magnetic bar provides a more convenient mixing system.
5. 50 mL buret — This is a long, thin, calibrated glass tube that is filled with base solution (titrant). The base is then dripped into the acid sample through a bottom stopcock, and the buret markings allow the volume of base consumed to be measured.
6. Buret stand and clamp — The buret must be clamped in a vertical position for accurate titrant delivery.

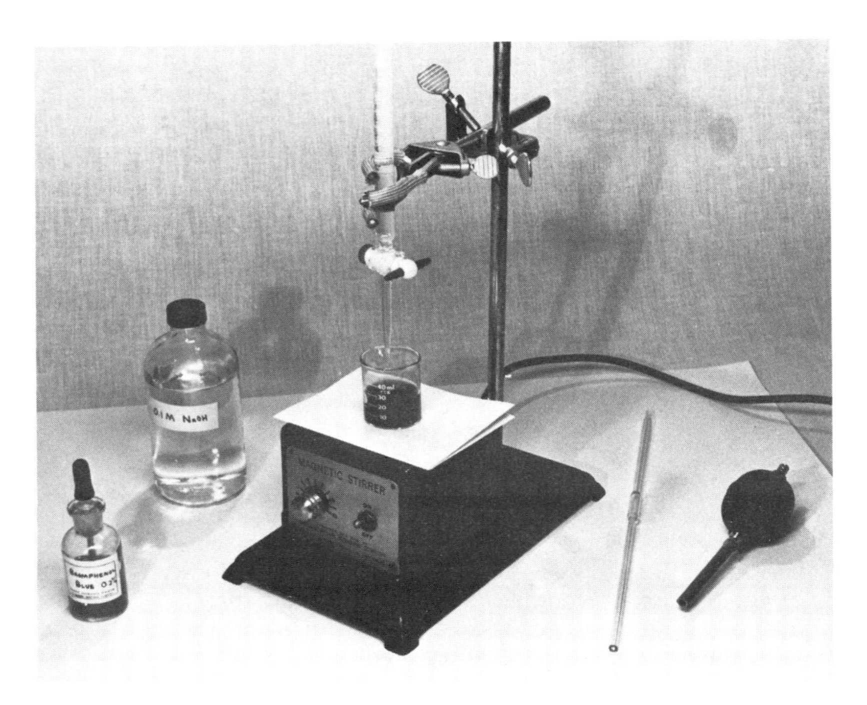

Fig. 6-8. Acid titration equipment.

7. Indicator solution — Recommended indicator is 0.2% Bromphenol Blue in isopropyl alcohol; 3–4 drops are added to each 1 mL acid sample.

8. Base titrant — Sodium hydroxide, 0.10 Molar (0.4%). This standard titrant solution can be prepared by dissolving 4.0 g of solid sodium hydroxide pellets in one liter of water.

The titration procedure consists of the following steps:

1. Measure out 1.0 mL of etchant solution with a pipet and drain it into a 50 mL beaker containing a small magnetic stirring bar. Dilute this sample with 20 mL of water.

2. Fill a clean 50 mL buret with 0.1 Molar sodium hydroxide titrant and note the starting mark.

3. Add 3–4 drops of indicator solution to the sample and place it on the magnetic stirrer base underneath the buret. The initial solution color should be yellow-green.

4. Add titrant to the stirring solution while observing the color. Slow down this addition as the endpoint approaches, as evidenced by the formation of blue color in the region where the drops hit.
5. Neutralization is complete when the solution becomes completely pale blue in color. Just one or two drops of titrant may be enough to cause the color to change at the endpoint.
6. Note the amount of titrant consumed. The etchant is calculated to contain 0.31% HCl per mL of titrant consumed. Therefore, a fresh etchant solution containing 3.7% HCl would require 12.0 mL of titrant to reach the endpoint.
7. As an example, assume that 8.0 mL of titrant was consumed for an unknown etchant sample. The acid concentration of the etchant would then be 8.0 × 0.31% = 2.5% HCl, which is low.

HCl Adjustment — Acid adjustment should be made as the last step in process control, because the addition of concentrated HCl (37.5%) will not appreciably change the density of the etchant; concentrated HCl has a density of 1.19 g/mL, very close to that of etchant. The actual adjustment is made as follows. Assume that the acid concentration is a low 2.5% from the titration test. We have three gallons of etchant in the tank and want to bring the acid level up to 3.7% HCl. The necessary amount of concentrated (37.5%) HCl to be added is calculated from the following equation:

$$Y = \frac{X(3.7 - A)}{37.5 - 3.7}$$

where,

X is initial etchant volume,
A is initial etchant HCl concentration in %,
Y is volume of 37.5% HCl to be added.

For the 3 gallon @ 2.5% example, the amount of acid to be added would be:

$$3(3.7 - 2.5)/(37.5 - 3.7) = 0.1 \text{ gallon, or } 450 \text{ mL}$$

Wear rubber gloves and safety glasses, and be extremely careful when handling concentrated acid. Never pour any material directly into concentrated acid, which can cause splattering; always add the acid to the other material. A well-ventilated area is necessary to avoid acid fumes, which are both a corrosion and a breathing hazard.

Etchant Disposal • The interesting characteristics of cupric chloride etchant result in the fact that the bath will never contain materials other than cupric chloride, HCl, and water following air oxidation; this statement assumes that no significant contamination occurs, and no metals other than copper are etched. Therefore, waste etchant is never generated. The solution volume will gradually grow as periodic water and acid adjustments are made to counter the increasing copper content, a result of etching circuit boards. Excess etchant can be sold or donated to other persons interested in printed circuit board fabrication. If the bath becomes contaminated with undesirable materials (such as organic solvents) it must be disposed of using the same precautions as with ferric chloride. Neutralization and mixing with concrete or mortar mix is the safest way to render a chemical etchant immobile for landfill.

Post-Cleaning • As with ferric chloride, cupric chloride will leave undesirable deposits on circuit board surfaces following etching. The most common residue is cuprous hydroxide, which is water-insoluble and appears as a yellow film on laminate surfaces. Cuprous chloride can also appear as a white residue if the etchant is low in acid content. Both of these contaminants are efficiently removed by soaking the board in 5% HCl following etching and resist stripping, and then rinsing well under running water.

ETCHING PROBLEMS • Various etching problems have been noted throughout the chapter, but the most common problems will be summarized in this section, together with possible causes and solutions. It will be assumed that KPR 3 photo resist is used for image transfer, and that cupric chloride solution is used for bubble etching.

Resist • Resist peels, degrades, or otherwise breaks down during etching — this is usually due to improper copper surface cleaning prior to photo resist coating, resulting in poor adhesion of the resist. Poor adhesion can also be due to lack of a post-bake step for the final resist coating. Some other possible causes for resist degradation are too thin of a coating, etchant temperature too high, acid concentration too high, excessively long etch times, and the presence of solvent in the etching tank.

Long Etch Times • Exhaustion of the bath is the most likely reason for long etch times, requiring chemical rejuvenation or etchant replacement. If etchant becomes too concentrated due to water evaporation, etch time will also increase; a density measurement will reveal this condition. Low HCl acid concentration is another process control situation that results in long etch times. Assuming that the bath chemistry is properly balanced, etching time can usually be shortened by increasing agitation and by raising the operating temperature of the bath.

Uneven Etching • Uneven etching is illustrated in Fig. 6-9 with a partially etched board that has some areas completely resolved, while other areas (light-colored) contain substantial amounts of copper yet to be removed. Uniform agitation will prevent uneven etching, and this is the reason why an even pattern of air bubbles should be generated in the bubble etching tank. If certain areas of the bath are observed to produce consistently faster etch action than others, it may still be possible to obtain overall even results by moving the board around in the solution every few minutes. The situation shown in Fig. 6-9 is undesirable because it requires that some areas of the board be exposed to etchant longer than necessary, causing pattern undercut and resist degradation. Unevenness in etching will always occur to some extent. For example, board edges and areas with many closely spaced traces will generally etch faster than open copper areas. However, as a general rule, once etchant breaks through to the PCB laminate in one spot, the rest of the board should be completely etched within an equal additional time period.

Fig. 6-9. Uneven etch patterns.

Etchant Does Not Attack Copper • Etchant does not attack copper in some areas where it should — the culprit is usually a thin film of photo resist or dye scum remaining after image development. Rubbing the board with a soft cloth soaked in isopropyl alcohol may remove the offending film, but this may not help if the pattern has been well-baked. In such a case, it will be necessary to remove the board from the etching bath, wash and dry it, and scrape off the undesirable scum using a razor knife or other sharp metal object. Very thin films of photo resist may not be noticed after dyeing the resist pattern, because this coating is not thick enough to retain dye. Therefore, it is good practice to stop and examine all circuit boards after the first minute of etching; copper areas that are properly attacked will change from shiny bright to dull reddish-brown in appearance.

Excessive Undercut • Pattern undercut occurs due to lateral etch action instead of the desired perpendicular attack. Long etch times in an exhausted bath tend to promote this problem. Undercut is minimized by fast etching and good solution agitation at the surface of the board. Spray etching is the optimum method to avoid undercut because it produces perpendicular etch action and fast results. Bubble etching is also a significant improvement over simple immersion in a solution, which can produce almost as much lateral as perpendicular attack.

Sludging in the Etch Tank • Sludging in the etch tank should not ordinarily occur with cupric chloride solutions containing excess chloride ion in the form of HCl. Some users add 10% sodium chloride or ammonium chloride to help complex and dissolve cuprous chloride solids. Sludge will precipitate if the bath density becomes too high, however; the presence of solids interferes with etch action and leaves residue on the surfaces of the finished board.

Corrosion of Circuit Board Traces • Extensive copper corrosion is almost always a result of poor surface cleaning following PCB etching. Etch undercut promotes trace corrosion because it produces nooks and crannies which retain etchant residues. To prevent this problem (which shows up as discoloration at the edges of metal traces) use HCl at 5% concentration with steel wool scrubbing. When acid is used for cleaning, it must also be completely removed with a good water wash, because mineral acid will attack solder joints. Even properly cleaned copper circuit traces will eventually discolor through air oxidation, making the metal surfaces difficult to solder. This is the main reason that a final solder coating should always be applied to copper PCBs, as detailed in Chapter 7.

CHAPTER 7

Electroplating and Coating

INTRODUCTION • Edge connector fingers provide the most common system for connecting printed circuit board lines to external signals, and are the standard mounting/connect method used in microcomputer systems. A typical microcomputer board will have 100 signals brought to the tab finger contacts, and it is imperative that these mechanical connections remain trouble-free; a circuit board may be re-inserted many times in its socket, particularly if it is a prototype in the design and testing stages. Poor electrical/mechanical contact is the most common problem found in electronic circuits, so the industry has adopted *gold-plated contacts* as a standard approach to ensure reliability of edge connector systems.

Prototype circuits will have enough problems without adding the aggravation of poor connections; this chapter therefore presents information on the gold/nickel electroplating of PCB tab connectors so that professional results can be obtained with low-volume equipment. Because plating is somewhat of an art as well as a scientific skill, the basic principles and techniques in electroplating will be discussed.

Another important step in PCB fabrication which has been stressed throughout this book is the deposit of a final protective solder surface coating to cover copper traces and preserve them in a readily solderable state. Because prototype boards are hand-assembled and will require a great deal of careful soldering, both during assembly and during circuit alterations, proper surface preparation is an important consideration to the experimenter. Circuit pads that can be easily soldered several weeks after PCB fabrication may very well make the difference between a successful project and one that is marred by burnt components, lifted traces, solder bridges, and short tempers.

OBJECTIVES • The objectives of this chapter are as follows:

- Explain the general principles of electroplating.
- Describe specific equipment for low-volume laboratory plating.

- Recommend procedures and baths for plating PCB edge connector tabs with gold/nickel.
- Present a simple manual procedure for coating PCB copper traces with a smooth layer of protective solder.

GENERAL PRINCIPLES OF ELECTROPLATING •

Electrolysis • In *electrolysis,* two conducting electrodes known as the *anode* and the *cathode* are suspended in a solution known as the *electrolyte,* and current is passed between them through the solution. Various chemical changes occur in this process, usually at the electrode surfaces; electrons enter the solution at the cathode, causing reduction reactions, and leave at the anode, causing oxidations. When electrolysis is used for metal plating, the aqueous plating solution contains metals in the form of ionic salts. Metal is dissolved at the anodes, and electrodeposition of metal occurs simultaneously at the cathode. As shown in Fig. 7-1, if the article to be coated has a conductive surface and is connected as the cathode, it will receive a metal deposit. A metal bar of the same composition is generally used as the anode, and metal is dissolved into solution from this bar; the net effect is a transfer of metal from anode to solution to the workpiece, a process known as electroplating. Nickel is used in the example of Fig. 7-1, with an electrolyte containing nickel sulfate.

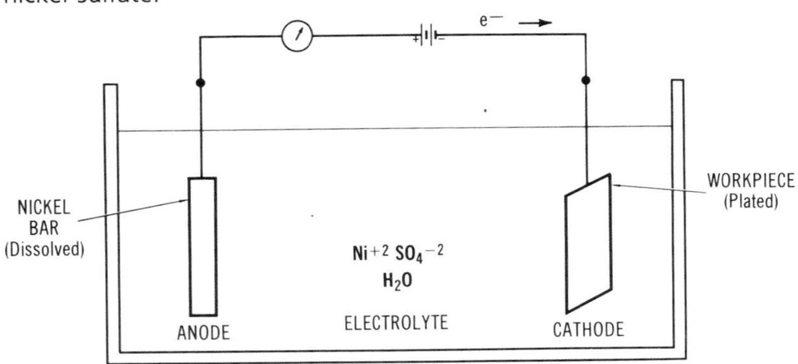

Fig. 7-1. Electrolysis with nickel and nickel sulfate.

Faraday's Law • Faraday's law is the fundamental principle of electrolysis, and can be generally stated as follows:

> *The quantitative extent of chemical reactions occurring at either the cathode or anode during electrolysis is directly proportional to the amount of electric current passed through the solution.*

This principle allows the electroplater to estimate how much metal is being deposited with time during plating, based on the measured rate of flow of electrons. Unfortunately, however, the different reactions occurring at the electrodes may not be completely straightforward, as we shall see.

Electrode Potentials • The accepted parameter that measures the relative tendencies of various metals to dissolve and form their positively charged ions is known as the *electrode potential*. Simply defined, the single electrode potential for a particular metal element is that potential voltage difference that occurs when a piece of the metal is dipped into a solution of its ions under standard conditions. (Electrode potentials are always relative values. When they are standardized versus the hydrogen electrode, they become known as *standard reduction potentials*.) Zinc shows a negative potential, and it tends to produce zinc ions and electrons by dissolving; on the other hand, copper has a positive potential, with a tendency for copper ions to deposit on the metal surface and consume electrons.

When the standard electrode potentials for many various metals are arranged in quantitative order, they form what is known as an *electromotive series*. Metals with positive values, such as copper and gold, are termed *noble* metals, while the negative ones are referred to as *base* metals. In general, the noble metals are the easiest to deposit electrolytically on a cathode, but are the most difficult to dissolve into solution at the anode. The main objective in metal plating is always to control the voltage difference that exists around the cathode so that the proper chemical reduction reaction occurs, and to minimize undesirable side reactions. For example, copper can be plated out in preference to zinc existing in the same solution, because copper is more positive in the electromotive series. However, if the cathode voltage becomes too high, zinc will begin to co-deposit, and hydrogen gas will also start evolving as another less positive side reaction.

The ordering of metals in the electromotive series assumes that variables such as ionic concentration (or more exactly, ionic *activity*) are identical in the electrolytes. However, concentrations can be effectively manipulated to control the tendency for various cathode reductions to occur. Thus, it is possible to actually plate zinc in preference to copper by adding complexation reagents (such as cyanide ions) to the bath, which significantly reduce the activity of copper ions. As another example, tin and lead have about the same electrode potentials, so they are plated simultaneously as an *alloy* whose relative composition depends directly on the respective metal ion concentrations in the electrolyte.

Polarization • The potential voltage difference that exists across an electrode immersed in solution is normally dependent upon the electrode composition and the chemistry of the electrolyte; quantitatively, it is on the order of a few tenths of a volt. Most metal electrode reactions are reversible, so the direction of the reaction can be controlled by superimposing an external voltage, as is done in electroplating. Ideally, this external voltage does not affect the electrode potentials, but only influences the direction and magnitude of electron flow based on the inherent resistance of the solution. However, in actual fact, as electrolysis reactions proceed, changes occur near each electrode that alter electrode potentials significantly. This effect is known as *polarization,* and is defined as the change in potential of an electrode as a result of electrolysis. The potential of the anode always becomes more positive and that of the cathode more negative than the initial potentials, so the net effect of polarization is to *oppose* the passage of current, requiring additional external voltage to maintain a given electrolysis rate.

Polarization is caused by three types of changes that may occur near the electrode region as a result of current flow. The first and most obvious change is in the local concentration of metal ions, which become depleted near the cathode and more concentrated at the anode surface; this effective change in ionic activities has a direct effect on electrode potentials. The second change is in the surface composition of the electrodes themselves, and is most commonly observed as the formation of insoluble products on the anode. The third category of changes is termed chemical polarization, and is used to explain effects not adequately covered by the first two categories. Chemical effects can be viewed as the variations in actual chemical processes occurring at the electrode/electrolyte junction, whose reaction rates may be limited by factors other than simple ionic current flow.

Polarization is important in electroplating because it has direct effects on the composition, quality, and rate of metal deposition. Cathode polarization in particular must be controlled to achieve the desired plating results. In many cases polarization is undesirable because it limits the maximum rate of current flow, and increases the time needed to obtain a required coating thickness. This is the reason that many plating baths are operated at higher temperatures and with vigorous agitation, because the resultant improvement in ionic mixing tends to prevent polarization due to concentration gradients. In other cases polarization is beneficial to the deposit characteristics.

Current Efficiency • The weight of metal actually deposited on a cathode surface is usually somewhat less than that predicted by Faraday's law, because other chemical reactions are occurring simultaneously at the cathode, consuming their share of the electric current. For example, hydrogen gas is commonly liberated from the cathode by the reduction of hydrogen ions (H^+) from aqueous plating solutions. *Current efficiency* is a useful measure of the extent of the *desired* reaction. If, in the example of Fig. 7-1, 90% of the current passed through the nickel sulfate bath were used to deposit nickel metal, and 10% to produce hydrogen, the *cathode efficiency* for nickel plating would be 90%. Much of the work that goes into formulating and testing chemical plating solutions is geared towards maximizing the current efficiency, because side reactions can lead to some difficult plating problems as follows:

1. Hydrogen gas bubbles forming on the workpiece can lead to pitting and plating voids in the deposit.
2. Undesirable materials may be reduced at the cathode and may become incorporated into the metal coating, changing its physical properties.
3. If the anode reaction does not match the cathode reaction, the chemical composition of the solution will gradually change during plating. For example, the metal anode may be attacked to form oxides and basic salts instead of the desired metal ions. Oxygen may also be liberated at the anode in place of metal dissolution.

Low cathode efficiency is a particularly troublesome problem in the printed circuit industry, which uses photo resist or screened resist to direct metal deposition in additive PCB processes. Hydrogen gas is the principal side reaction, and its evolution exerts a scrubbing or lifting action that tends to dislodge resist particles during plating. Therefore, copper and solder baths used in electroplating PCB patterns almost always have cathode efficiencies approaching 100%.

Throwing Power • In addition to wanting metal deposits with the right physical properties and appearance, the electroplater wants to apply coatings in such a way that the workpiece is covered with a uniform metal deposit. *Throwing power* is a measure of the ability of the plating bath to produce a uniform metal thickness across the entire surface of the object. An even coating is always difficult to deposit on an irregularly shaped object, but the goal still remains to make it as even as possible. The main factors that influence throwing power can be manipulated effectively, and will be briefly discussed.

Current Distribution — When electric current is passed between two electrodes in solution, the current density will not generally be the same at each point on the electrode surfaces. For example, if a printed circuit board edge connector tab finger is immersed in a solution as the cathode surface for gold plating, more gold will be deposited on the surface facing the anode than on the opposite side, which is at a greater distance. The initial current distribution is determined by variations in electric potential across the electrode surfaces, being strictly a function of their geometric shapes and relative positions. Uniform plating of flat objects (such as PCBs) is not difficult because anodes can be easily positioned so that the current distribution will be even. In the case of edge connector tabs, two parallel anodes in the shape of rods extending the length of the tab would be positioned in the bath, one on each side of the workpiece; this example illustrates the general objective of making anodes *conform* to the shape of the surface to be plated.

Irregularly shaped objects are difficult to plate evenly because electrode geometry cannot be adjusted to give uniform current distribution. Separating anode and cathode by greater distances may help somewhat, but protruding surfaces will always be at a higher potential than recessed areas due to unequal charge distributions. It is not unusual for current density to vary as much as 10 to 1 in various areas of an irregular surface. Therefore, techniques other than simple electrode geometry are needed to increase throwing power of plating baths.

Polarization Effects — As previously discussed, electrode polarization tends to oppose the passage of current through a bath, and usually increases with an increase in current density. Therefore, when plating an irregularly shaped object, the effect of cathode polarization is to *equalize* current density and increase throwing power. Areas with initially higher current densities will become more polarized, making it more difficult for current to reach those areas, and thus evening out the *secondary current distribution* that is in effect at the cathode surface. This is a simple principle, but it allows the electroplater a great deal of freedom in controlling throwing power by adjusting the variables that affect cathode polarization, including both chemical composition and physical operating characteristics of the bath. An example of this principle in action occurs in copper plating when two standard baths are compared: the acid copper sulfate solution, and the alkaline copper cyanide or pyrophosphate formulations. The acid bath is cheaper to prepare and allows much higher plating rates, but polarization is small and the system shows poor throwing power; it is widely used in large-scale industrial processes for coating flat, regularly shaped objects. When good throwing power is needed, the more expensive alkaline copper baths are used with their appreciable polarization, which is probably due to the chemical effects of hindered complex ion dissociation near the cathode. The alkaline copper complex baths are widely used in the printed circuit industry to plate the recessed surfaces found in PCBs with plated-through-holes, where an even deposit of copper in the recessed hole walls is of prime importance.

Effect of Electrode Efficiency — Many baths show a decrease in cathode efficiency as current density increases. This characteristic tends to level out metal deposition rates by counteracting the effects of unequal current distribution on the surface of irregularly shaped objects. Although electrolysis occurs faster in some areas than others, increased hydrogen evolution prevents excessive metal deposition.

Effect of Conductivity — The total voltage drop across electrodes in a bath is a sum of the electrode potentials, polarization effects, and the IR voltage drop across the solution itself acting as a resistor. If cathode polarization is having a positive influence on throwing power, then an improvement in solution conductivity will further improve the throwing power, because with a smaller IR drop, the polarization will represent a larger portion of the total voltage drop, and thus will have a relatively greater effect on the secondary current distribution. In addition to its possible help in improving coating uniformity, good conductivity is almost always desirable in electroplating baths because it allows a lower voltage to be applied to the electrodes, and hence reduced power requirements and less tendency to heat the solution.

ELECTROPLATING PARAMETERS • Electroplating baths are composed of metal salts dissolved in water. Many other reagents may then be added to affect physical properties (such as conductivity), chemical characteristics (such as pH and stability), and specific properties of the desired metal deposits (such as appearance and hardness). In practice, most commercial formulations are quite complicated, and their analytical control and upkeep can become a difficult task. When all of the variables associated with operating conditions are added to the possibilities for chemical composition, the formulation of a plating bath suitable for a particular application may become more of an art than a science. This is in fact the case throughout the electroplating industry, which is dominated by hundreds of "proprietary" baths sold by companies who extoll the virtues of their particular products, and promise to provide technical support after the customer buys.

Due to extensive research and advances in electrochemistry, the mechanisms for electrolytic reactions are now fairly well understood, and the effects of most variables in the composition and operation of plating baths can be explained. However, the final effects of changing several variables in combination are still difficult to predict due to their complicated interactions. This is the reason that plating laboratories will buy a proprietary formulation intended for a specific use, rather than mix their own chemicals from various standard bath compositions published in the literature. In some cases, the mechanisms for certain plating effects are still unclear; for example, organic addition agents are commonly added to nickel baths at low levels to increase brightness, but their exact chemical function is often unknown.

The chemical constituents of electroplating baths can be divided into four general groups: primary metal salts, other organic/inorganic salts, acid/base reagents, and low-level additives with special purposes.

Metal Salts • Water-soluble metal salts are dissolved in the bath to provide a source of metal ions during electroplating. The most common materials used are simple salts, such as nickel chloride and copper sulfate, which are very soluble, show high conductivities in solution, and dissociate to give high metal ion concentrations. Complex metal salts are also used frequently, such as metal cyanides; other examples include fluoborates, pyrophosphates, and sulfamates. These latter metal salts are used in place of the simpler compounds when some advantage results in terms of solubility, conductivity, metal ion activity, or compatibility with operating conditions. For example, if copper plating is to be carried out in a basic pH range, copper sulfate is unsuitable, but copper cyanide is readily soluble due to the formation of metal complexes. As with all reagents used in plating solutions, it is important that metal salts be highly pure; contamination from other metals or certain anion impurities can drastically affect the results of metal deposition. All chemicals used should carry the label "Reagent Grade," "CP" (Chemically Pure), or "Plating Grade" to avoid introduction of unknown materials into the bath.

Additive Salts • For any particular type of plating bath there is usually a wide range of useful compositions and reagent concentrations. However, for specific operating conditions, an optimum level usually exists for each chemical species to give the best results. For example, very concentrated metal solutions tend to give large, granular deposits, but if the bath is operated at a higher temperature that permits increased current density, these changes may give fine metal deposits from a concentrated solution. When formulating a bath, a standard composition is chosen, salts are added to suit a particular purpose, and then small changes are made to modify or correct the characteristics of deposits actually obtained.

A good example of the use of additive salts to influence plating results is in nickel plating with the famous Watts bath, or its numerous variations. The Watts bath uses nickel sulfate as the primary source of metal ions, but a certain amount of nickel chloride is commonly added to increase conductivity and to improve the anode efficiency, or *anode corrosion* as it is known. Without the presence of chloride ion, high purity nickel anodes will not dissolve easily to form metal salts, but instead will exhibit low current efficiency with the production of oxygen and basic metal compounds at the anode surface. This problem results in excessive sludge buildup and the bath becomes quickly depleted of its desirable metal content. Various formulations use different amounts of nickel chloride because this additive also affects the properties of deposits; for example, an all-chloride nickel bath produces harder, finer-grained deposits, but they are highly stressed. (Internal stress in plated nickel is undesirable because it means the coating is under a constant tension, contractile in nature, that tends to cause cracking and separation of the coating from its basis material. Another metal salt often used in place of nickel sulfate for Watts-type baths is nickel sulfamate, which considerably reduces the stress inherent in nickel deposits if correctly used.)

Concentrated metal solutions give good conductivities and show high cathode efficiencies at fast plating rates. However, there are often advantages in lowering the metal ion activity, which can improve throwing power, increase anode efficiency, and decrease crystal size to give deposits with finer grain. A dilute solution of a simple metal salt can also produce these results, but it has limitations — the metal supply will be quickly exhausted, and the maximum current density will be very low. Therefore, other compounds are routinely added to concentrated metal solutions with the objective of lowering the metal ion activity while maintaining a constant reservoir of metal in solution. Ionic activity can be reduced in two ways: with *common ions* that hinder dissociation of the original metal salt, or with the formation of *metal complexes* using both organic and inorganic complexation reagents. The common ion effect is simply illustrated in nickel plating with the addition of sodium sulfate to solutions of nickel sulfate; this technique can easily reduce the nickel ion activity to 10% of its normal equilibrium value.

When using complexation additives, the typical anions are cyanides, fluoborates, pyrophosphates, tartrates, and citrates; in some cases, the cation portion of the salt may also be important, as with ammonium salts in a nickel bath, or sodium and potassium in double cyanide metal complexes. Because of the greater throwing power imparted to complex baths, some of these formulations are extremely important in the printed circuit industry. Lead/tin fluoborates are the standard form of solder plating baths, and copper pyrophosphate is widely used for copper plating.

Hydrogen Ion Concentration (pH) • The acidity or alkalinity of a plating bath is usually an important parameter of its chemical composition and operation, and must be periodically adjusted with acidic or basic reagents. The pH scale is normally used for measurement of hydrogen ion concentration, and extends from 0 to 14; 0 is strongly acidic, 7 is neutral, and 14 is strongly alkaline. Practical pH measurements are made with a glass electrode, with colored pH paper, or by adding pH-sensitive dyes directly to a sample of the solution. (This latter approach is known as colorimetric analysis.)

Hydrogen ions are important in electroplating because they are positively charged ions and migrate directly to the cathode during electrolysis. They can then be reduced by the negative electrons and are evolved as hydrogen gas, the main cathodic reaction in addition to metal reduction. Cathode efficiency is determined largely by the ratio of metal deposition to hydrogen evolution, and hydrogen ion concentration has a direct influence on this process. It can also have a direct influence on the character and quality of metal deposits, so most plating baths are specified to operate within a narrow range of pH limits. The vast majority of baths operate under acidic conditions, but some, like the concentrated cyanide baths, operate above pH 7. (Cyanide ion is inherently basic in chemical action, and is always in equilibrium with hydrogen cyanide; this toxic gas is evolved from cyanide solutions that become acidic, either intentionally or by accident.)

The pH of a plating solution must be monitored and adjusted fairly often because electrolysis reactions tend to either reduce or strengthen the H^+ concentration during plating, and these side reactions are rarely balanced. A low cathode efficiency will produce hydrogen gas by the following reduction, causing pH to slowly rise:

$$2\,H^+ + 2\,e^- \rightarrow H_2\,(gas)$$

On the other hand, a low anode efficiency will produce hydrogen ions and oxygen gas, which lowers the pH as follows:

$$2\,H_2O \rightarrow 4\,H^+ + 4\,e^- + O_2\,(gas)$$

In many plating baths, such as gold, an insoluble anode is used such as platinum or carbon, and the only appreciable anode reaction occurring is the oxidation of water to O_2. In these baths both the pH and metal ion concentration must be monitored and periodically adjusted with additives.

When a pH adjustment must be made to a plating bath, acid is added to lower the pH and alkali is added to raise it. These acid/base additives are chosen in each case so as to least disturb the composition of the bath, and specifically to avoid introducing any new ions into the solution. For example, in a nickel sulfate bath, nickel hydroxide would be added as the base, and sulfuric acid as the acid, causing only slight changes in the overall nickel and sulfate concentrations.

To avoid the need for frequent pH adjustments, many plating baths include *buffer* compounds to stabilize the pH near a desired operating point. A typical buffered solution contains a weak acid in combination with its salts, and acts in such a way as to oppose pH changes: if acid is being consumed, the weak acid dissociates to replace lost hydrogen ions, and if acid is being produced, the weak acid dissociates to replace lost hydrogen ions and make more weak acid. A typical example of buffer action is the use of boric acid as an additive in nickel plating baths operated with pH values ranging from 1.5 to 6.0. Boric acid is actually only effective as a buffer in the pH range from 5 to 6, but it is interesting that this compound is a useful additive in nickel sulfate baths operated at much lower pH's, such as 1.5–3. Such a situation serves as a good example of how complex plating action can be, and emphasizes that plating effects can often be explained but are rarely predictable. Boric acid is actually beneficial in nickel baths because it prevents the pH from rising above 6, where basic compounds such as nickel hydroxide begin to precipitate. These compounds can become included in the metal deposit and lead to dark, or *burnt* coatings. Due to polarization near the cathode, hydrogen ions become depleted and the actual pH inside the *cathode film* is usually above pH 5; however, unionized boric acid opposes further pH increase by exerting its buffer action inside the film, and thereby prevents burnt deposits from occurring.

Low Level Additives • These materials are often organic compounds, and are added to the plating bath in small amounts to achieve specific desirable characteristics of the metal coating. Additives are most often used to control hardness and brightness of the deposit, but they are also useful for producing smoothing action (*leveling*) of rough surfaces, and for controlling *pitting* effects. Organic substances are widely used in nickel plating to control brightness, and often become included in the deposit to a small extent. Examples of nickel brighteners include aldehydes, ketones, and sulfonic acid derivatives. If these organic additives are not present in proper proportions, they may also cause nickel deposits to become brittle; however, analytical determination of low-level additives is extremely difficult. Therefore, when metal deposits are observed to have poor characteristics, the plater may filter the entire bath through an activated charcoal bed to remove all organic material, after which the proper materials are added back to the bath at known concentrations. The precise mechanisms for producing bright deposits are often unknown, but the results have been connected to specific chemical compounds in many cases.

Additives to produce smooth, refined deposits in acid copper plating have included proteins, such as glue and gelatin, and sugars, such as molasses and dextrin. Thiourea is often added as a brightener. Many of these materials are extremely active and will show an effect at a concentration of 10 parts per million.

Pitting is a common problem in some plated metal coatings. It is usually a result of hydrogen evolving and sticking to the metal surface as small gas bubbles during electrolysis. If a bubble stays around long enough for metal to be deposited around it, it will result in a pit, or void in the coating. This type of defect is particularly troublesome in acid nickel baths, which must be kept free of most impurities and suspended particles that seem to encourage pitting action. Assuming the bath is clean, two other approaches are used to prevent pitting. The first idea is to prevent hydrogen evolution from ever occurring by adding oxidizing agents that are more readily reduced at the cathode than H^+; hydrogen peroxide is a common additive because its reduction product is water, a harmless substance. However, peroxide is an active chemical and may react with other materials in the bath, such as organic brighteners. Therefore, a second more general approach to discourage pitting in any plating bath is to change the *surface tension* of solutions so that gas bubbles will have less tendency to stick to metal surfaces. It was early observed that some baths, such as chromic acid, show vigorous hydrogen evolution with very low cathode efficiency (15%), but very little pitting occurs because of the low surface tension of this solution. The same results can be achieved in nickel baths by the addition of *surface-active* materials, or *wetting agents.* Typical wetting agents are salts of long-chain organic sulfonic acids, such as sodium lauryl sulfate, added at concentrations as low as 0.1 g/liter.

Properties of Plated Metal Deposits • An electrodeposited coating is used because it has some set of properties more desirable than those of the original substrate material. The metal composition itself may be important, but numerous other properties of the deposit can be varied by choosing the appropriate bath composition and operating it under suitable conditions. Some of these properties have already been mentioned; they will be summarized in this section, with particular notes as to their importance in the plating of PCBs.

Thickness — More than any other factor, thickness is the most important property that determines the value of a metallic coating. Even when the stated purposes for plating are fairly different, such as (1) protection against corrosion, and (2) improved wear resistance, most of these practical requirements can be boiled down to a particular metal type at a specified thickness.

The plater controls metal thickness by adjusting the current density and the plating time. By knowing how much current has passed through the solution, the cathode efficiency, total area of the plated object, and the density of the metal deposit, an average final thickness can be predicted. Examples of thickness calculations will be presented later in this chapter. Unfortunately, it is not the *average* thickness, but the *minimum* thickness that is important, because most coatings are only as good as their weakest point. This is the reason that throwing power is so important in electroplating. Uniformity is especially important in precious metal plating because of economy; the objective is to obtain the minimum required coating thickness across the entire surface without wasting metal.

As an example of minimum thickness requirements, copper contact fingers for PCB edge connectors are routinely plated with nickel followed by gold to provide wear resistance, tarnish resistance, and low electrical contact resistance. These desirable features are achieved by specifying a nickel deposit of 0.0002–0.0004 inch (0.005–0.001 mm) thick, and a gold deposit of 0.00005–0.0001 inch (0.0013–0.0025 mm) thick. A maximum roughness of the final deposit is also commonly specified, since a smooth surface allows the contacts to slide easily in and out of the connector.

Wear Resistance — When a metal surface will be subjected to mechanical forces, its resistance to wear may be the key factor in choosing a coating. *Hardness* is important when the surface must resist *abrasion,* such as scratching or tearing out of metal fragments. But by itself, hardness may not be sufficient; chromium is one of the hardest metals, but its usual *brittleness* makes it undesirable for use on surfaces that receive impact forces. In such a case, a more *ductile* (flexible) deposit is needed, with considerable *toughness.*

Some metals are inherently harder than others: nickel is hard, copper is soft. This is the reason that nickel is chosen to protect PCB copper edge connector contacts, which receive considerable abrasive wear during insertion or removal. Different nickel deposits may also show varying degrees of hardness depending upon the type of bath used and its operating conditions; this is a result of the final crystal structure, or of the impurities co-deposited with nickel. For example, the Watts bath gives harder nickel deposits at higher pH's, but hardness decreases with a rise in plating temperature. The nickel sulfamate bath gives harder nickel deposits with less stress than the Watts bath, but they are also less ductile.

Adhesion — To be useful, a plated coating must be strongly adherent to the basis metal. Otherwise, externally applied stress may separate the layers, as often occurs when a plated article is deformed or subjected to sharp temperature changes. The most common reason for poor adhesion of a plated deposit is *improper cleaning* of the original surface. Dirt and grease can cause the problem, but a more likely culprit is metal oxide that forms as a thin layer on most metal surfaces exposed to air. Foreign materials can also become chemically absorbed on active metal sites. In any case, an objectionable, weakened layer usually exists on the surface of metal objects that have not been specifically prepared for plating.

Electroplaters go to great pains to clean and prepare metal surfaces for electrolytic deposition. In many ways, this pre-cleaning operation is similar to the steps taken in cleaning blank copper-clad PCB laminates before applying photoresist coatings, as described in Chapter 5. The first step in plating is to remove all dirt and grease with an alkaline cleaner. The object is then soaked in dilute acid to neutralize any residue remaining from the cleaning agent, and also to dissolve metal oxides. Acid treatment is known as *pickling* in the plating industry. The acid chosen is usually a strong mineral acid that is compatible with the actual plating solution; if the solution contains chloride ion, for example, then dilute (5–10%) HCl would be used as the pickling treatment. Care must be taken to avoid contamination of the plating bath with foreign ions. Copper PCBs are commonly pickled in 10% sulfuric acid before solder plating in a lead/tin fluoborate bath, which will not tolerate sulfate. It is, therefore, necessary to add an intermediate water rinse and a soak in dilute (10%) fluoboric acid before proceeding from pickling to solder plating.

Another step that is often included in the metal preparation sequence is a slight etch of the metal surface. Etching has two purposes: in addition to removing any contamination in the top surface layer, it serves to slightly roughen the surface, which promotes good adhesion of subsequent deposits. Etching is accomplished by two methods. First, common etchants such as cupric chloride can be used for a brief treatment. In this case, another cleaning dip in pickling acid is needed after etching to prevent contamination of the plating bath. The second method used for etching is known as *electrolytic polishing, anodic pickling,* or *anodic etching.* The article to be plated is placed in an acid pickling bath (or sometimes an alkaline cleaner) made anodic with respect to a second, dummy cathode electrode. Current is passed through the solution and metal is effectively dissolved by electrolytic action. This procedure is particularly useful for preparing nickel surfaces for plating, because nickel quickly forms oxide films that are difficult to remove by other methods; the nickel is said to become *passive.*

Corrosion Resistance — Some of the most important applications of electroplating have the objective of providing a corrosion-resistant coating to common metals, such as steel and iron, which would otherwise oxidize (rust) during exposure to air. Tin, zinc, and nickel are plated coatings that fall under this classification. Nickel is actually one of the highest volume plated protectors of steel in the automobile industry, because of its use in conjunction with decorative chromium plating. Although chromium is highly desirable as a bright metal finish, its porosity is such that little protection is provided to the underlying steel; therefore, most chrome finishes are largely nickel, with a thin final coating of tarnish-resistant chromium.

Several types of metal are used on printed circuit boards to provide corrosion protection for copper patterns. The most important ones are 60/40 tin-lead solder, nickel, tin-nickel alloy, and gold. All of these coatings are effective as *etch resists* during copper etching with appropriate solutions. In addition to functioning as resists, solder and gold are particularly useful for more permanent protection — solder for preserving copper solderability, and gold for preventing tarnish on contact fingers. Despite its expense, ready replacements are not being found for gold plating on electrical contacts because of its excellent conductivity, and because it is one of the few metals that does not readily form oxides upon exposure to air. Rhodium (one of the platinum group metals) is a possible gold substitute for contact protection, but it is also an expensive metal.

Brightness and Appearance — The noble metals, such as silver, gold, and platinum, are valued for use in decorative coatings, but their popularity is due to more than just monetary value; these metals produce coatings with definite physical appeal due to their brightness and luster. Other more common metals can also be plated as bright deposits, including copper, zinc, brass, tin, nickel, and chromium; however, of these only nickel and chromium can maintain their luster following exposure to air. Nickel, like aluminum, becomes passive very quickly with the formation of a very thin, transparent oxide layer that protects the metal surface from further corrosion. (Aluminum cannot be plated from aqueous solutions and is, therefore, not considered as a decorative coating.)

The appearance of any plated deposit depends mostly on its crystal structure. In general, bright deposits are obtained when the deposit is fine-grained and smooth, and this type of deposit is encouraged as the cathode polarization increases. Plating conditions that give bright deposits are usually ones that are associated with polarization, such as lower metal concentration, lower temperature, and higher current density. Sometimes low-level additives will increase the brightness of deposits through their effect on crystal structure, and this is the reason for adding specific organic contaminants to many baths: gelatin and peptone in solder fluoborates, sulfonic acids and aldehydes in the Watts nickel formulations, and thiourea, proteins, and sugars in acid copper. However, the operating conditions are very critical for baths with additives, and very stringent controls are necessary to maintain the desired quality of results. Fortunately, the appearance of most plated coatings can be improved by buffing and polishing, particularly if the entire surface is accessible. In a small plating operation, the tight controls needed to produce bright deposits directly from the plating bath are unwarranted, and mechanical polishing is a routine necessity.

Bath Operating Conditions • The operating conditions for an electroplating bath can be separated into both chemical and physical parameters. Chemical conditions include such things as pH and reagent concentrations, which have already been discussed. This leaves the physical/mechanical variables, the most important being temperature, current density, agitation, and anode characteristics.

Temperature — Operating temperature is usually raised when a higher current density is needed, because at higher temperatures the solution becomes more conductive and ion dissociation/mobility increases. However, if current density is not increased along with temperature, deposits will become coarse-grained and rougher due to a decrease in cathode polarization. One other advantage of higher temperatures is a general improvement in electrode efficiencies. Some possible disadvantages of a high temperature are excessive water evaporation, decreased throwing power, and more tendency towards undesirable side reactions, such as hydrolysis and oxidation of reagents.

Current Density — The average current density is defined as the total plating current divided by the cathode surface area. In combination with cathode efficiency, this parameter directly determines the rate of metal deposition during electroplating. All baths have a certain range of permissible current densities, such as 25–100 amp/ft^2 for Watts nickel baths. A certain minimum current is needed to establish polarization, which gives dense, fine-grained deposits. At the upper limit, deposits begin taking on a rough, spongy appearance, until they eventually become dark or "burnt," due to the precipitation of metal hydroxides and other basic compounds that form in the H^+ depleted, high pH region around the cathode. It is generally desirable to use the highest practical current density in order to save time during plating.

Agitation — Agitation is like temperature in that it increases ionic mobility, which allows a higher operating current to be used. Agitation also helps maintain more uniform conditions throughout the tank because it mixes the solution, prevents stratification of concentrated salt layers, and keeps temperature uniform. An important advantage of any form of agitation is that it sweeps hydrogen gas bubbles away from the cathode surface, thus discouraging pitting. Typical methods used to agitate plating baths include stirrer paddles, physical movement of the electrodes, pumping of the solution through a recirculating loop, and introduction of air bubbles. Air agitation is often risky because small amounts of oil (from compressors, etc.) can ruin the bath, and the presence of air itself may promote oxidative decomposition in certain sensitive solutions.

One frequent disadvantage of agitation during plating is that it tends to separate insoluble residues that form on anode surfaces, and spreads this sludge throughout the system. It also prevents free metal particles and sludge from settling out of the way to the bottom of the tank. Fine metal particles are continuously lost at the surface of most anodes, and if these reach the cathode, they tend to cause rough deposits. In practice, highly agitated plating tanks use continuous solution recirculation through filters to eliminate sludge, and anode residue is contained by the use of *anode bags,* which are made of permeable cloth tied securely around the anodes. Cloth compositions may be cotton, glass, plastic, or polyester, depending upon the type of bath.

Anodes — The anodes used in electroplating are usually bars or rods made from the same metal that will be deposited at the cathode. It is advantageous to keep the anode surface area as large as possible to maintain uniform current density throughout the bath. The use of several anodes wired in parallel and positioned on all sides of an object to be plated is common practice, since this evens out the current flow. High purity anodes are important to prevent unknown contaminants from entering and spoiling the bath. However, in some cases, certain impurities are purposely included in the metal anodes with the purpose of improving anode efficiency. Nickel is an example — nickel anodes containing small amounts of carbon or nickel oxide are said to be "depolarized" and dissolve easier under electrolytic action. In general, the anode efficiency must approximately equal the cathode efficiency to maintain a steady concentration of metal ions in solution. In the case of solder alloy plating with 60% tin and 40% lead, both tin and lead anodes are hung in the plating tank to replenish both types of metals being lost.

Some plating baths do not use anodes of the same composition as the metal deposit; this is usually because of some extreme differences in the reactions occurring at the cathode and anode electrodes. The best example is chromium, which is plated from a bath of chromic acid using lead anodes. If chromium anodes were used, they would dissolve at a rate 5–6 times faster than the inefficient metal deposition rate, so inactive lead anodes are used and the bath is periodically replenished with chromic acid. Another example is gold plating baths, which tend to have a higher anode than cathode efficiency. Gold baths are operated fairly dilute because of the high cost of the metal; a dilute bath represents a less-expensive loss when problems occur in plating. Rather than let the gold concentration build up with gold anodes, platers prefer to use insoluble carbon, titanium, or stainless steel anodes, and replenish the bath with gold salts based on use or analysis.

ELECTROPLATING EQUIPMENT • The equipment and materials needed for electroplating include a suitable tank (usually plastic or plastic-lined), a power supply, cleaning and plating chemicals, metal anodes, and various accessories such as a rack for hanging objects in the bath, heaters, pumps, filters, and agitation devices. When preparing prototype PCBs, the electroplating equipment requirements for gold/nickel plating of edge connector contacts are fairly simple, with the main items being a small bath and a low-output power supply. A commercial prototype tab-plater system from Kepro was previously illustrated in Chapter 2, Fig. 2-18; Kepro also supplies chemicals and supplies needed to operate this unit.

As the volume of PCB fabrication increases, prototype shops will also want to apply solder coating by electroplating instead of manual coating. Solder plating will not be detailed in this chapter, but a suitable prototype unit from Kepro for this purpose is shown to the right of

Fig. 7-2. (Other equipment in the same figure is used for plating copper to form plated-through-holes in PCBs.) Tin/lead fluoborate electrolyte is fairly touchy to operate because of the instability of stannous ion, and because of the need for additives to get a fine, smooth coating. Also, when considering the move to solder plating, keep in mind that solder deposition must be carried out *before* etching in order to have a completely conductive substrate for plating. The result of this requirement is that subsequent etching must be carried out in a solution compatible with solder as a resist, such as ammonium persulfate.[20] (Ferric chloride and cupric chloride etchants attack solder.) Plated solder coatings will also require that solder fusion equipment be used to treat the final deposit, as is commonly done with an infrared reflow oven.

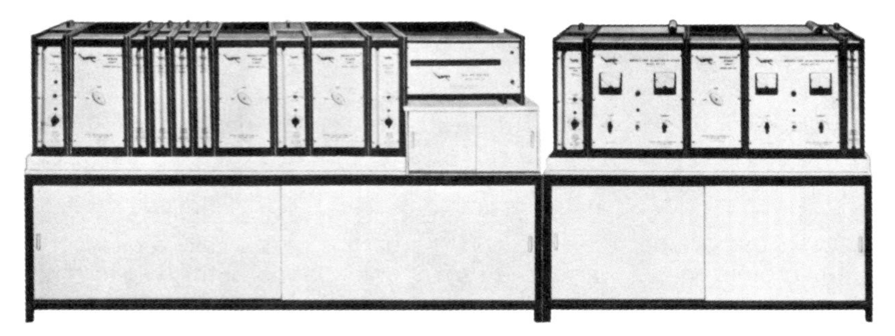

Fig. 7-2. Prototype PCB plating equipment for copper and solder. ***(Courtesy Kepro Circuit Systems, Inc.)***

Plating Power Supplies • A schematic diagram for a typical 10-amp plating power supply is shown in Fig. 7-3. The circuit is basically an adjustable, unregulated dc supply with a large filter capacitor across the output to smooth the voltage. A small (<5%) ac ripple component on the dc line is not objectionable in plating. The component values in this circuit should be adjusted for the particular power requirements needed. In general, a 6.3-V CT 10 amp filament transformer should supply enough voltage and current for simple electroplating needs, and will isolate the output circuit from line voltage. An autotransformer is used ahead of the primary for control; the author uses a 10 amp Powerstat variable transformer of the type used to control electrical heaters in the laboratory. Both an ammeter and a voltmeter are indicated in the output circuit, but it is the current that is adjusted (using the autotransformer) to give the proper current density for plating objects of varying surface areas. The output voltage should normally be less than 4 volts if highly conductive plating solutions are used. If a higher output voltage is needed, as in anodizing aluminum (20 V) then a higher voltage filament transformer may be substituted, with an appropriate current

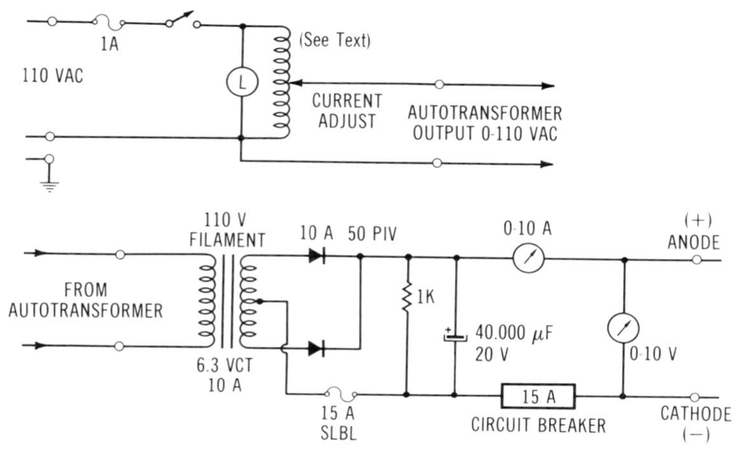

Fig. 7-3. General-purpose plating power supply schematic.

rating. A fuse or circuit breaker should always be placed in the output line to protect the power supply against short circuits, and the fuse value will depend on the maximum current capabilities of the circuit design.

In plating tab contacts, the small surface area will rarely require more than 0.5 amp of total current during plating. For precisely adjusting and maintaining small currents, a *constant current source* is very useful. A schematic diagram for a simple design is shown in Fig. 7-4. This power supply is inherently short-circuit proof, and will float the output voltage up to about 30 V to maintain a constant output current based on the control set-point. The circuit shown has two possible ranges, 0–100 mA and 0–1000 mA, and a user-calibrated meter to monitor current. This constant current power supply is shown assembled in Fig. 7-5. (The device is also very useful for charging rechargeable batteries at a low, constant, adjustable rate.)

Tab Plating Tank • A simple tab plating tank is shown in operation in Fig. 7-5. The tank is actually a rectangular plastic ice holder used in refrigerator freezers to hold ice cubes. Such a small, narrow tank is desirable because it minimizes the volume of plating solution needed. In the case of gold, cost savings are significant. The plating solution is poured into the tank until its level is high enough to completely cover PCB tab fingers, usually about 2 inches deep. When not in use, the solution is stored in a plastic quart bottle to prevent evaporation. The anodes in this system consist of two rods (nickel for nickel plating, carbon for gold plating), one on each side of the tray, whose ends are brought out of the solution for electrical connection. The circuit board to be tab-plated is stood up inside the tray with the tab

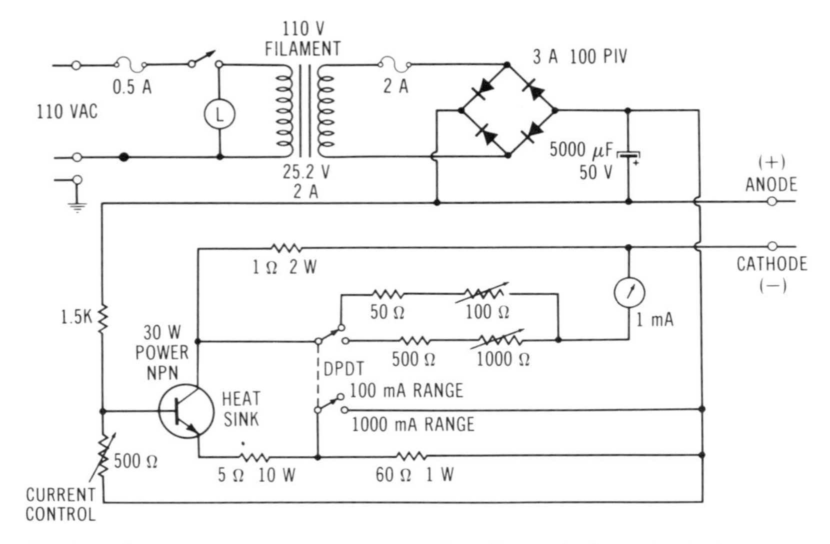

Fig. 7-4. Constant current power supply schematic for tab plating.

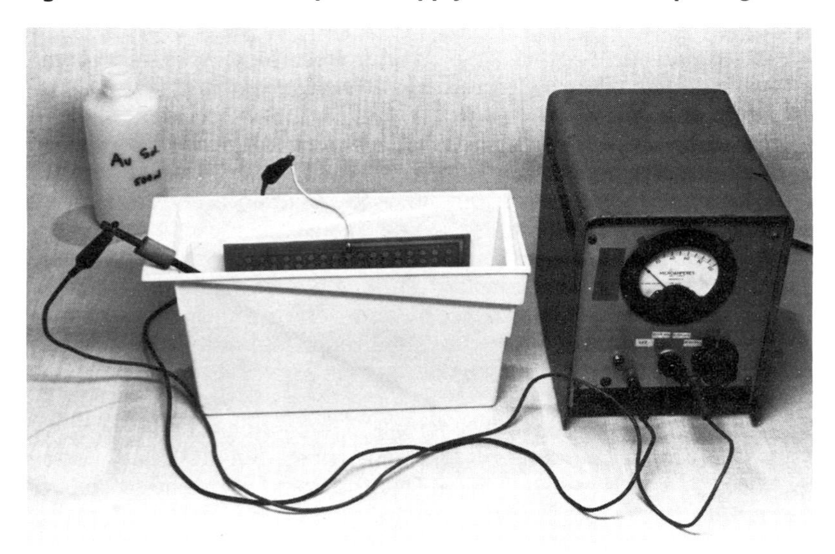

Fig. 7-5. Set-up and equipment for tab plating.

connector submerged in the bath, and held vertically with two blocks of wood glued at one end of the plastic tank. No provision for agitation is made in this simple plating system, but the author has obtained good results with both nickel and gold using a still bath, and occasional movement of the circuit board during plating.

BATHS FOR TAB PLATING ● Distilled water and high-purity chemicals should be used to make up the following electroplating solutions. Unpredictable plating results will result if contamination occurs, the worst type being extraneous metal ions and organic materials. Consult the Metal Finishing Guidebook-Directory[10] for plating material sources.

Nickel Plating • Because of its simplicity and ease of operation, the following bath[21] is recommended for plating nickel on printed circuit tab connectors. The formulation is basically a modified Watts bath with equal amounts of sulfate and chloride ions. Although the deposits may be somewhat stressed as compared to more recent sulfamate baths, this should not be a problem if plating thickness is held fairly thin. The advantages of this bath are smooth, fine-grained, ductile deposits with little tendency towards pitting, and these qualities are obtained without the need for any additives.

Nickel Sulfate ($NiSO_4 \cdot 6H_2O$)	195 g/liter
Nickel Chloride ($NiCl_2 \cdot 6H_2O$)	174 g/liter
Boric Acid (H_3BO_3)	40 g/liter
pH (adjust with HCl)	1.5
Temperature	25°C
Current Density	20 amp/ft^2

Both the temperature and current density are lower than normally recommended for this bath, but such operation is consistent with the simplicity of room temperature operation and little agitation. High-purity nickel rod should be used as the anodes, and the cathode efficiency is assumed to be 90% when calculating plating times and thicknesses. The only periodic bath adjustment needed is HCl addition in order to maintain the specified pH; otherwise, the bath should operate well for a long time before any problems occur, at which point it can be disposed of and a new batch prepared. If any pitting due to hydrogen bubbles is observed, this can be remedied by occasionally stroking the deposit with a plastic brush to remove bubbles during plating. Current flow should never be interrupted once nickel plating is begun; this leads to nonadherent deposits. Remember that the solution is acidic and highly corrosive.

As an example of calculating plating current and time, assume that a 0.0002 inch-thick nickel deposit is wanted. The finger area to be plated is estimated to be 2.4 in^2, including both sides of the tab and the exposed plating strip. From the bath specifications, a target current density of 20 amp/ft^2 translates into about 140 mA/in^2, so the *total plating current* is:

$$2.4 \times 140 = 336 \text{ mA}$$

The question now remaining is *how long* should the current be allowed to flow to get the desired thickness. From a plating textbook[22], the standard rate of nickel buildup from Ni^{+2} sulfate baths is found to be 0.0053 in/hr at 100 amp/ft², assuming a 100% cathode efficiency. At 20 amp/ft² and a cathode efficiency of 90%, the buildup rate is 1.7×10^{-5} in/min, so the total required plating time for a 0.0002 inch deposit is:

$$\frac{0.0002}{1.7 \times 10^{-5}} = 12 \text{ minutes}$$

(The same result can be calculated knowing that the density of nickel is 8.9 g/cm³, and that 2 electrons are required to reduce each atom of nickel, which has an average atomic mass of 58.7.)

Gold Plating • The classic hot alkaline cyanide gold plating bath is not used in the printed circuit industry because of its tendency to attack resist coatings. The bath is also not welcome in the prototyping laboratory because of its extreme toxicity, and high operating temperature; the high cyanide content makes this bath hazardous to use in areas where concentrated acids are routinely handled. Fortunately, some new gold plating baths have recently been developed which operate in the acid pH range and make use of more innocuous conductivity salts. The following formulation[23] is an example of this new (patented) technology, and the author has used it with good success to deposit gold over nickel on PCB tab contact fingers. Operation of the bath is simple, the chemistry is stable, and deposits show good adherence to previous nickel coatings.

Potassium Gold Cyanide ($KAuCN_2 \cdot 2H_2O$, 61% gold)	4 g/liter
Citric Acid	90 g/liter
pH (adjusted with KOH or citric acid)	3–6
Current Density	5 amp/ft²
Temperature	25 °C

The anodes used in this bath should be carbon rods or platinum-coated metal rods, since the cathode efficiency is only about 50%. Gold is replenished (based on cumulative use) by dissolving more salts, and the gold concentration is not very critical. Occasional movement of the circuit board will suffice for agitation. Several advantages make this bath ideal for plating gold on prototype PCBs as follows:

1. Dilute solution — Low gold concentration means the bath represents a minimum investment. At today's prices ($400/ounce in 1983), a 500 mL solution can be prepared using less than $20.00 worth of gold metal. The salt is available from the following recommended sources listed in Appendix B:

    Engelhard Corp.
    D. F. Goldsmith Chemical & Metal Corp.

2. Strike capability — Gold will often deposit on less-noble metals by simple immersion displacement before electrolysis begins. This type of deposit is weak, and leads to a nonadherent plated coating. Special dilute "strike" gold baths were formerly necessary to overcome this problem; they were used to electrolytically deposit a thin layer of adherent gold before the regular bath was used. However, the new acid bath described here does not require an initial strike on nickel or copper if it is used in dilute form, as was recommended previously.
3. Nickel activation — The low cathode efficiency of this gold bath means considerable hydrogen is evolved at the nickel surface when plating begins. Hydrogen tends to act as a reducing agent and thereby activates the passive oxide film that forms readily on nickel surfaces. The result is a more adherent gold deposit over nickel.
4. Deposit characteristics — Gold deposits from the citrate bath are smooth, fine-grained and bright, requiring little polishing. The deposit can be made very thick and does not become stained with time. Electrical conductivity is excellent.

Due to the lack of agitation and the high rate of hydrogen production, it is recommended that the deposit surface be brushed occasionally during plating to liberate hydrogen bubbles, which might otherwise stick and cause pitting. A clean plastic brush should be used on the submerged tab without interrupting current flow. Remember that this bath does contain a small amount of cyanide due to the nature of the gold salts, so handle it carefully.

A sample calculation will now be used to show how plating current and time are determined for the gold bath. Assume, as in the previous example for nickel, that the area to be plated is 2.4 in^2. The desired gold thickness will be 0.00005 inch. From the bath specifications, the current density should be 5 amp/ft^2, or 35 ma/in^2. The *total plating current* is therefore:

$$2.4 \times 35 = 84 \text{ mA}$$

Referring to a plating textbook[22], the standard rate of gold buildup from a bath containing the Au^+ aurous ion is found to be 0.0162 in/hr @ 100 amp/ft^2. At 5 amp/ft^2 and a cathode efficiency of 50%, the buildup rate is 6.8×10^{-6} in/min, so the required *total plating time* for a 0.00005 in-thick deposit is:

$$\frac{0.00005}{6.8 \times 10^{-6}} = 7.4 \text{ minutes}$$

In this example, the gold weight deposited is about 0.037 g, which is worth about $0.50 at the current price of $400/ounce. It should be possible to plate 15–20 tab connectors before a 500 mL solution must be replenished with gold salts.

TAB PLATING PROCEDURE • The selective plating of PCB edge connector contacts with nickel and gold consists of the following steps using the baths and equipment described:

1. Electrical preparation — The procedure starts with a clean, etched copper circuit board having an edge connector tab. During plating, all of the individual finger contacts on both sides of the tab must be electrically connected in parallel to the negative side of the power supply. These common connections are normally etched into the board using a *plating strip* (see Fig. 7-6) that is included in the initial artwork design. The plating strips extend around to the top of the board, where they are connected to a single wire soldered to both sides. After plating is complete, the plating strips are cut away when the board is sheared to its final dimensions. It is also possible to plate an edge connector tab whose contacts are not initially connected together, but the procedure is somewhat tedious. A wire must be soldered to the top of each contact or to a connecting trace so that all contacts may be electrically connected during plating. The connections are easier on a PCB intended for wire-wrap assembly, because each contact will then have an associated pad for installing a wire-wrap pin.
2. Surface preparation — Copper is one of the easier metals to activate for plating, and deposits are usually strongly adhesive. The tab contacts should first be scrubbed with medium-grade steel wool and an alkaline cleaner, followed by a water rinse. This polishes and cleans the copper surface. The board is then dipped in 5–10% HCl to remove all traces of alkalinity and copper oxide. Another water rinse should now reveal any problem areas by the *water break test* previously described in Chapter 5, and if the surfaces look good, the board is wiped clean and dried.
3. Masking — Certain areas of the board that are electrically connected to the tab contacts, but which are not intended to receive plating, must be *masked* with liquid resist or enamel paint. It is only necessary to apply this coating in areas that will be submerged in the bath during plating. Fig. 7-6 illustrates how resist is applied around the contacts with a small brush, and then allowed to dry.
4. Nickel plating — Just prior to dipping the masked PCB in a nickel plating solution, another brief dip in 5–10% HCl will ensure activation of the copper surface. The board can be transferred directly from the acid pickling bath to the plating solution because

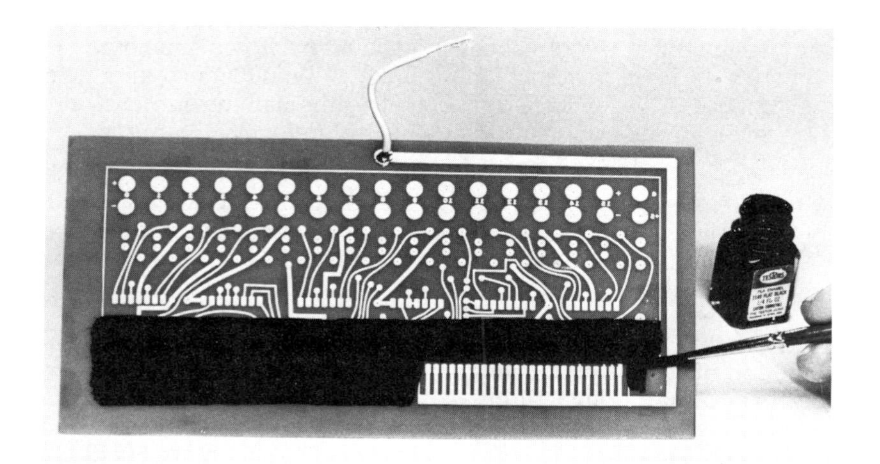

Fig. 7-6. Tab plating requires a plating strip and selective masking with resist.

both contain chloride ion. A 0.0002 inch-thick nickel deposit is obtained by operating the bath as previously described for about 12 minutes at a current density of 140 mA/in². A separate nickel anode rod should be positioned on each side of the tab to get uniform current flow and equal deposit thickness on both sides of the board. Do not interrupt current flow at any time until the entire plating is complete.

5. Gold without nickel — The purpose for nickel plating is to give the sliding contacts *wear resistance*. Both copper and gold are relatively soft metals, and will not withstand abrasion. By providing a more durable, hard coating under the thin gold deposit, contact wear is greatly improved, preventing loss of gold particles through mechanical gouging. However, the nickel deposit is not necessary as far as low electrical contact resistance, if the connector does not receive much wear. It may be acceptable in some prototype PCB applications to skip the nickel plating step and deposit gold directly on copper contacts. Such a coating will provide good electrical connections until gold is rubbed off to expose the underlying copper, which might tarnish through air oxidation and create problems. If gold is plated directly on copper, be sure to rinse off the HCl pickling solution with water before transferring the board to the citrate bath, which should not receive chloride contamination.

6. Gold plating — Following nickel plating, the board should be rinsed with water and transferred to the gold bath without delay

to avoid deactivation of the nickel surface. If an unavoidable time delay occurs, the nickel coating should be stored submerged in 10% HCl until the gold bath is ready. A 0.00005 inch-thick gold deposit is then obtained by operating the bath as previously detailed for about 8 minutes at a current density of 35 mA/in^2.

7. Polishing — After water rinsing, the mask resist is removed with solvent, and the gold coating can be polished to a bright finish with fine-grade steel wool. The final gold/nickel deposit should be as bright and as smooth as the original copper surface. The printed circuit board is now ready for solder coating.

PCB PROTECTIVE COATINGS • Copper PCB traces require some sort of coating protection to preserve solderability and to prevent metal corrosion through air oxidation. (A company that specializes in PCB coating chemicals is the London Chemical Co. (Lonco) listed in Appendix B.) The most common coating materials used are 60/40 tin-lead solder, pure tin, and lacquers.

Solder • Of these three materials, solder is the most desirable, and a simple manual procedure therefore, will be presented for solder coating. As was discussed in Chapter 2, solder is commonly coated on an industrial scale by electroplating as the 60/40 tin-lead alloy, which serves as an etch resist during persulfate etching. The coating is then leveled and fused to the underlying copper with heat to leave a shiny, bright layer of protective solder. However, this approach is unattractive to the electronic experimenter because it requires the complications of maintaining a fairly large metal fluoborate electroplating system, with inherently sensitive and unstable chemistry. Fortunately, a nice solder coating can be obtained by simpler manual methods. The other two coating materials are even easier to apply and are in common use; they will be mentioned, but are not generally recommended because of certain disadvantages.

Immersion Tin • Pure tin can be coated on a clean copper surface using an inexpensive commercial solution known as *immersion tin*. The circuit board is simply dipped in the solution at room temperature (or slightly higher) for about 5–10 minutes, and tin metal is deposited on the copper surface through a chemical oxidation/reduction process. Immersion plating is a surface phenomenon that does not require the use of electrodes or passage of current to produce a deposit. The mechanism is actually a displacement reaction, where a metal ion in solution with a higher reduction potential exchanges with the surface metal, which is oxidized into solution. Tin deposits from immersion baths are fairly thin, because once the base copper metal is completely covered, no more reaction can occur. Typical thickness is 50 μin of tin, which will gradually oxidize upon exposure to air.

With proper application immersion tin provides excellent protection for copper traces, and gives a bright coating of excellent appearance. Unfortunately, however, immersion tin does not usually produce an *easily soldered* surface. In fact, after a few days the coating may be noticeably difficult to wet with normal rosin-flux solder. The author would much prefer soldering to a clean copper surface, or even one that is slightly oxidized. The nice appearance of an immersion tin coating does not compensate for the difficulties it can create if PCB assembly is delayed or if solder alterations are necessary at some later date.

Lacquer Coatings • Another approach for long-term protection of PCB traces is to brush or spray the board with a clear, nonconductive lacquer after solder assembly is complete. Since this coating does not affect the initial solderability, some other pre-cleaning steps are necessary to ensure that bare copper pads can be readily soldered before the final lacquer protection is applied. An etched circuit board that has been stored for longer than a few days will develop a thin layer of copper oxide across the exposed surfaces. (The oxide is often seen as discoloration or as a streaky appearance.) Therefore, the purpose of pre-cleaning is to remove oxides and activate the copper surface just prior to solder assembly. The following steps are sufficient:

1. Scrub the board with soap and fine steel wool, and water-rinse. This should remove any oil or dirt.
2. Dip the board in dilute acid (5% HCl or sulfuric acid) to dissolve oxides.
3. Wash and dry the board without exposing it to excessive heat, which encourages re-oxidation. The resulting metal surfaces will now be easily solderable, and should remain active for the time needed to assemble the complete board.

Once the board is assembled, a lacquer coating can be applied to seal and preserve all surfaces permanently. Typical lacquers available from electronic distributors are based on acrylic plastics, or silicone resins. Both types can be removed with solvents when a component is replaced or some other solder modification is required; otherwise, the protective coating interferes with the desoldering and resoldering operation. The lacquer approach is effective for low-cost commercial boards, but the following two disadvantages are apparent for prototype circuits requiring frequent alterations:

1. Removal of the coating with solvent is inconvenient.
2. Coating the component side of double-sided PCBs may present problems. Sprayed lacquer will not reach some areas shielded by

components, and it will penetrate other undesirable areas, such as sockets and potentiometers.

As a solution to these problems, another type of protective lacquer has been developed for use on prototype PCBs, and is based on rosin materials. An example of a rosin-based protective coating is Lonco Seal-brite No. 230-10 that is applied to the etched, clean, unassembled PCB before any soldering is begun. Protected boards can be stored for many months before assembly, and the coating is then soldered directly through without requiring solvent removal. The rosin material is designed to enter into the soldering process and act as a thin coating of flux.

Flow-Coating • The most simple approach to forming a solder coating is to manually flow solder onto each and every trace with a small soldering iron. This may sound like a tedious operation, but it can in fact be accomplished very rapidly with proper technique, as illustrated in Fig. 7-7. The key step is to obtain an extremely clean copper surface, activate it with dilute (5%) HCl, wash and dry it, and immediately apply a coating of liquid rosin flux. The isopropanol flux solution available from GC Electronics is ideal for this purpose. With proper surface preparation it is remarkable how fast and smooth solder can be flowed across the open circuit board traces, covering an entire board in 5–10 minutes. A small, 25-watt pencil iron used with normal 60/40 tin-lead rosin-core solder is effective for spreading the solder. To obtain optimum results, circuit board holes should not be drilled until after

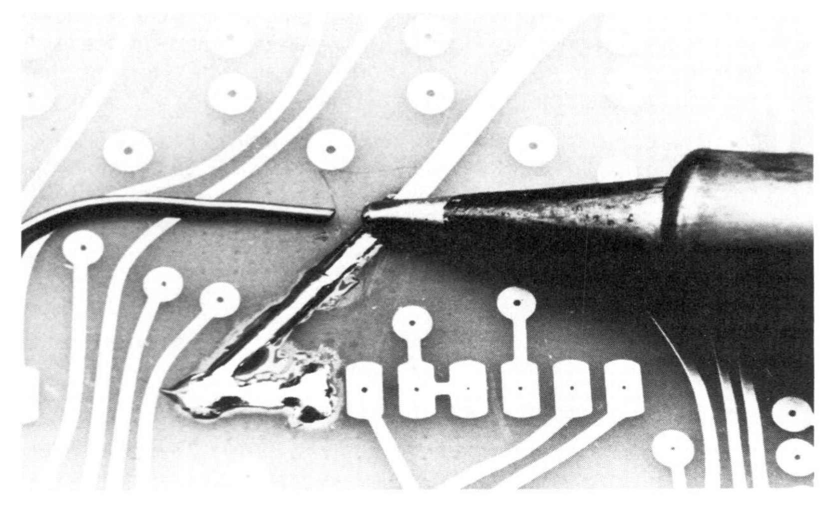

Fig. 7-7. Manual flow coating with a soldering iron.

solder coating has been completed; this avoids ridges and depressions that make the board more difficult to clean, and may make the solder more difficult to spread smoothly.

If solder does not readily bond or "wet" the copper traces, the traces have not been properly cleaned. An example of this problem is shown in Fig. 7-8, where solder "beads up" instead of spreading out. If poor wetting occurs, the alloying process can be forced by using extra heat and flux; however, traces may become delaminated from excess heat, so careful pre-cleaning of the board is more desirable. Never immerse a printed circuit board in mineral acid once it has received a solder coating or soldered joints, because acid (particularly HCl) will attack tin-lead, turning it into a gray, brittle material. A better approach for improving the solderability of a partially coated board is to scrub it vigorously with a steel wool pad soaked in water. If it is necessary to completely remove a solder deposit from copper, the following stripping solution can be used at room temperature:

Acetic acid, glacial	1 gal.
Hydrogen peroxide, 30%	20 fl oz
Water	3 gal

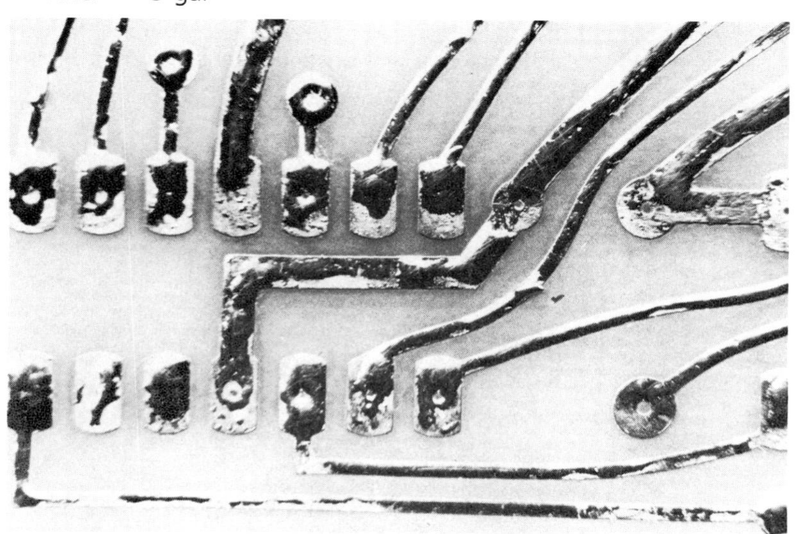

Fig. 7-8. Poor solder wetting of copper PCB traces.

Plating / Coating Considerations • As a precautionary measure, PCBs having edge connector fingers should always be plated with gold/nickel *before* a solder coating is applied. There are two reasons for this order in the coating steps. First, it avoids the problem of exposing solder to acidic cleaning and plating solutions, which will cause discoloration or substantial attack to the solder alloy. It is certainly possible to mask the entire soldered surfaces with resist during plating, but this is an unnecessary procedure if the plating operation is carried out first. The second problem with applying solder before plating is that any solder spots occurring on the contact finger surfaces will prevent adherence of nickel and gold deposits, which should be applied directly to copper. For best results, plate the fingers first and let the gold overrun into the trace area intentionally. Then, mask the fingers with paper tape to control the flow of solder away from these areas during solder coating.

Reflowing • With care and attention, a thin, even layer of solder can be applied on traces and pads across the board that will suffice for a final coating. However, for a thinner coating and more professional appearance, the board should be *reflowed* and *leveled* in hot oil. This operation is illustrated in Fig. 7-9. A steel pan is filled with oil and heated on a hot plate to 200–210°C (392–410°F), which is 10–15°C (50–59°F) higher than the melting point of 60/40 tin-lead solder. Do not exceed 210°C (410°F) because epoxy-glass circuit board laminates may show darkening and possible damage at higher temperatures. Once the proper temperature is reached, the board is gently lowered into the hot oil with metal tongs, and excess molten solder is quickly brushed away with cardboard or a fine-bristled steel brush. It is helpful

Fig. 7-9. Solder reflow with hot oil.

to position the oil bath on a slight incline to prevent excess solder from running back under the board. Double-sided boards are then turned over to brush away solder on the second side. The entire reflow operation at 200 °C (392 °F) should be completed in less than 1 minute to avoid damage to PCB laminates.

Reflowing with hot oil is a fairly safe operation if high-temperature fluid with low volatility is used. Avoid splashing hot oil while manipulating the circuit board; cotton gloves should be worn to guard against burns. The oil recommended for solder reflow is RO-1G from Kepro Circuit Systems, a 1-gallon container of high-temperature water-miscible polyethylene glycol-based fluid. This type of oil makes final cleanup easier, because it can be removed with water and a little soap. Other types of oil (such as pure lard) are messier to remove, requiring solvent-washing or extensive scrubbing of the board with detergent. Although lard can give successful results, it also tends to smoke at 200 °C (392 °F) and should be used in a fume hood; lard is more flammable than polyethylene glycol.

Following reflow, the hot board is removed from the oil bath and placed on a stack of newspapers to cool. Use a flat resting surface to prevent any warp from occurring as the laminate cools.

Once it has cooled to room temperature, the solder-coated PCB is ready for cleaning. If water-miscible reflow oil was used, this can be removed with soap and water. However, the use of lard as the reflow oil will require an initial solvent dip in xylene or paint thinner to dissolve oil before a final scrubbing with soap and water completes the cleaning. The circuit board should now exhibit an attractive bright metallic finish. If solder deposits are dull and gray in some areas, the appearance can be improved by dipping the board in 20% phosphoric acid and polishing the wet surface with extra-fine steel wool, grade 000. Do not use a coarser grade of steel wool or sandpaper, because the soft, thin solder layer might be removed with rough buffing. (Phosphoric acid is commonly available as the rust remover Naval Jelly.)

Chapter 8
Screen Printing

INTRODUCTION • Screen printing is the most versatile of all printing methods because of its unique ability to deposit many types of ink on almost any surface. It is used for printing precise images on articles made of paper, wood, glass, plastics, metal, ceramics, cloth, and leather; the objects may be as diverse as wallpaper, or the front panel of a stereo receiver. The printed pattern may be 8 inches wide, or it may be 8 feet long. Not surprisingly, such a flexible technique has been refined and developed for use in the electronics industry to transfer images for fabrication of printed circuit boards and intricate front panel displays.

The simple idea of using a *stencil* for image reproduction was known to early civilizations, and is now being applied in automatic screening machines capable of turning out several thousand impressions per hour. However, screen printing remains a relatively simple and inexpensive reproduction method for manual printing applications, and in combination with photography will yield precision images with a high degree of dimensional accuracy.

The basic stencil principle is to form a pattern of openings in a flat surface corresponding to a positive image. Ink is then forced through the openings to reproduce the pattern on an underlying surface. For a long time, stencil printing was severely limited because to retain its dimensions, the stencil had to form a continuous pattern with no isolated elements. It was in the United States, between 1901 and 1906, that the continuity problem was first overcome by attaching stencils to finely woven, tightly stretched silk, providing support for the pattern but allowing essentially unobstructed penetration of inks through the open mesh holes. Silk is still widely used for this purpose, but the term *screen printing* is now preferred over *silk screening* because newer fabrics, such as nylon, polyester, and stainless steel, are widely used in place of silk. As applied in the printed circuits industry, wire mesh can reproduce extremely fine detail with metal weaves having up to 250,000 uniform openings per square inch of surface area! Other properties, such as durability, tensile strength, chemical resistance, and resistance to distortion have made synthetic or metal fabrics preferable to natural silk in industrial screen printing.

For many years the stencil patterns attached to silk screens were applied using primarily manual methods, such as knife-cut films glued to the fabric. Graphic arts designers can become quite skilled with sharp instruments used to cut intricate patterns in an adhesive stencil film attached to a temporary backing sheet. These and other manual methods are still used extensively by artists to imbed their designs in silk screens. However, it was *photography* that revolutionized screen printing methods, as it did for most fields of graphic communications. The perfection of durable light-sensitive emulsions that could be imbedded in screens and developed to form photographic stencil images was the major advance that made screen printing of value in the electronics industry, allowing fine detail to be reproduced from original high-contrast artwork.

Modern improvements in screen printing technology can thus be summarized as (1) photographic techniques, (2) better fabrics, and (3) automated process machinery. Of these, the last two are only important in high-volume industrial printing. Photography can be used with good results by everyone, including the hobbyist; processing of photographic screen stencils does not even generally require a darkroom. The basic procedure in screen printing remains simple: fabric is stretched across a frame, a stencil is attached to the mesh, and ink is forced through the screen to print images.

Screen printing is a valuable technique in the electronics prototyping laboratory because of its usefulness in PCB fabrication and in labeling panels and enclosures with printed information. In the case of PCBs, white ink might be used to print information on circuit boards; in the case of enclosures, brightly colored enamel paint can be used to create labels or designs on panel plates. Results are only limited by the imagination of the designer.

OBJECTIVES • The objectives of this chapter are as follows:

- Show the value of screen printing in electronic prototype construction.
- Discuss the general principles of screen printing.
- Detail the various elements that comprise screen printing methods, and relate them to specific applications in electronic construction.
- Give detailed instructions for a simple manual screen printing procedure using high-resolution photographic techniques.

MATERIALS SOURCE • The following sources listed in Appendix B are recommended for equipment and supplies related to screen printing:

Cincinnati Screen Process Supplies Inc.
International Printing Machines Corp.
The Naz-Dar Co.

ELECTRONIC APPLICATIONS OF SCREEN PRINTING • Before proceeding to more detailed discussions of the elements that make up screen printing techniques, it would be helpful to outline some specific areas in electronic fabrication where this versatile printing method can be of value to the hobbyist, experimenter, or prototype designer. Such a summary will give the reader a better perspective when considering how various approaches and materials are chosen for specific printing objectives in the area of electronics. Screen printing covers a wide field, and although basic principles will be presented, we are mainly interested in a few well-defined applications. Other books[7,24] are available that cover the complete range of screen printing techniques in detail.

The four areas of immediate potential interest are as follows:

1. Chassis and panel designs/labels
2. Component labeling information on PCBs
3. PCB solder masks
4. PCB etch and plating resists

Screen printing parameters of most variability in these four areas are usually the image resolution, latitude for distortion, multipattern registration requirements, ink qualities, and process volume. Assuming that photographic stencil films are used to prepare printing screens, the original artwork will be similar in each case, and is prepared using methods previously described in Chapter 3. Note that a *positive* transparency is always used as the photofabrication mask (see Chapter 4) in screen printing because the final screen emulsion is a negative image of the ink pattern to be printed, and the photographic stencil is a negative-acting film when exposed to light. In other words, the initial transparent photo mask will always look like the final ink pattern to be screen-printed.

Chassis and Panel Printing • A good example of a screened electronic panel design is illustrated in Fig. 8-1A. This is the original photo mask, and the final assembled panel is shown in Fig. 8-1B. Some of the most important advantages of the screen printing process are realized when a control panel with detailed graphics and professional appearance is produced. Inks may be chosen to suit any type of surface, such as brushed aluminum, finished wood, or smooth plastic. Many combinations of colorful inks, both transparent and opaque, can be printed in registration to give striking designs; this approach requires that a matched set of screens be prepared, one for each color pattern to be printed, and that the colors be printed/dried in separate, successive steps. Another important advantage of screening methods is that cabinet and chassis shapes are not critical to the printing procedure. As long as the surface to be printed is reasonably flat, an arrangement can be devised to position the object underneath a screen frame. The front side of a formed, rectangular metal box can thus be printed as easily as a flat nameplate to be attached with screws.

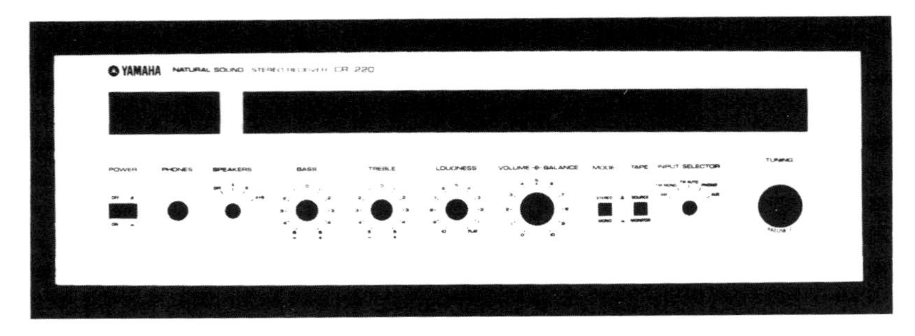

(A) *Positive photo mask.*

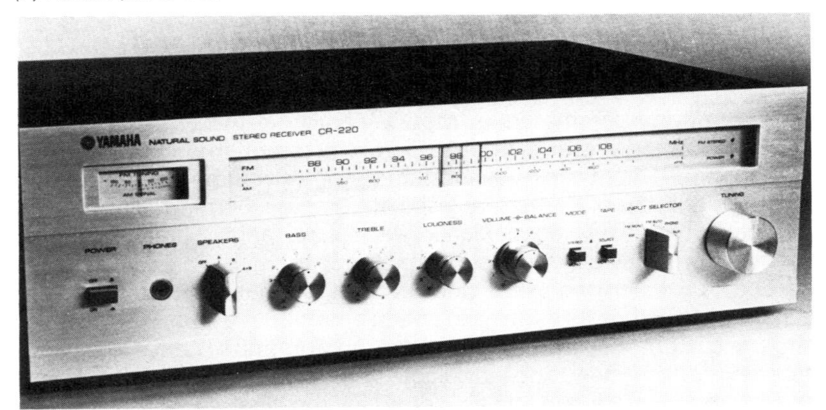

(B) Final assembled panel.

Fig. 8-1. Screen printing electronic panels.

PCB Component Side Labeling • Labeling information is commonly printed with white ink on the component side of printed circuit boards as an aid to assembly, calibration, and troubleshooting. A typical positive photo mask for this purpose was shown in Chapter 3, Fig. 3-24, and an assembled board with printed labels was shown in Chapter 2, Fig. 2-19. The ink used for PCB labels is usually an epoxy-based formulation capable of withstanding both heat and solvents, to which boards are exposed during solder assembly and final solvent cleaning. Viscous, chemically active epoxy inks are difficult to use with printing techniques other than screened application. The flexibility of a screen and squeegee system also allows ink to flow smoothly across the irregular surfaces created by etched copper traces covering the board. PCB labels will usually consist of letters, numbers, and component symbols whose image sharpness is not particularly critical, so long as the information is readable. Registration of the label pattern with the underlying board surfaces also is not critical. Therefore, inexpensive, easily handled silk material can be used for the screen fabric in label printing. It should be noted that inks used for PCB labeling must be electrically nonconductive, because they are permanent coatings that may directly bridge circuit board traces. Even a slight conductivity may produce some unexpected results in high-impedance, high-frequency circuits.

PCB Solder Masks • The transparent green coating widely seen on recently manufactured printed circuit boards is a solder mask applied as an ink, using screen printing methods. *Solder masks* are designed to direct molten solder to specific areas of the board during assembly, primarily in combination with automatic wave-soldering equipment. An example of solder mask artwork was shown in Chapter 3, Fig. 3-31. The pattern usually consists of a circular opening around each terminal area to receive solder; when hot solder is applied uniformly across the entire board, either by hand-dipping or wave technology, small solder fillets form around each lead and wick up into the holes. Without the presence of a mask, solder would stick indiscriminately to metal traces across the board in large globs, bridging conductors and forming a mess. Because they must repel hot solder at temperatures up to 250°C (482°F), solder resist inks usually contain tough, permanently cured resins, such as epoxy or melamine formulations. These resins are difficult to remove after screening, and are not necessary for prototype boards where component leads will be individually hand-soldered. However, solder masks are mentioned here because the thick ink is easily applied by screening, with noncritical printing tolerances. Even manual assembly is neater with the use of solder masks on high density PCBs.

PCB Etch and Plating Resists • Screen printing is a good method for forming etch-resistant patterns on copper-clad PCB laminates, unless extremely fine detail is needed. Screened resists can be applied as 0.015 inch-wide (0.381 mm) lines, which should be sufficient resolution for most PCBs. (By comparison, photo resist is easily capable of 0.005 inch-wide (0.127 mm) lines.) Screens are particularly well suited for producing "Print and Etch" boards in high volume, especially if the boards are single-sided and do not require pattern registration during printing. Once a screen is set up and in operation, a single stroke of the squeegee will complete the printing process for each board. Double-sided boards can also be fabricated with screen printing methods, if dimensionally stable screens and precise registration equipment is available. A particular advantage of screen-printed plating resists exists in the *pattern plating* process for forming plated-through-holes (discussed in Chapter 2), because no resist is deposited in predrilled holes, which can obstruct the subsequent metal coating step. On the other hand, when photo resist is used for image transfer, residual monomer coating is often difficult to remove from the hole walls during the developing step.

In any case, no particular advantage exists for the use of screened resists in low-volume PCB fabrication. As seen in Chapter 5, photo resist is a more practical approach for PCB prototyping, and has the advantages of better resolution and less difficulty in registration of double-sided boards. Working directly with negative masks on photo resist is the simplest manual approach for achieving registration; screen printers must worry about screen distortion and precise positioning of the stock material during squeegee printing. In the case of single-sided boards, the purchase of precoated photo resist laminates is a time-saving approach compared to screen preparation.

If a screen printing process is justifiable for PCB fabrication in higher volume runs (where the process becomes efficient), then the properties of the resist ink are of most importance. First, the ink must have the right viscosity to be transferrable to a smooth metal surface without running; line definition must not become distorted after printing. A thick, gel-like, greasy resin is usually the best consistency. Second, the ink must be strongly adherent to the metal substrate without peeling or flaking after drying. And finally, a chemically resistant ink is necessary to withstand all of the highly active etching and plating solutions in common use. Vinyl-based inks are usually the most suitable because they fulfill these basic requirements, and can also be removed easily from the board when the plating or etching operation is complete. The UV-curable inks are a more recent development in electronic screen printing.

GENERAL PRINCIPLES • Screen printing differs from other methods of printing in that ink is transferred *through* a plate instead of *from* the surface of the plate. The screen plate consists of a fine mesh fabric that has been *masked,* or *blocked-out* with various materials imbedded or attached to the fabric. The final mask is known as a *stencil,* and the fabric is called the *screen.* Since the open mesh screen is essential for supporting the stencil pattern, another term for screen printing is *mitography*[25], meaning to *print* with the use of *fibers.* Many techniques have been developed to make mitography a suitable process for achieving results impractical by any other printing method, principally due to the wide variety of inks that can be used and the possibility for printing on almost any surface while maintaining high image resolution. Both hand procedures and mechanized equipment will produce equally fine results.

A basic screening operation is illustrated in Fig. 8-2 with a simple, complete hand printing unit that can be constructed of wood

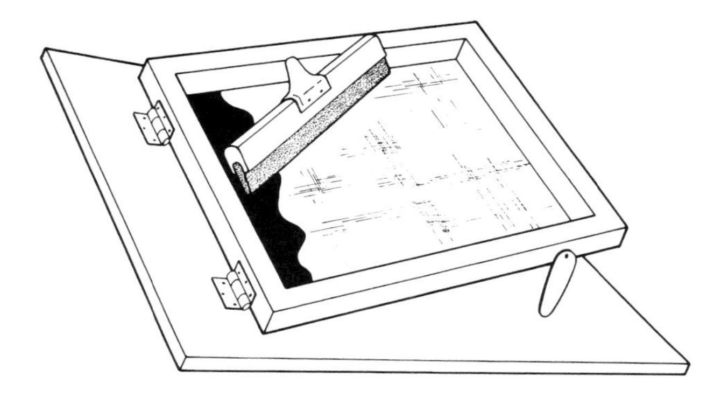

Fig. 8-2. Basic screen printing setup.

materials. The frame is mounted on a sturdy flat base with hinges, or it may be directly attached to a table top. The frame is a rigid rectangular border over which the fabric has been tightly stretched and fastened, sometimes simply with tacks or staples. Silk is the traditional fabric used in mitography, and it is still effective for the majority of low-volume printing needs. After imbedding an appropriate ink-resistant stencil mask into the open screen areas, the frame is mounted on its base hinges so that it can be swung up and down over the work material which is positioned underneath. Guides are often attached to the base to align the work with the screen as it swings down for printing. Another helpful addition is a prop bar or counterbalance to hold the frame up when material is being inserted or removed. During actual printing, the screen is swung down over the work surface, sometimes touching, or sometimes slightly offset above the surface. Ink is poured across one end of the screen, and a flexible *squeegee* is used to draw ink across the stencil in a smooth, sweeping motion which forces it through the screen pores to deposit on the work. The screen is then carefully raised, the printed material is removed, and the apparatus is ready to make another print.

Printing screens always have a usable life-span that limits the number of successful prints that can be made from any one stencil/screen combination. Useful life is limited by the toughness of screen fabric, its resistance to ink and cleaning solvents, deformation or loosening of the screen material, and durability of the imbedded stencil. When a particular stencil is no longer useful, in many cases it can be removed and another one attached, depending upon the value and condition of the screen fabric. Silk is considered to be an expendable fabric by most printers and is discarded following each series of runs; a variety of methods have been developed to allow new fabric to be quickly stretched and attached to the frame in preparation for a new stencil.

Preparation of the stencil itself was a major undertaking at one time, requiring the hand assembly of knife-cut films or other ink-resistant screen coatings. However, the present use of *photographic stencils* has opened the screen printing field to persons with little artistic skill or manual dexterity. It also means that images as detailed as actual photographs can now be screened using masks consisting of fine dot patterns, or *halftones*. In general, any black-and-white image with good contrast can be converted into a working screen stencil using photographic techniques. Light-sensitive emulsions imbedded in the screen behave similar to the photoresist polymers discussed in Chapter 5. After being exposed to light through a mask, the emulsion can be developed with a solution that dissolves unexposed areas and leaves the stencil pattern imbedded in the screen as a tough coating.

Although the apparatus shown in Fig. 8-2 appears simple, all screen printing techniques are merely variations on this approach. The most complicated commercial equipment is designed to speed up the printing process through mechanization, but no substantial changes are made to the nature of the system. High-volume screening equipment was previously illustrated in Chapter 2, Fig. 2-13; it allows material to be rapidly fed under the screen, accurately positioned, an automatic squeegee gives controlled ink transfer across the surface, and the material is withdrawn and moved automatically into a drying device. Except for differences in the rate of production, an experienced printer can achieve the same quality results using hand-operated screens that machines can produce automatically.

ELEMENTS OF SCREEN PRINTING • The type of fabric used in screen printing is chosen for each particular job based on the resolution required, the type of stencil to be used, the number and quality of prints needed, and other variables in the printer's process. The fabric must be woven in a uniform mesh, it must be durable and capable of being tightly stretched, it must stand up to all of the inks, solvents, and cleaning solutions employed in printing, it must be permeable to ink, and it must be capable of retaining stencil materials during the normal wear and abrasion of the squeegee process. Many types of fabrics have been tested, but the only important ones used today are silk, Nylon (polyamide), Dacron polyester, and stainless steel.

Silk • Silk is the traditional fabric used in screen printing because of its resilience, strength, and chemical resistance. Silk is the strongest natural fiber known, and a good grade of silk cloth can be used to make several thousand screen prints. Although synthetic fibers are coming into greater use today, silk is still the easiest one to use for general screen printing because of its good stretching characteristics and its ability to retain stencil materials securely. These qualities are due to the strength and fibrous nature of silk threads, which are composed of many smaller entertwined filaments. *Multifilament* fabrics give the most "bite" and surface area for retaining gelatinous photographic stencils, although they have the disadvantage of less open pore area available for ink transfer through the mesh.

Silk fabric used for screening is sometimes known as bolting cloth, and is classified according to the type of weave, the weave density, and the overall fabric strength. The best grade of silk is still imported from Switzerland.

Weave — The most common weaves used for silk and other fabrics are illustrated in Fig. 8-3. Coarser cloth maintains better strength and uniformity if some of the multiple thread weaves are used, but for most screen printing work the plain *taffeta weave* is chosen because it allows the greatest aperture area relative to the total screen surface area. The plain weave is strong and uniform when high density mesh is used (greater than 120 threads per linear inch). Cloth is available up to 80 inches (2 m) wide and in bolt lengths as long as 60 yards (55 m).

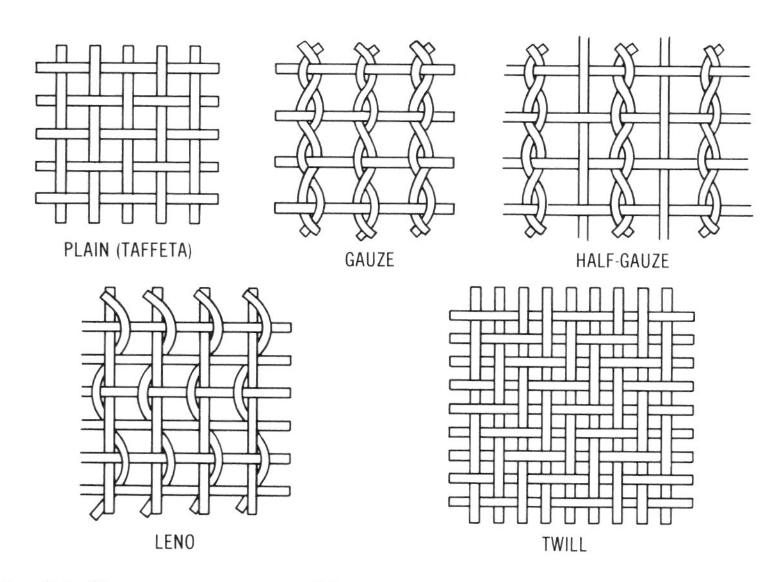

Fig. 8-3. Weave patterns used in screen printing fabrics.

Weave density — Number classifications are used to specify mesh count information for silk cloth, as seen in Table 8-1. A typical silk used for general-purpose screening is No. 12, having about 125 openings per inch. For fine detail printing with photographic stencils, such as in printed circuit applications, higher number silk is used ranging from No. 16–20 (157 to 173 mesh count). Silk is available up to No. 30, which has about 220 openings per inch; finer mesh is capable of finer printing detail, but the ink deposit also becomes progressively thinner, due to decreasing open area. Finer silk is more difficult to use because the smaller pores make it harder for ink to pass through and screen clogging is more likely to occur. Therefore, fabric should be chosen as fine as is necessary to hold detail, but as coarse as possible to maximize open area.

Table 8-1. Number Classifications for Taffeta Weave Silk Fabric

Mesh Number	Mesh Count	Mesh Opening (Inch)	% Open Area
2XX	54	0.0132	54
4XX	65	0.0107	49
6XX	74	0.0092	46
8XX	86	0.0077	43
10XX	109	0.0054	35
12XX	129	0.0040	27
14XX	139	0.0037	26
16XX	157	0.0032	26
18XX	168	0.0030	25
20XX	173	0.0028	24
25XX	196	0.0026	24
30XX	220	0.0024	23

Strength — The quality of silk is designated with X's following the number, with more X's indicating a stronger weave. Double X quality is almost always used in screen printing, such as No. 16XX. Although triple X silk is available for producing more durable screens, the thicker threads in such material result in smaller pore size.

Silk screens are still used considerably in electronic printing applications because they are easily attached to frames, indirect photographic stencils adhere well to silk, the material is unaffected by organic solvents, and silk has considerable resilience (the ability to withstand temporary pressures without deforming). However, several disadvantages of silk should be noted as compared to newer synthetic materials:

1. Abrasive cleaning is difficult without damaging silk fibers. This makes it difficult to reclaim silk screens by complete removal of old stencils.
2. Silk tends to absorb moisture and lose tautness. If the screen is then re-stretched, images become distorted. When low toler-

ances and close alignment of patterns are required, most printers use polyester or wire cloth.

3. The multifilament, fibrous nature of silk threads gives less open pore area as compared to single-stranded (*monofilament*) materials. This limits passage of ink. Silk meshes above No. 20 are therefore prone to ink-clogging, and the problem is intensified by the *wicking* action of multifilament fibers that tend to absorb ink.
4. As silk threads wear during printing, fine filaments break and separate from the main fiber. Since these strands tend to wick ink, the result is low-quality print: poor line definition, ragged edges, and smeared patterns. For this reason, silk is not highly recommended for printing PCB conductor patterns where a bridged trace can be troublesome. However, it is often used for printing letters, numbers, and symbols in electronic applications. Silk wear is less of a problem when a low number of prints is intended and the fabric is not reclaimed from previous screens.

Silk must be cleaned before use, but the rough fibrous nature of the threads makes stencil adherence less critical than with other fabrics. Do not use abrasive powders or a stiff scouring brush, which will damage silk threads. Instead, silk is prepared by washing in warm water and a mild, nonbleaching detergent, followed by a thorough water rinse and drying.

Synthetic Fabrics • Silk cloth is being replaced in the screen printing industry by synthetic fibers, particularly nylon (polyamide) and dacron polyester. In addition to being less expensive than high-quality silk, the synthetics have some real process advantages as follows:

1. Synthetic fibers are formed as smooth, single filament threads with high tensile strength. These threads can be made very thin in diameter, allowing a very high mesh count with a large percentage of open area. The result is that finer detail can be printed. For example, 196-mesh nylon is equivalent to No. 25XX silk in threads per inch, but the nylon percent open area is almost 40% compared to 24% for the silk. Since this allows higher mesh screen to be used, nylon and dacron polyester are available woven with more than 400 threads per inch.
2. Smooth monofilament threads do not soak up ink during printing, allowing smooth, even printing, and preventing screen clogging problems.
3. Synthetic fabrics are resilient like silk, but they also have better strength, durability, and abrasion resistance. Manufacturers claim that 100,000 prints can be made from a single well-prepared nylon screen and stencil. Because abrasive scouring agents and

harsh chemicals can be used to treat the fabric, cleaning procedures are effective and screens can be reclaimed for attaching new stencils.

4. Polyester holds tension and dimensional stability extremely well and is suitable for close-tolerance printing. Nylon absorbs moisture like silk, but this problem can be minimized by stretching nylon fabric while it is wet.
5. Both nylon and dacron polyester are resistant to organic solvents normally encountered in screen printing. There are some solvents that will attack synthetic fabrics, and these liquids can be used to good effect for cleaning. For example, nylon is attacked by acetic acid and phenols, which become good cleaning agents when used at the proper dilution.

Synthetic fabrics have two particular disadvantages relative to silk that should be noted:

1. The smooth nature of monofilament thread makes it less adherent for some stencil materials, such as indirect photographic films. It is, therefore, desirable to use direct photographic stencils that are actually imbedded in the fabric as an emulsion before exposure and development of the stencil pattern. Cleaning of nylon and polyester is also extremely important; no trace of grease or grime is permissible. Strong alkaline cleansers may be used and fine scouring powder is helpful for introducing some "bite" into the smooth fiber surfaces.
2. Synthetic fabrics are more difficult to stretch on frames than silk, which is often hand-stretched. Nylon in particular is stiff and fairly difficult to attach to frames. Mechanical stretching aids and tension gauges are used to properly prepare synthetic screens. Manufacturer instructions should be followed carefully when stretching nylon or polyester screen fabric.

Cleaning procedures for nylon and dacron polyester are similar, although specific instructions accompanying fabric and stencil materials should be considered. A typical cleaning cycle consists of the following four steps:

1. The fabric is slightly roughened with an abrasive.
2. Fabric is dipped in an alkaline solution of about 10% sodium hydroxide or tri-sodium phosphate (TSP).
3. Fabric is neutralized with 5% acetic acid (vinegar).
4. Fabric is thoroughly washed with water to remove all previous cleaning materials.

Reclaimed fabric should not be scrubbed with abrasives again because this step is only necessary once, and it will weaken fibers.

Stainless Steel • Metal screen fabric (also known as *wire gauze*) offers the most durability, uniformity, and strength of any material. It is used where extreme precision and long wear resistance is needed. In the hands of an experienced printer, wire screen can produce prints within tolerances of 0.0005 inch (0.0127 mm). A common screen fabric used in printed circuit production is 270-mesh, 0.0014 inch-diameter (0.0356 mm) type 304 stainless steel wire cloth. Because of the ability to control wire diameter and strength, manufacturers can produce metal screens with the finest mesh and largest amount of open area of any screen fabric. A properly prepared wire screen can be reclaimed and used many times while remaining perfectly taut. However, several disadvantages (including its high cost) make the use of stainless steel screen impractical for inexperienced users — they are as follows:

1. Metal screen is the most difficult fabric to correctly stretch on a frame. Special frames are built for this specific purpose.
2. Metal screen is not resilient — this makes it undesirable for printing on highly irregular surfaces. Also, the metal screen must be carefully handled. Dropping an object on a tight metal screen will produce kinks that are impossible to remove.
3. Metal fabric must be extremely clean before attaching the emulsion, coating, or film. At one time, cleaning steps included solvent degreasing, strong alkaline solutions, abrasive scouring, and even torching with a gas flame. A more recent approach is to degrease the metal and introduce a slight surface etch with etchant solutions, such as dilute ferric chloride. Once it is properly cleaned, stainless steel mesh is intermediate between silk and nylon for encouraging the adherence of transfer stencils. Direct, imbedded photographic emulsions are most reliable, but also the most difficult to prepare.

In summary, the reader who is inexperienced with screen printing techniques should start with silk fabric, which permits simple hand-stretching and the use of indirect (transfer) photographic stencil films. The reader can then progress to synthetic or metal fabrics that produce better quality prints and more detail, but that require more careful cleaning, better quality frames, and often the use of direct photographic emulsions for durable stencils.

Frames • The frame used for screen printing is a square or rectangular rigid border to which the mesh fabric is attached and stretched to "drum-tight" tension. The frame should be several inches wider than the largest stencil pattern to be attached, giving some working freedom during cleaning, stencil transfer, and squeegee inking. The frame material thickness may range from 1 inch to 5 inches (25.4 mm to 127 mm) wide depending upon the total frame width and strength desired. Most screen printing frames are made from wood or aluminum.

Wood Frames — Low-cost printing frames are made from kiln-dried white pine, which has the qualities of light weight and minimal tendency to warp. These frames are usually used with silk, although synthetic fabrics may also be attached. Wooden surfaces should be oiled or varnished to prevent warp due to moisture absorption. The following four different types of wood frames are commonly seen:

1. Plain wooden frame — The simplest possible system, a plain wooden border can be used to attach silk fabric by stapling or tacking down the hand-stretched material. A thin strip of tape, paper, or cardboard is usually positioned between silk and staples to prevent tearing of the material. Starting at one corner of the frame, fabric is tightly stretched and fastened along one side. The fabric is then pulled toward the opposite side and fastened along its entire edge, keeping threads parallel to the frame at all times to maintain even tension and maximum fabric strength. Fabric is attached to the other two sides in similar fashion. A little practice will produce an even, "drum-tight" printing screen. Two offset rows of staples or tacks should be sufficient to hold the silk in place, and the frame is usually sealed with strips of gummed paper tape attached to both sides of the fabric and the wooden borders. This tape gives the screen strength where cloth attaches to the frame, and it prevents ink from leaking out of the stencil/screen area. Wooden frames prepared in this manner can be reused many times by tearing out the staples, but eventually they must be discarded due to the steady damage inflicted on the wood.
2. Groove and cord — An improvement over the plain wooden frame is one containing a recessed, round, machined groove in its border, into which the fabric and a waxed cord are stuffed with a blunt tool to hold the silk firmly stretched in place. Such a frame is shown in Fig. 8-4; Cincinnati Screen Process Supplies furnishes this type of frame stock in the form of interlocking pieces ranging from 8–72 inches (203–1829 mm) in length and 1–2.75 inches (25.4–69.9 mm) in width. The cord and groove system is very fast and easy to use, and is nondestructive to the frame. Screens can be quickly removed for restretching or replacement by merely pulling the cord out of its groove. The stretching technique is essentially the same as with a plain frame: starting at one corner, the fabric is steadily stretched and attached around the perimeter,

Fig. 8-4. Wooden frame with groove and cord system for attaching fabric.

keeping fabric threads parallel at all times.

3. Groove and cleat — Similar in principle to the groove and cord method, the frame has a rectangular groove cut along each side. The fabric is positioned over the grooves one side at a time, and cleats are pushed into the grooves and screwed in place. The system is illustrated in Fig. 8-5A. When all four cleats have been installed in their grooves, the screen fabric should be wrinkle free but not completely taut. The cleats are then carefully screwed down deeper in their slots to develop proper uniform tension across the screen surface.

4. Floating bar frames — This type of frame is illustrated in Fig. 8-6A. Fabric is initially attached to the four inside floating bars by one of the previous methods, wrinkle free and under slight tension. Further tension is then developed on each side of the frame by steadily tightening the bolts with wingnuts. One advantage of this system is that strong, uniform pressure can be developed, even sufficient for stretching metal fabrics if the frame is strong enough to hold tension without distorting. Floating bar frames are often made of aluminum for this application. Another advantage of the floating system is that the degree of stretch may be easily measured with a ruler. For example, nylon fabric is often specified to be stretched to 105% of its wrinkle-free dimensions. The fabric is attached to the bars, and extended to the exact required final length on each side to satisfy this specification.

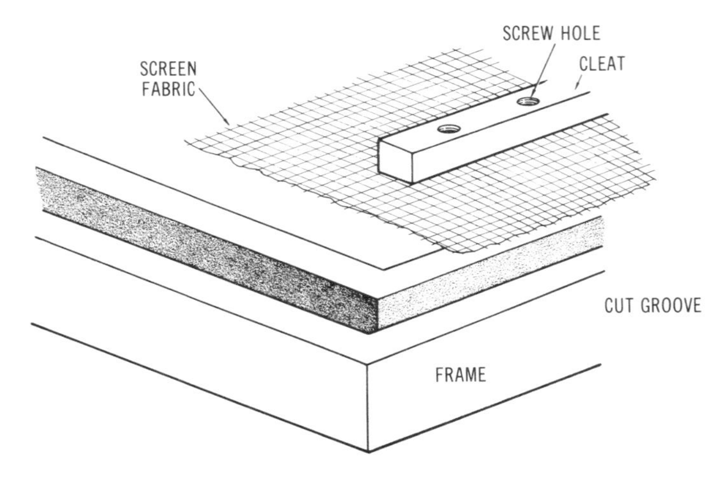

(A) Wooden groove and cleat design.

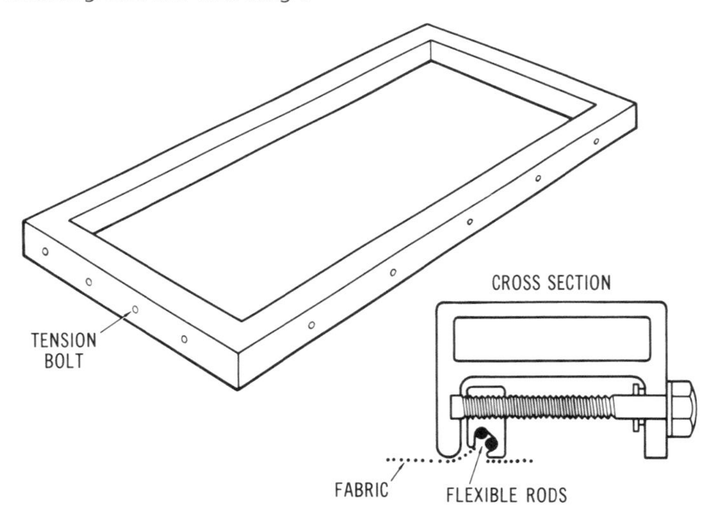

(B) Diamond Chase machined aluminum frame.

Fig. 8-5. Construction of screen printing frames.

Metal Frames — Fabricated, welded, machined aluminum frames are a definite improvement over wooden frames because they give better dimensional stability and may have built-in devices to aid in the attachment and uniform stretching of screen fabrics. One of the simplest approaches in industrial high-volume screen preparation is to use simple, flat aluminum frames made from smooth welded stock. An elaborate, free-standing stretching device (see Fig. 8-7, the Harlacher Model H-23) prepares the fabric to proper dimensions and tautness;

the simple frame is then brought up from below the tight screen and glued in place with a quick-setting adhesive.

Once the reader gains some experience in screen printing, he or she will probably want to invest in a good machined aluminum frame, such as the Diamond Chase frame illustrated in Fig. 8-5B. A cross section of this frame is also shown in the figure. Screen fabric is pushed into a recessed groove inside a drawbar on each side of the frame, and clamped in place with two flexible nylon rods. A bolt is located every six

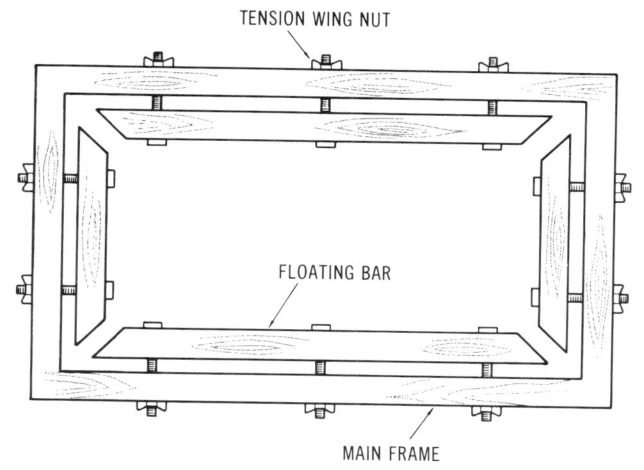

(A) Floating bar frame.

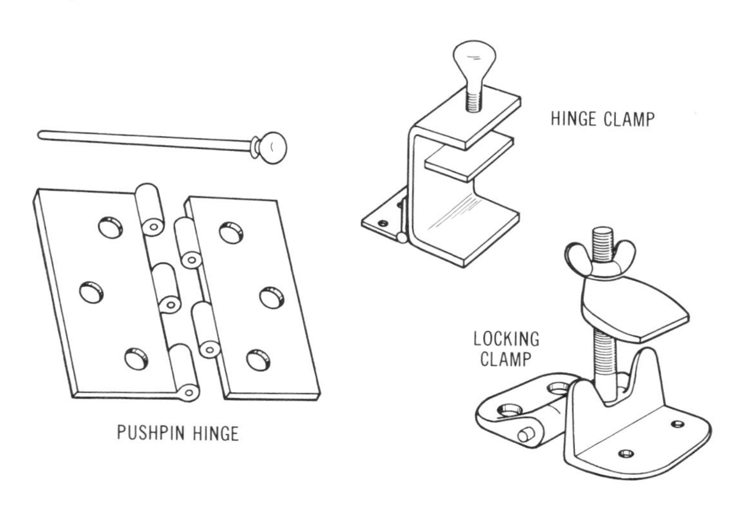

(B) Hinges for mounting screen printing frames to a base.

Fig. 8-6. Floating bar frame and hinges.

Fig. 8-7. Commercial precision fabric stretching device. ***(Courtesy the Naz-Dar Co.)***

inches along the length of the sides, and is manually screw adjusted to develop the proper screen tension and pattern registration. In principle this system is similar to floating bars, but more versatile adjustment is possible with the evenly spaced bolts; also, screen fabric is held firmly against the smooth inside frame border at all times. Diamond Chases can be made to any specified size, and represent a very convenient method for preparing screens with any type of fabric material. Although they require a high initial investment, the printer will obtain longer life, better results, and less aggravation by using such a patented system. (A 14 by 18 inch ID, 2 inch-thick Diamond Chase sold for about $90.00 in 1982.) Complete inertness to organic solvents, dimensional stability, and the ability to take extreme tension are other advantages of this type of frame. One caution on aluminum frames is to avoid exposing them to strongly basic cleaning solutions, such as sodium hydroxide, which attack aluminum metal.

Photographic Screen Printing Films • Once a printing screen has been assembled by attaching tightly stretched fabric to a frame, a stencil must be adhered to the screen to form the desired pattern of openings for ink transfer. Hand prepared stencils are still being used by artists and commercial designers of large posters, but photographic methods are most efficient for preparing screen stencils having fine detail. In general, the fabrication of a photographic screen involves the following steps:

1. Master artwork is prepared using high-contrast materials, as described in Chapter 3 for printed circuit artwork. Typical artwork is in the form of a black-and-white drawing, or black transfer patterns adhered to a transparent sheet of polyester film. Artwork is produced on the largest scale needed to hold detail. If the final print will consist of multiple inks printed in sequential registration, then artwork must be prepared so that the pattern for each different ink can be processed into a separate screen, and they can all be aligned with some type of registration system.
2. The artwork is photographed and converted into a high-contrast, positive transparency of the pattern to be printed. Continuous-tone photographs can be screen printed if the original photograph is converted into a *halftone,* which means the image is converted into a transparency having a distinct pattern of high-contrast dots. The individual dots must be of sufficient size to be resolved by the screen stencil and fabric mesh pores. Photography with high-contrast films is discussed in Chapter 4, although halftone work is not mentioned; for more information on this technique, the reader is referred to a text[26] on graphic arts photography for screen printing.
3. A contact print is made using a light-sensitive emulsion exposed through the positive transparency. The emulsion is developed to produce a stencil having open areas and blocked-out areas corresponding directly to the light and dark areas of the transparency. Some film emulsions are attached to the screen *before* exposure, and some are attached *after* both exposure and development. The final stencil emulsion must be resistant to ink and solvents, and must remain adherent to the screen fabric during squeegee printing.
4. In some cases photographic emulsions can be removed from the screen after use so that a new stencil may be attached to the same fabric.

The substances or coatings used to produce photographic emulsions for screen printing may be natural materials, such as gelatin and glue, or they may be synthetic polymers, such as polyvinyl alcohol and polyvinyl acetate. In most cases the emulsion must be *sensitized* with special chemicals, such as ammonium bichromate or diazo compounds, which render it light sensitive. The coating is then exposed to light through a transparent mask, and areas that receive radiation become hard, while other areas remain soft and are washed away in the developing, or *wash-out* step. Three distinct methods are used to prepare photographic printing screens, each having certain advantages and disadvantages depending upon the particular application. These three methods are termed *direct, indirect,* and *direct-indirect* photographic screens.

Direct Screens — In the direct approach, a liquid emulsion is coated directly on the screen by brushing, dipping, or squeegee coating. In most cases the emulsion is sensitized in the liquid form, while in other cases a sensitizer may be applied to the coating after it is dry. The complete screen is then exposed to light, and processed. In any case, the direct type screen is the most durable of all photographic stencils because the coating completely covers and penetrates the screen fabric. Because of this inherently strong adherence, direct emulsions are most often used with fabrics such as stainless steel or synthetic monofilaments that do not retain flat films well. Disadvantages of the direct approach include difficulties in producing smooth, uniform coatings, inconvenience in processing the entire screen/frame assembly, and loss of image resolution due to the influence of the screen mesh inside the emulsion layer.

Recent emulsion formulations and improved coating methods have eliminated the problem of ragged edges in lines printed from direct screens. The direct process produces very accurate registration of multiple prints, because no image distortion occurs due to transferring a soft, limp, processed stencil to the screen, as is done in indirect photographic film methods. An example of a modern process is the Encosol direct emulsion from Naz-Dar Company. The use of a diazo sensitizer in this product eliminates many of the problems found in earlier bichromate-sensitized emulsions. For example, the Encosol liquid emulsion is not sensitive to heat and is considerably more stable than bichromate products. Once coated on both sides of the screen with a squeegee and dried, the light-sensitive screen may be stored for many months under darkroom conditions before use.

Exposing a direct screen to light presents somewhat of a problem because the entire screen must be held in close contact with the photo mask. Commercial contact printing frames for this purpose use vacuum and a flexible blanket to encapsulate the entire frame and transparency during exposure (see Chapter 2, Fig. 2-12). Two simpler methods are illustrated in Fig. 8-8. In Fig. 8-8A, the screen frame is inverted over the transparency and a rigid glass plate; the sandwich is held firmly down with weights resting on the inside (squeegee side) screen surface. Ultraviolet light is then introduced from below the glass plate to give the proper exposure. In Fig. 8-8B, buildup material is placed underneath the frame to support the screen from the inside surface, and the transparency is placed between the top outside surface of the screen and a heavy glass plate to hold them in close contact. Ultraviolet light is then positioned above the glass plate for exposure. It should be noted that

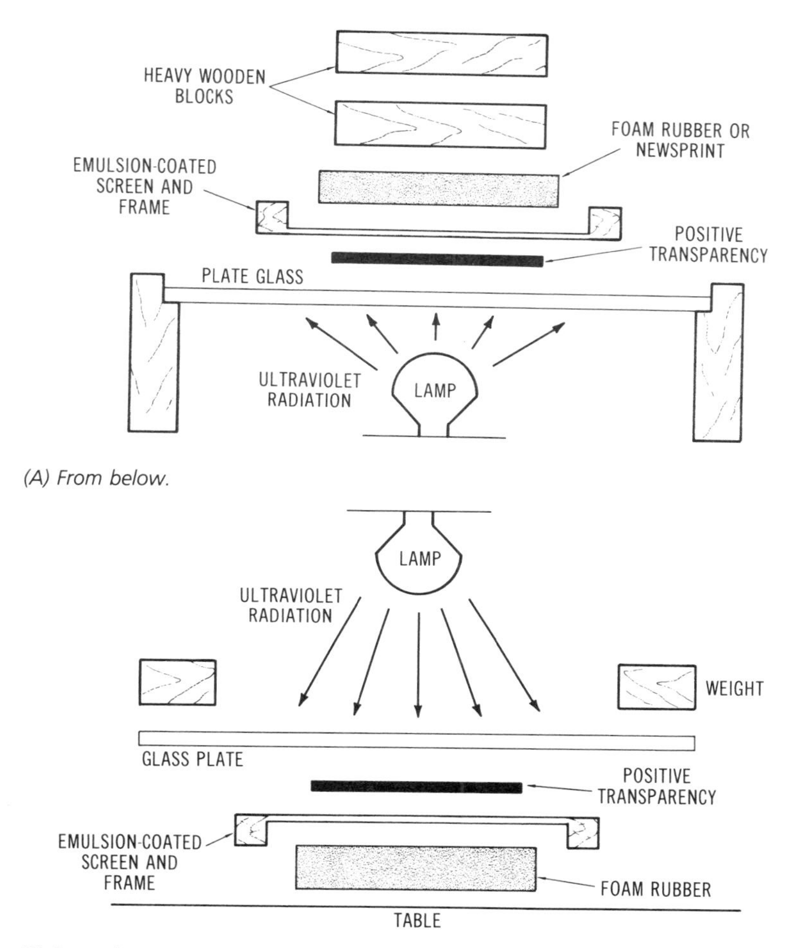

(A) From below.

(B) From above.

Fig. 8-8. Exposing a direct photographic screen.

the exposure and sensitivity of photographic screen printing emulsions is similar to that of photo resists (discussed in Chapter 5), and similar light sources are used for polymerizing both types of materials. Depending upon the type and strength of light, and distance from the film, exposure time is typically 3–15 minutes.

Developing, or washing out of the Encosol emulsion is fairly simple. The screen is not particularly light sensitive in the visible region, so it can be processed under normal room lighting. After exposure, the screen and frame assembly is placed in a sink and washed with a low pressure spray of water from a hose — warm water may be helpful. After a few minutes, unexposed emulsion separates clean and sharp from the screen, leaving the desired stencil image. Excess water is removed by blotting with paper towels, and the screen is air-dried in front of a fan.

Synthetic and metal screen fabrics can be reclaimed after use with Encosol emulsion coatings. However, natural silk cannot be cleaned without damaging fibers. To reclaim a screen, printing inks are first completely removed with solvents. The screen is then soaked on both sides with strong bleach for about 20 minutes, and washed with a strong spray of warm water. Before applying a new emulsion, the screen must be thoroughly cleaned to remove all traces of bleach or old emulsion particles. Degreasing and detergent solutions are important for ensuring good adhesion of the next emulsion coating.

Indirect Photographic Stencils — Indirect films for screen printing are also known as *transfer* type films, and represent the earliest method of preparing screen stencils by photographic techniques. The method is known as *indirect* because the film is processed to form a stencil *before* it is attached to the screen. One of the first successful transfer films was called *carbon tissue,* composed of a gelatinous coating on a backing sheet of paper; it was developed for screen printing in the 1920s, and similar films are still in common use today. Carbon tissue has the advantage of being easily reclaimed from silk or other fabrics after use because the gelatinous base is easily dissolved by mild enzyme-type cleaners and warm water. However, carbon film must be processed on a temporary support during exposure and wash out, which makes it more difficult to use as compared to newer transfer films that are processed directly on their clear plastic backing sheets. Also, less image distortion is likely to occur when the original backing sheet stays with the film until it is securely attached to the screen and dried. Recently developed indirect films are based on mixtures of gelatin, polyvinyl alcohol, and polyvinyl acetate; they are formulated to give characteristics such as presensitized stability, better photographic resolution, and better adhesion to synthetic monofilament screen fabrics.

Indirect photographic films are often used in preference to direct types because of the following advantages:

1. Resolution — Indirect screens give exceptional detail in electronic printing applications because the emulsion is not imbedded in the screen, but is attached more by surface adhesion. This provides a screen printing surface as smooth as the original photographic film surface.

2. Processing — Processing is very convenient. A suitable-size film sheet is cut from the roll and exposed to light, in the same manner as making a contact photographic print. The image is then washed out in an aqueous developing solution, and the wet film

backing is directly adhered with pressure to a clean screen. After drying, the backing sheet peels away to leave the stencil mounted in position for printing. (Open screen areas around the film are filled in with ink-resistant block-out material.)

3. Reclaiming — Indirect screens are easier to reclaim because the film does not encapsulate the fabric. This is the only practical photographic stencil that allows silk screens to be used for more than one stencil.

The adhesion characteristics of indirect film also lead to its main drawback, which is a lack of toughness and, therefore, decreased printing life as compared to direct, imbedded films. Synthetic and metal screen fabrics may be used with indirect film, but the mesh surface must be correctly prepared and extremely clean. Finer mesh screens (at least 150 lines per inch) are preferred because they offer more points for film fabric contact and thus better adhesion. This lack of stencil durability should not discourage the printer who only intends to make a few prints, as in electronic prototyping. Silk cloth is recommended in these applications because its multifilament nature ensures secure film attachment.

A recommended indirect photographic screen printing film is the Blue Poly 2 material from Ulano, which has become well accepted in the trade as a general purpose film for all types of printing and for use on any screen fabric. Also recommended is Ulano Super Prep Foto Film, an improved version that is said to have better resolution and better adherence to synthetic fabrics. Both of these films are presensitized and mounted on a 0.002 inch (0.05 mm)-thick polyester plastic backing support, available in sizes as large as 25-foot (6.35 m) rolls, 40 inches (1.02 m) wide. Wash-out development is accomplished with water and a proprietary aqueous solution of two packaged materials, the Ulano A and B Developers, which provide oxidative chemical action when mixed. More detailed information on the use of Super Prep transfer film will be given at the end of this chapter as part of a general procedure for manual screen printing.

Direct-Indirect Screens — Efforts to combine the best properties of direct and indirect photographic screen methods have resulted in the *direct-indirect* emulsion systems, also known as *direct film.* The idea is to provide a durable, adherent emulsion with the photographic resolution of a transfer stencil. When making direct screens, the printer forms his or her own film emulsions using liquid starting materials. Good photographic results are possible *in principle,* but one cannot really hope to equal the careful composition and uniformity inherent in prefabricated film sheets coming directly from the manufacturer. Although indirect transfer films do derive the benefits of a prefabricated emulsion layer, after wash out they cannot be deeply imbedded in the screen without producing image distortion. Therefore, the direct-indirect method seeks to combine both techniques. A prefabricated film layer is attached to the screen, and is then exposed and processed directly on the screen. The final film is well imbedded in the screen fabric, and its characteristics and thickness (typically 0.001 inch or 0.0254 mm) are not dependent upon the coating procedure or the printer's expertise.

Preparation of a direct-indirect screen consists of the following general steps:

1. Screen preparation — As with all photographic stencils, the screen fabric must be degreased and cleaned before use. Synthetic fabrics should also be roughened before initial use by rubbing a fine abrasive on the outside mesh surface with a wet rag. A fine silicon carbide grit (500 mesh) is ideal for this mechanical treatment because the particles are so fine that they are easily washed out of the screen after treatment. Manufacturers of screen printing films supply cleaning materials and give recommendations for surface preparation of any type of fabric used in combination with their products.

2. Secondary emulsion — The durability and adhesion characteristics of direct-indirect stencils are achieved by coating the screen with a liquid emulsion at the same time that the dry sheet film is attached. This secondary emulsion is the real key to success; it impregnates the screen, bonds to the dry film, and sensitizes the film. Although sheet film could be supplied presensitized, most manufacturers put the sensitizer in the secondary emulsion because this improves film storage stability. The secondary emulsion is mixed at the time of use, and typically contains a gelatinous emulsion combined with a bichromate or diazo sensitizing compound.

3. Attaching film — A suitable-size piece of film is placed emulsion side up on a glass plate, and the screen is placed in direct contact with the film. Liquid emulsion is then poured on the inside (squeegee side) of the frame, and brushed thoroughly into the screen. A dye is often added to the emulsion to clearly show when the liquid is evenly coated, at which point the assembly is allowed to air-dry. Some processes allow coating to be carried out under room lighting, while others require safelight conditions.

4. Exposing and processing — After drying, the film backing is peeled off and the screen is ready for exposure. Exposing and developing are carried out in the same manner as with direct film, with the inherent inconvenience of having to handle screen and frame together. The final screen/emulsion assembly is relatively tough because the stencil extends into the fabric mesh, unlike

indirect films. At the same time, the outside printing surface that will contact the work is uniformly thick and produces excellent line quality with sharp detail.

A recommended direct-indirect emulsion film is the Chromaline type A-100 system. Type B-100 is available for the finest detail and high precision required for commercial printed circuit work. The principal advantage of the Chromaline system lies in its ability to produce a predetermined stencil thickness, combined with easy application. As pointed out in vendor's literature, another advantage of direct-indirect screens is that good resolution can be obtained with coarser mesh screens. Fine mesh is not needed for adherence of the film, and the outside film layer provides the capability for detail. Screens coated with Chromaline stencils can be reclaimed by treating the emulsion with room temperature chemicals and a water spray.

Film Recommendations — To summarize the various approaches for preparing photographic printing screens, the reader is advised to start with indirect films and silk fabric, which should satisfy the demands of most low-volume electronic printing applications. If fine detail in combination with high dimensional stability are needed, then a natural step is the progression to monofilament polyester fabric. Problems with the adhesion of indirect stencils on polyester may then lead to the use of direct-indirect film emulsion systems. The need for direct emulsions will probably never be experienced unless high-volume output is necessary, and in this case considerable skill and experience may be needed to obtain good emulsion coatings.

Block-Out • *Block-out* mediums are substances used to fill in open areas in the screen fabric to prevent ink from being squeegeed through these areas during printing. Photographic transfer film stencils are usually adhered to the underside (printing surface) of the screen fabric, with some open border space around the outside edges of the film; this area is filled with block-out medium to complete the stencil pattern. Commercial fillers are generally water-soluble gel-like liquids that are brushed or squeegeed on either or both sides of the screen. After drying, the block-out forms a durable, flexible coating that is resistant to inks and solvents; coloring is often added so that pinholes will be readily evident after application. Water-soluble fillers are preferred because this formulation simplifies removal of the entire stencil if the screen will be reclaimed and used for another printing task.

An important step in screen preparation is to seal the fabric and frame with masking tape, gummed paper tape, or plastic tape. Tape should be attached to both sides of the screen and extended partly onto the frame surface, giving the fabric extra strength and preventing ink from leaking under the screen edges during printing. Block-out filler may be coated around the stencil either before or after the sealing tape has been attached.

Inks • Screen printing inks have been formulated for almost every application imaginable because of the versatility of this printing technique. In many cases, it is the chemical composition of the ink itself that largely determines the success of a screen printing process; in other cases, the color or appearance of the ink is most important. With the exception of decorative labeling, most electronic screen printing inks are chosen for physical properties other than their color. The following qualities are important for inks:

1. Squeegee action — The ink must deposit cleanly by squeegee pressure through the mesh, without running or producing ragged line edges. Consistency and viscosity are important.
2. Clogging — The ink must be composed of sufficiently fine particles to prevent clogging of screen pores. The screen can also become stopped up if the ink dries too rapidly during printing.
3. Solvents — Ink components must not dissolve or attack the screen, stencil, or block-out medium. Most screen materials are designed to withstand solvents, but not aqueous-based materials such as water-soluble inks.
4. Drying — After printing, the wet surface must be dried in some manner. In the case of PCBs, drying time may be important to prevent delays in subsequent fabrication steps; however, drying must not be so rapid that screen clogging occurs during the normal course of printing. Conventional inks are dried through two processes: oxidation, as in oil-based products, and evaporation, as in solvent-based inks. Either of these processes can be accelerated with hot oven curing. As an example of new developments in screen ink formulations, the new UV (ultraviolet) curing inks do not contain volatile solvents, but are hardened by polymerization under UV light in a matter of minutes. No ovens are needed, no solvent fumes must be swept away, and no screen clogging occurs during printing because the ink is a stable liquid under normal room lighting.
5. Adherence — Different printing surfaces will require varied ink formulations to obtain good adherence of the screened coating. Specific inks are chosen for printing on metals, plastics, wood, glass, paper, and cloth. For example, vinyl-based inks are good for vinyl plastics. On difficult, smooth surfaces such as glass, catalytic epoxy inks may be the only practical formulation for obtaining good adhesion; other types of ink will peel after printing.
6. Physical toughness — The final physical properties of dry inks are

particularly important in printed circuit work. Labeling inks must resist solvents and be completely nonconductive; etch-resist inks must be immune to attack from powerful etching baths; and solder-masking ink must repel molten solder and be stable at high temperatures. Where extremely durable inks are needed, the types that cure by chemical action (such as epoxy-based inks) are chosen.

7. Appearance and finish — In screening front panels, the final appearance will be important. Qualities such as opaqueness, transparency, gloss, and color are designed into the inks. For example, the following colors are available in the KC Enamel Plus line, an ink useful for printing on aluminum and other metal panels:

	Opaque	
Primrose Yellow	Brilliant Orange	Yellow
Lemon Yellow	Peacock Blue	Warm Red
Medium Yellow	Ultra Blue	Rubine Red
Emerald Green	Black	Rhodamine Red
Medium Green	White	Purple
Scarlet Red	Century 21 Brown	Process Blue
Fire Red	Super Opaque White	Green
	Transparent	
Primrose Yellow	Orange-Highway	Purple
Medium Yellow	Blue-Highway	Magenta
Green-Highway	Gold	Stop Sign Red

8. Solubility — Cleanup solvents must be available to remove excess ink from the screen while preserving the stencil. Most printers use solvent-soaked rags or blank newsprint to blot ink out of screens, either periodically to prevent clogging, or at the end of a printing operation when the screen must be completely cleaned. In some cases, screened inks are not intended to be permanent; they will be removed with a solvent some time after drying. A good example is the resist ink used to mask printed circuit boards during etching and plating. Because the use of organic solvents is being discouraged due to their volatility and toxic properties, special resist inks have been developed for printed circuit use that can be dissolved, or *stripped*, in an aqueous alkali solution.

When choosing a screen printing ink, the reader is advised to use a formulation that can be used directly out of the can, with minimal needs for mixing or thinning. (Epoxy inks are an exception, since they usually require the addition of a catalyst just before use.) Use the solvent or thinner recommended by the manufacturer for cleanup and follow specific directions for preparing the surface to accept ink. The following inks are recommended for electronic applications:

1. Etch and plating resist	Naz-Dar Vinyl Resist 221 Black
2. PCB labels	Wornow Cat-L-Ink 50-100R White Epoxy
3. Solder mask resist	Naz-Dar 242A Green Solder Resist
4. Front panel printing	KC Enamel Plus

MANUAL PRINTING TECHNIQUES •

Frame Mounting • After the screen has been completely prepared, it must be mounted in position for printing. This is usually done by attaching the frame to hinges screwed into a sturdy base or table top surface. For convenience, two types of hinges are useful: ones with removable pins, and *clamp-on* hinges, both illustrated in Fig. 8-6B. In any case the hinges should be sturdy and tight to prevent side-to-side wobble of the frame during printing, which will affect print quality and make pattern registration difficult.

During printing, the screen is swung down over the work so that it is parallel to the printing surface. Most objects to be printed are flat, but they may be of varying thickness. Appropriate spacer strips are inserted between the baseboard and the frame hinges to offset the screen to the correct distance above the work; another spacer is mounted at the front of the screen for support. It may be possible to accommodate objects of varying thickness if the spacers are thick enough to handle the tallest possible workpiece. Thinner objects can then be built up to the proper height by placing flat sheets of plastic, cardboard, or wood in between the baseboard and the work.

For printing the surfaces of very tall objects, such as boxes and cabinets, an extended, adjustable support must be devised to mount the screen frame at an elevated position above the work table. Such an arrangement is shown in Fig. 8-9. In this approach, drilled wooden block spacers of proper height are stacked on rods attaching the screen hinges to the baseboard. Another stack of blocks is positioned in front of the screen to support it in the printing position.

Off-Contact Printing • When printing on absorbent surfaces such as paper, cloth, or wood, the ink is usually applied with the screen resting in direct contact with the work surface. After printing, the screen is carefully peeled away from the object. However, a slightly different approach is needed for printing on smooth surfaces, such as aluminum plates or epoxy-glass circuit boards. Smooth surfaces tend to smear the ink when printed in direct contact with the screen, so *off-contact* printing is a preferred technique. The screen is swung down and supported very close, but slightly above and exactly parallel to the printing surface. A typical offset distance is one-sixteenth inch. Contact between the screen bottom and the surface of the object occurs only at

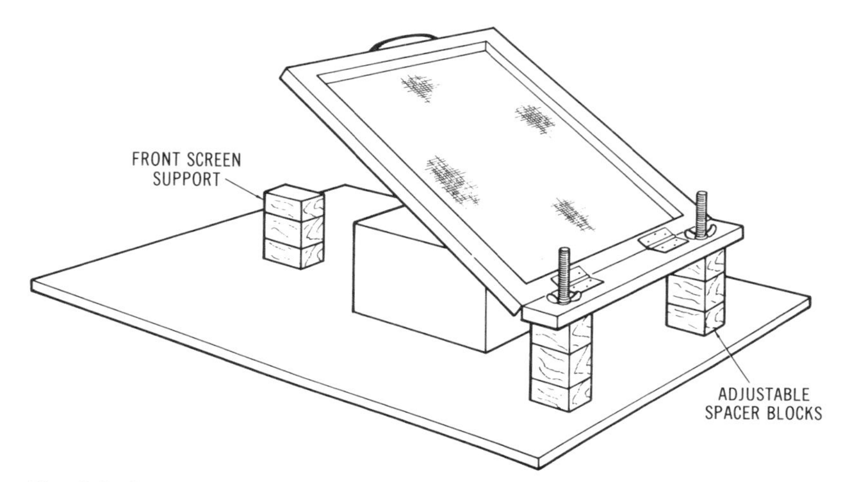

Fig. 8-9. Screen mounting system for printing tall objects.

the point where the squeegee is being pressed and pulled across the surface during printing; the trailing screen springs back away from the inked object and thus prevents smearing. The off-contact principle is illustrated in Fig. 8-10. An extension spacer should be placed at the far end of the workpiece, usually a piece of scrap material, to prevent the screen from being dragged down against the far edge at the end of the squeegee stroke.

Squeegee • The *squeegee* is the tool used to force ink through the screen and deposit it on the surface below. Squeegee construction is important to produce a clear, sharp print. The blade must be firm, but flexible enough to respond to the operator's touch; it is usually made from a solvent-resistant synthetic rubber. Typical squeegees are shown in Figs. 8-2 and 8-10. The wooden handle is slotted and, depending upon its size, is designed to be comfortably held with one or two hands. The rubber blade is mounted in the slot, and should be wide enough to completely bridge the stencil pattern. For example, a 9-inch screen frame would typically require a 7- or 8-inch squeegee blade. Standard blades extend about 2 inches from the handle, are three-eighths inch thick, and are made from a medium-hardness neoprene rubber. The blade edge must be completely straight with a flat end and square corners. Since this edge will wear with use, it must be sharpened periodically on a sanding block to prevent rounding of the corners. Once the operator becomes comfortable with a particular style of squeegee, the operator should stick with that design and vary only the width for different size screens. A squeegee blade that is kept sharp and square at the point where it contacts the screen will give sharper and cleaner prints.

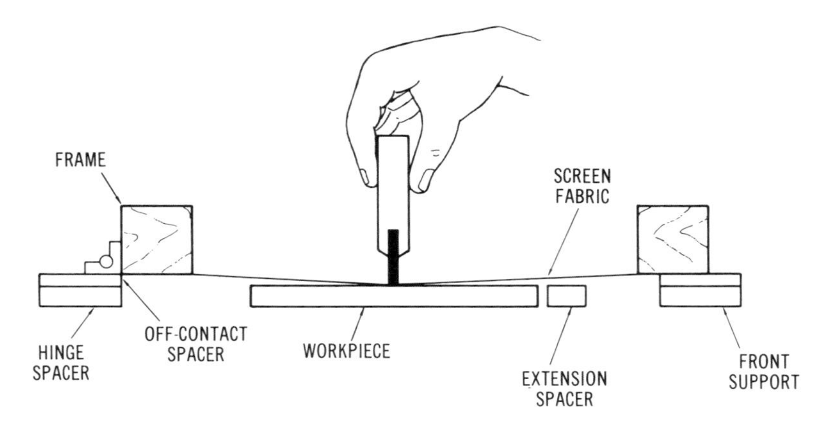

Fig. 8-10. Off-contact screen printing.

It may be possible to obtain the same printing results with a short, thick, soft blade as compared to one that is harder, longer, and thinner, but the printer must develop a feel for the squeegee by practicing on disposable material, such as blank newsprint. During actual printing, a pool of ink is deposited at one end of the screen outside of the stencil area. Rub the squeegee blade in the ink and distribute a small ribbon of ink in front of its entire length. Then, holding it firmly and evenly at a constant, slight angle to the screen, push or pull it steadily across the screen, feeling the blade force the ink through the smooth screen surface. Too much pressure is undesirable because excess weight causes the fabric to drag and results in a blurred impression, or *shadowing*. The amount of ink in front of the squeegee is not critical because the screen will only admit a thin layer of ink, regardless of how much is available on the top surface. In general, you will learn how much ink to use and how to manipulate the squeegee through practice and experience. After use, excess ink should be wiped off the blade and never allowed to dry, which will damage the rubber surface. Remove all traces of ink with solvent before storing the squeegee, and support it by its handle rather than resting its weight on the blade.

Registration • *Registration* in screen printing refers to the ability to make a print in exactly the right position and orientation on the surface of the object. Most electronic applications of screen printing demand accurate registration of the printed pattern. For example, labels for the component side of a printed circuit board must be in registration with the pads and traces already etched into the board. Electronic panel designs must register with the geometry of the panel-mounted knobs and switches, and if multiple colors are used for the design, they must register with each other, since a different screen is required to print each color ink.

The ability to obtain pattern registration depends on close attention to each step in the printing process as follows:

1. The original positive transparency used to mask the photographic stencil must have proper dimensions. This depends on careful artwork preparation and accurate photography. In many cases, it is helpful to include specific registration marks in the artwork, such as targets and dots, which aid in final screen alignment and show how well the pattern is registered after printing. Registration marks should be positioned outside the printed pattern area so that they can be cut away or wiped off with solvent following printing.
2. Screen fabric must be tightly stretched to obtain dimensional stability. Also, some fabrics are more stable than others; polyester and stainless steel are better than nylon and silk for extremely fine tolerance work.
3. Distortion must not occur during preparation of the screen stencil. In this regard, direct photographic films give better results than indirect transfer stencils, but the latter are adequate for most work.
4. The object to be printed must be correctly positioned under the screen. Loose hinges must be avoided so that the frame can be brought down in exactly the same position each time it is lowered for printing.

Various methods are used to position the stock material accurately underneath the screen. If many objects are to be printed and they are all the same size, guide rails or short pins may be attached to the printing base as an aid to feeding and quickly adjusting the material for each print. In general, it is difficult to visually align the stencil with the surface of an object simply by looking down through the screen from above. If the off-contact printing method is used, stencil patterns will shift slightly in the direction of squeegee pull. The following approach (see Fig. 8-11) is recommended for achieving good registration of flat surfaces during screen printing, and is ideal for low-volume work:

1. A transparent sheet of plastic or tracing vellum is taped to the printing base along one edge so that it can be folded back out of the printing area. A piece of blank stock is then placed under the sheet to bring it up to the proper height, and a test print is made on the surface of the transparent material.
2. The object to be printed is placed underneath the transparent, printed sheet, and carefully aligned with the fresh pattern to achieve proper registration. Once positioned, the object should be taped down securely to prevent it from moving around during squeegee action.

Fig. 8-11. Method for achieving pattern registration.

3. The alignment sheet is folded back out of the way and a print is made on the registered object.

SCREEN PRINTING PROCEDURE • The following procedure will demonstrate the important steps in screen printing by manual methods, using an indirect photographic stencil prepared on silk fabric. Although the instructions are fairly explicit, the user should use them primarily as a guide to learning basic techniques, because considerable practice and experimentation are necessary to become adept at the art. Screen printing is versatile because so many variations are possible in frames, screen materials, stencils, inks, and surfaces to be printed; the printer must choose the materials that are applicable to his needs, and develop a suitable process that meets his goals and with which he can feel comfortable.

Screen Preparation •

1. A general-purpose frame adequate for most printed circuit work will have 10 by 12 inches (254 by 305 mm) inside dimensions. Printing large front panel plates may require a larger frame; be sure to maintain an extra 1–2 inch (25–51 mm) border around the stencil for ink deposit and squeegee freedom. A wooden frame with the cord and groove system for screen attachment is a good starter frame.
2. Use a No. 20XX silk screen fabric for good resolution. This fabric

has 173 threads per inch. Attach the dry silk to the frame drum-tight, keeping the weave parallel to the frame at all times.

3. Wash the screen using warm water, a mild, nonbleaching detergent, and a soft bristle brush. Thoroughly rinse it in warm water and set it aside.

Stencil Preparation •

1. Ulano Super Prep indirect film is a good photographic transfer stencil for low-volume screening of sharp patterns and detail. The film has a slow photographic speed, allowing it to be used under subdued lighting conditions. Avoid direct sunlight or fluorescent lighting, which both contain a substantial UV component. The film is presensitized.
2. A suitable size piece of film is exposed using a photographic contact printing frame (see Chapter 4, Fig. 4-17). Super Prep film is exposed through the transparent backing sheet; therefore, place the emulsion (dull) side down on a white background, position the positive transparency above it, and hold the two in firm contact with the glass plate of the printing frame. Because the film is exposed from the back, the positive pattern should appear *reversed* when viewed from above the frame.
3. Expose the film sandwich to a UV-rich light source, such as a No. 2 photoflood lamp, or long-wave UV fluorescent lamps. The UV lamps are preferred because less heat is generated, which can distort the film material. Exposure time will vary from 5 to 15 minutes with the light source positioned 12–15 inches (305–381 mm) away from the film surface.
4. A test film sheet should be prepared to determine the optimum exposure time for your particular setup. This is done by placing 10 parallel strips of paper over the printing frame and removing them one at a time, every 2 minutes, over a 20-minute exposure period. This operation will produce a test pattern with exposure times ranging from 2 to 20 minutes time in 2 minute increments. Develop the film and observe which strip area produces the optimum stencil pattern. Underexposed areas will wash out excessively, and overexposed areas will not wash out fine detail.
5. The developer for Ulano Super Prep film is prepackaged in "A" and "B" envelopes containing powders that are mixed in 16 ounces of 65–70°F (18–21°C) water just before use. Place the exposed film in a tray of developer, emulsion side up, for about 2 minutes, with gentle rocking agitation. Then transfer the film to a tray of warm water at about 100°F (38°C) and wash out the emulsion with gentle agitation for 1–2 minutes. The wash out

step can also be accomplished under gentle running warm water if the film is placed emulsion side up on a glass plate for support. Remember that film emulsion is very soft after being in the developer solution, so handle it carefully to avoid smearing the pattern. If wash out is done in a tray, it is still a good idea to rinse the stencil under running water for a short time to completely clear the pattern.

Adhering Stencil to Screen •

1. After washing, the film should be immediately attached to the screen while it is still soft. Place the film, emulsion side up, on a glass plate on a flat surface. The screen frame is then positioned over the film and gently pressed down, so that the film attaches to the outside screen surface. Lightly blot the screen from above (squeegee side) with blank newsprint paper to soak up excess moisture, and then blot the stencil side by turning over the frame. Excessive pressure may distort the stencil pattern.
2. Before drying the stencil, apply a block-out material to the screen open areas and boundaries of the stencil with a small brush. A recommended filler is Ulano No. 60 Water Soluble Fill-In. The only openings in the entire screen surface are now the ones formed by the stencil pattern.
3. The screen containing stencil film and block-out should now be allowed to dry completely. Natural air-drying is recommended, which will take about 1 hour. Once the screen is dry, the plastic backing sheet takes on a milky appearance and may be peeled away from the stencil. If the film shows any dark patches, it is not sufficiently dry, and the backing should not be removed. After removing the plastic sheet, inspect the stencil. Sometimes, a residual film of glistening, translucent material will show up here and there. It can be removed with a solvent, such as mineral spirits, and a brush. The film emulsion and block-out materials are resistant to organic solvents, but they should not be exposed to water that will soften or dissolve them.

5. Seal the frame and screen on both sides with gummed paper tape if you have not already done so. Other tapes that resist inks and solvents may also be used.
6. A cross-section of a completed screen is shown in Fig. 8-12.

Using the Screen •

1. The frame is attached to the work surface with sturdy hinges. It must be offset above the surface so that when stock is positioned for printing, the screen will be at the proper height above the

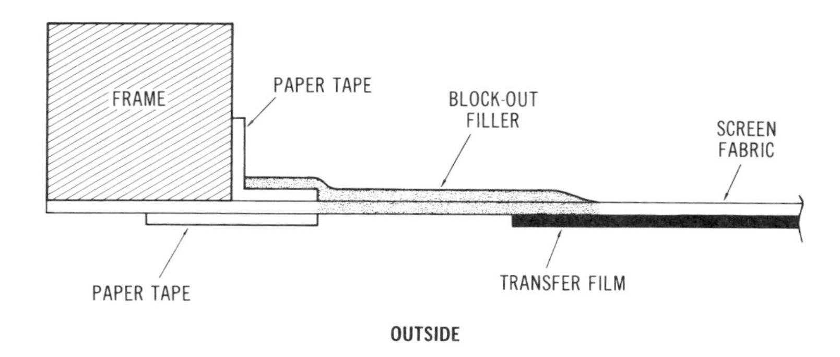

Fig. 8-12. Cross-section of completed indirect photographic screen.

work surface. This means that spacers are inserted between the hinges and the platform base. A cross-sectional view of a mounted screen is shown in Fig. 8-10.

2. In general, the screen height consists of three combined thicknesses: the off-contact distance (one-sixteenth inch), the thickness of the material to be printed (one-sixteenth inch circuit board, for example), and the thickness of any support placed under the work for registration adjustment. The object to be printed is often attached to a sheet of one-eighth inch flat plastic using double-sided adhesive tape; this plastic sheet is known as a *register platform,* and allows for easy positioning of the work. The platform can be moved around to achieve proper pattern registration, and is then taped to the base on all four sides to hold the work steady. This approach avoids having to attach tape to the top surface of the workpiece.
3. A swinging prop leg attached to the screen frame is handy for supporting the screen when it must be raised above the work surface (see Fig. 8-2).
4. Pattern registration is achieved using the transparent test print technique previously described and illustrated in Fig. 8-11.
5. To make a print, the screen is propped up and 2–3 tablespoons of ink are deposited at the lower end. The workpiece is positioned, and the screen is lowered. A band of ink is gathered in front of the squeegee, and holding at a slight angle, the squeegee is drawn firmly across the stencil in a smooth, sweeping motion. Pick up excess ink on the squeegee blade and return it to the lower end of the screen.

6. Gently raise the screen and support it on the prop leg. The work can now be removed, inspected, and dried. It is generally best to make a number of test prints on blank newsprint before attempting the actual printing of real stock, to reveal any problems in the setup and to acquire some technique and feel for the squeegee.
7. After all parts have been printed, or if clogging of the stencil occurs, the screen must be thoroughly cleaned. Remove excess ink in the frame with a couple of pieces of stiff cardboard. Then, with pieces of soft cloth soaked in ink thinner or other suitable solvent, wipe the top and bottom surfaces of the screen until all ink is removed and the stencil image is completely clear.

Reclaiming Silk Screens • When a stencil is no longer needed, the printer often wants to remove it from the fabric so that a new stencil can be prepared on the same screen. Unfortunately, silk is fairly fragile and will not stand up to harsh treatment, such as abrasive scouring or strong chemical cleaning. Sodium hypochlorite (bleach) is effective for removing hardened stencils from synthetic and steel fabrics, but it will damage silk threads and cannot be used for reclaiming silk screens. In general, it is not possible to reclaim silk screens made with direct photographic films or direct emulsions. However, *indirect* film, such as the Super Prep, is not deeply imbedded in the screen mesh and can often be removed with mild treatments, particularly if the stencil has not been attached to the screen for a long time — see the following:

1. All traces of ink should be removed with solvent-soaked rags.
2. The screen should be soaked in hot water for 5–10 minutes. A soft brush and hot water are often successful in completely removing all traces of stencil and block-out filler from the silk. The fabric can then be washed with detergent and water as it was initially prepared.
3. If hot water leaves some material behind on the screen, an enzyme powder is often effective for dissolving or loosening the film. Film manufacturers supply specific enzymes for this purpose. The powder is sprinkled on the wet screen and covered with a wet cloth for 5–10 minutes. Hot water is then used to clean the screen. If this approach is successful, the fabric should be neutralized with 5% acetic acid (vinegar) to render inactive any remaining traces of enzyme before finally soaping and rinsing the screen.

Chapter 9

PCB Machining, Soldering, and Assembly

INTRODUCTION • After a circuit board has been etched and final coatings have been applied, it must be cut to finished dimensions and drilled. Epoxy-glass laminates are easily cut and shaped with hand tools, but the hard, abrasive nature of this material will take its toll on drill bits. A drill press is almost always required for machining holes; accurate placement and drilling of hundreds of tiny holes in etched patterns is a tedious task, demanding the use of proper tools and techniques.

Assembly of components on the PCB will require a considerable amount of hand soldering, which is also necessary to complete external point-to-point wiring connections. Soldering skills are easily learned if the process is understood and appropriate equipment is used. In general, PCB soldering requires a fairly small, low-wattage iron and noncorrosive materials to protect delicate, miniature electronic components. Excessive heat must be carefully avoided to prevent damage to circuit board traces, particularly after having invested the considerable time and effort needed to design and fabricate your own prototype boards. The appearance and final quality of printed circuit assemblies can be further improved by removing solder residues with appropriate solvents.

Finally, this chapter includes a discussion on the basic types of modifications and repairs that may be necessary after PCB assembly is complete, including desoldering techniques and circuit alterations.

OBJECTIVES • The objectives of this chapter are as follows:

- Recommend tools and explain techniques for cutting and shaping PCB laminates.
- Discuss tools and techniques for drilling holes in circuit boards.
- Explain the basic principles of soldering.
- Discuss materials, tools, and techniques for successful soldering of PCBs and hookup wire.
- Explain the problems that can result in poor solder joints, and

show how to prevent them.
- Provide tips on desoldering electrical connections and PCB components.
- Outline procedures and recommend solvents for cleaning circuit boards after solder assembly.
- Present specific methods for making repairs and modifications to PCBs.

SHAPING • The printed circuit board is generally cut to size twice during fabrication: once as a blank copper-clad laminate, and once after etching is complete. The initial cut is rough and slightly oversize, giving an extra one-half to 1 inch border which simplifies handling and gives some leeway during image transfer. Grease and fingerprints are usually found near the edges of an object, so all of the critical cleaning steps during PCB fabrication are more effective when a border is maintained around the actual pattern area. The final cut is made before PCB assembly, and establishes the exact final board dimensions; it must be accomplished with care because the board can easily be ruined at this point. The author prefers to cut the PCB to its final shape before drilling holes, because drilling is the most tedious operation in forming a circuit board, and considerable time will be wasted if the board is ruined after holes have already been drilled.

Three common methods are used to cut PCB laminates: *shearing*, *slicing*, and *sawing*. All of these cuts will leave rough edges, so filing and sanding are necessary to smooth the board edges and make small cuts in tight spots. *Routing* and *blanking* are two other machining operations used to cut PCBs to close tolerances, but they require expensive machinery and will, therefore, not be discussed.

Shearing • A blade shear designed for cutting PCB laminates is illustrated in Fig. 9-1. Blade shears have a bottom stationary blade attached to the table, and a top blade that descends into the work through manual, foot-operated, or hydraulic pressure. The top blade is not parallel with the bottom, but is at a slight angle known as the *rake angle*. This feature allows the blade to cut steadily from one side of the workpiece to the other as it descends under constant pressure. Typical rake angles are 0.5–1.0 inch (12.7–25.4 mm) vertical drop per cutting foot. A uniform gap exists between the upper and lower blades, typically adjusted to 0.001–0.002 inch (0.025–0.05 mm) clearance. Note that this clearance is less than the gap used to cut sheet metal, approximately 0.003–0.005 inch (0.076–0.127 mm). Too wide a gap will produce a damaged edge; paper-based laminates will show feathered cracks, and epoxy-glass laminates will become deformed. In general, epoxy-glass PCB materials produce the cleanest edges by any cutting method because of their hardness and great flexural strength. However, this hardness also causes rapid dulling of cutting blades.

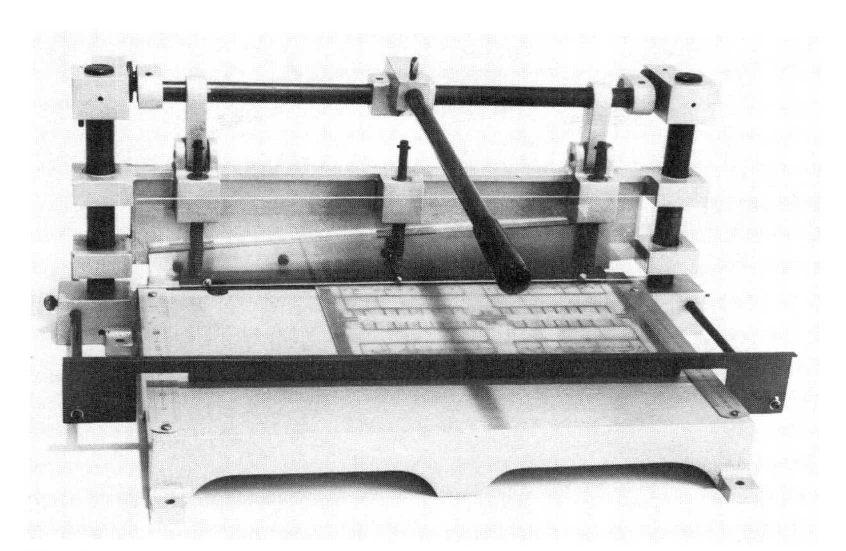

Fig. 9-1. A small hand-operated shear for cutting PCB laminates. ***(Courtesy Feedback Inc. and Circuitape Ltd.)***

Other component parts of a shear include a hold-down bar, and front/side guides to align the workpiece with the blade at adjustable angles. The hold-down bar must furnish even pressure across the laminate surface to prevent it from moving during shear action. Hand-operated shears are ideal for cutting small pieces of PCB stock, because they are simple to operate and allow close visual adjustment during cutting.

Slicing • Printed circuit boards can also be cut with a large paper cutter, as shown in Fig. 9-2. This method is similar to shearing, but it is termed slicing because the blade descends at a steep angle relative to the workpiece. Epoxy-glass laminates can be cut accurately with a paper cutter, although paper-based laminates tend to crack excessively due to the large rake angle. Blades must be sharpened often, particularly if the cutter is still needed to cut paper, since the hard epoxy-glass material will quickly dull sharp blade edges. The most difficult problem in using a paper cutter for slicing PCBs is to hold down the work securely during cutting. Rather than attempting to cut through the entire board in one fast motion, it is more effective to use a series of short, chopping blows while closely watching the cutting path. If the outside of a ruled line is followed, rough edges can be smoothed later with a file.

As shown in Fig. 9-2, the use of etched guidelines is a good practical approach for obtaining close-tolerance final PCB dimensions when the board will be cut and shaped using hand tools. The grid system

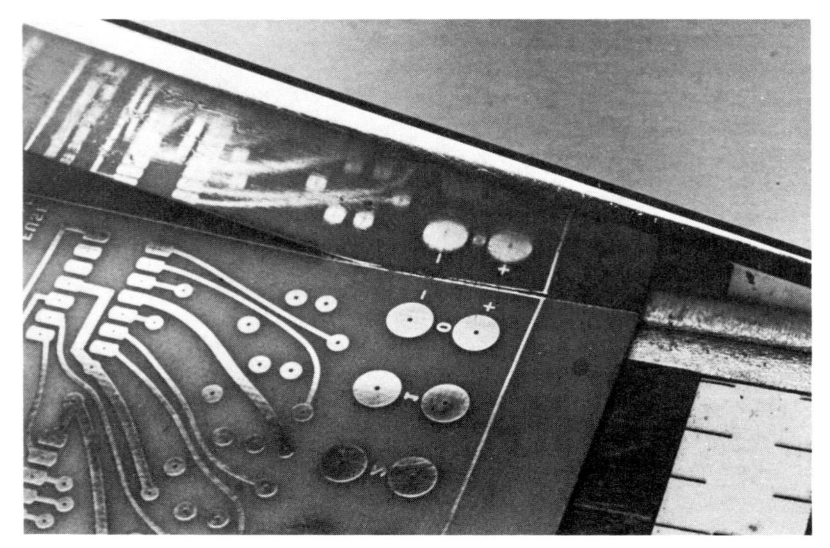

Fig. 9-2. Etched guidelines are helpful when cutting PCBs with a paper cutter.

used for artwork preparation allows perimeter lines to be carefully established during PCB design. These lines are taped up with very narrow printed circuit tape, and are positioned so that the board has proper final dimensions when the etched guidelines have been just barely cut or filed away. During shearing, slicing, or sawing, rough cuts are made along the outside edges of etched guidelines. Edges are then filed or sanded smooth to completely remove the etched lines and leave a board with exact final dimensions. This approach is particularly effective when special shapes, such as notches or edge connector fingers, are part of the PCB dimensions. Border outlines can be accurately established with fine printed circuit tape during artwork preparation, and allow the final board to be hand-tooled to exacting specifications without the need for error prone ruled lines drawn after etching is complete.

Sawing • High-speed circular table saws are used commercially to cut PCB laminates. The teeth may be made of carbide or diamond steel bonded because of the abrasive nature of laminate materials. In the hobbyist workshop, a *hacksaw* can also be used to cut PCBs; when blades become dull, they are simply replaced. In general, use of a hacksaw will give the roughest edge of any cutting method, but if some tolerance is allowed during cutting, board edges can be filed to give smooth final results. Work must be held securely in a vise or table clamp during sawing to prevent vibration and to allow a straight cut to be made. Paper-based laminates are more likely to feather-crack as saw teeth exit the work, but acceptable results can be obtained with a fine-toothed blade. Hacksaw blades with carbide teeth are available to give the smoothest cuts and longest life.

Filing • Filing is primarily a finishing operation to remove rough edges, smooth corners, and improve the overall appearance of circuit boards. However, filing is also an effective method for cutting and shaping printed circuit laminates because considerable material can be removed fairly quickly. Despite its hardness, epoxy-glass material crumbles easily into a fine powder during filing.

To remove large amounts of material, a fine-toothed double-cut flat file is used across board edges as illustrated in Fig. 9-3. This method is known as *cross-filing.* The work is mounted securely in a vise, with cardboard strips protecting board surfaces from the vise jaws. The file handle is held in one hand, while the thumb of the other hand rests on the file end to control direction and pressure of the stroke. The file is pushed in a smooth cutting stroke away from the operator and simul-

Fig. 9-3. Cross-filing to shape circuit board edges.

taneously at a 45° angle across the work surface. Cutting force is applied during the forward stroke only, and not on the return; to avoid curvature of the work surface, the file must be held level during the entire stroke and must not be allowed to rock.

Another filing technique used to obtain a completely straight edge is to lay a long, flat file on a table, and stroke the board edge firmly along the face of the file. After the desired board shape has been achieved with a double-cut file, final smoothing and finishing can be achieved using a finer-toothed single-cut mill file. Fine sandpaper or steel wool can also be used to obtain a polished final surface on PCB edges. In general, the edges are sufficiently smooth when they appear to be the same color as the flat surfaces of the board.

Extremely delicate filing work is accomplished with the aid of miniature files known as *Swiss pattern files,* which come in sets of many useful sizes and designs. An appropriate file shape (round, rectangular, flat, triangular, etc.) is selected to suit the contour of the surface to be worked. Swiss pattern files are illustrated in Fig. 9-4, where a miniature rectangular file is being used to cut a notch in a printed circuit edge connector tab, so that the board can be keyed to fit a particular edge connector receptacle.

DRILLING • Accurate drilling of holes is extremely important in printed circuit fabrication because the etched pads are small, and tolerances are close. If holes are drilled off-center, sides of pads will be torn out, and soldering of component leads may become a problem.

Fig. 9-4. Swiss pattern files for delicate work.

Extra care must be taken when drilling double-sided boards, because mistakes started on one side of the board tend to worsen when the bit exits on the other side. An example of poor side-to-side hole alignment is illustrated in Fig. 9-5. Similar results can occur due to sloppy drilling, poor pattern registration, or off-vertical drilling. To ensure 90° vertical drilling of double-sided boards with critical alignment needs, a *drill press* is necessary (as opposed to hand-held drill). Using a good manual drill press, the operator should be able to visually position holes to within 0.005 inch (0.127 mm) of pad center.

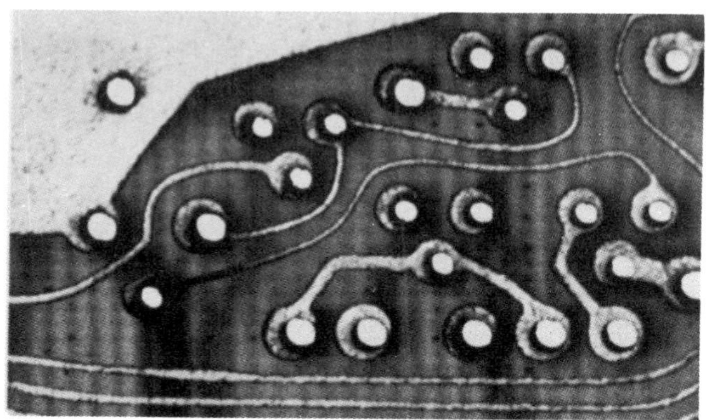

Fig. 9-5. The results of poor side-to-side registration in double-sided PCBs.

Drill Bits • Small diameter twist drill bits suitable for making holes in printed circuit boards are classified according to a *numerical* system consisting of 80 sizes, No. 1 through No. 80, ranging from 0.228 inch to 0.0135 inch in diameter. These bits should be adequate for most PCB work since they range from almost one-fourth inch to the size of a pin. If large diameter holes are occasionally needed (as for corner mounting holes) the *fractional* bit system can be used, which extends up to one-half inch in diameter in increments of one-sixty-fourth inch.

Twist drills are commonly made from three materials: *carbon steel,* high-speed *tungsten-molybdenum,* and *tungsten-carbide.* The carbon steel bits are used for drilling soft metals, such as aluminum and brass, and will dull quickly if used in harder materials. They are generally unacceptable for drilling printed circuit laminates. High-speed bits are used for drilling hard metals such as stainless steel, but they, too, will dull fairly rapidly when used on abrasive materials such as epoxy-glass circuit boards. Carbide bits are best for epoxy-glass; because they dull so slowly, entry and exit *burrs* are minimized and exceptionally clean holes are produced. Unfortunately, however, carbide bits are very brittle and fragile. The problem is compounded in printed circuit work by the very small diameter of bits needed. Carbide bits are very expensive, typically 5 times the cost of high-speed bits.

High-speed and carbide twist drills are pictured in Fig. 9-6. These bits are No. 69, or 0.0292 inch in diameter, the most common hole size used for ICs, resistors, diodes, transistors, and disc capacitors mounted on printed circuit boards. Two types of bit construction are illustrated: the common *straight shank* configuration of most high-speed twist drills, and the special one-eighth inch *enlarged shank* used for carbide drills. Because of the small size of printed circuit drill bits, they are difficult to rigidly secure and align in conventional chuck holders. Although high-speed bits can withstand some vibration due to off-center rotation, the fragile carbide bits cannot. The slightest misalignment of these brittle drills will lead to immediate seizure and breakage in the hole, so their shanks are carefully machined up to a size that can be accurately gripped and centered in the normal three-jaw Jacobs chuck used on most standard drill press heads.

The author has used both high speed and carbide drill bits for drilling holes in prototype PCBs, and has come to the conclusion that the carbide bits are difficult to work with. Unless the user has a high quality drill press ($2000–$5000) to allow vibration-free drilling, the long life of carbide bits will never be realized due to breakage. Workshop-style presses ($200–$300) are simply not smooth enough in their operation to prevent the slight seizure that will immediately destroy a carbide PC bit. When a miniature carbide bit does break off in a hole, it is often

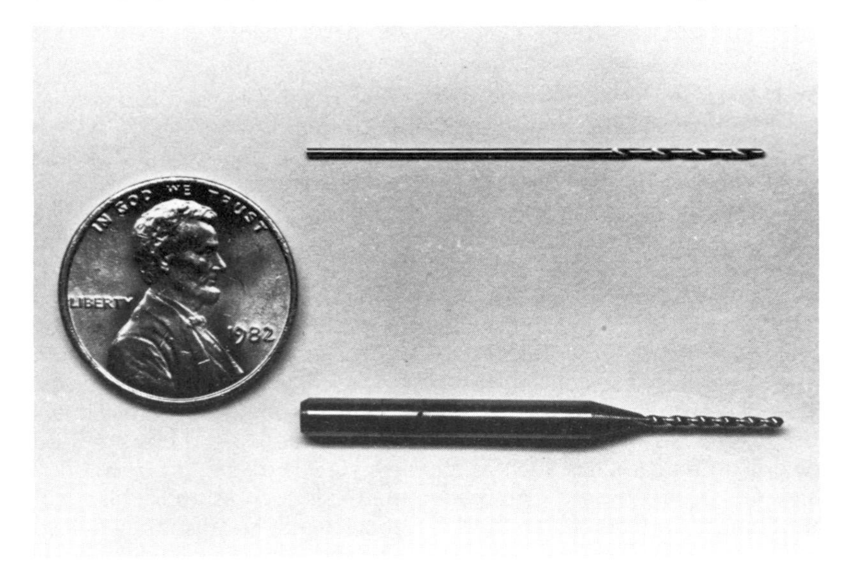

Fig. 9-6. Miniature high-speed (HS) and carbide twist drill bits for PCB drilling.

very difficult to extract without enlarging the hole to the point where the etched pad is ruined.

High-speed bits, on the other hand, are fairly flexible and therefore forgiving. The price paid is in drill life with respect to sharpness; high-speed bits will become dull after drilling about 100 holes in epoxy-glass, and entry/exit burrs will start becoming noticeable as the bit is used more. Since it is doubtful that the average user can obtain more than 500 holes with a carbide bit before shattering it, high-speed bits will give practical results and are economically favored even if they are discarded after every 100 holes. In practice, the author uses high-speed bits for about 200–300 holes before replacement; sharpening of these tiny bits is not practical.

Burrs • When drilling soft metals such as copper or aluminum, *burrs* are created as the bit pushes material aside instead of cutting it away. Burrs are commonly seen on the exit side of the hole, but they can also occur upon entry if the drill bit is extremely dull or if the operator forces the bit in too fast. To minimize burring during PCB drilling, the operator can use sharp bits, lower the feed rate, and employ *backup* materials underneath the board, such as aluminum or scrap PC laminate. The effects of burring are illustrated in Fig. 9-7 and discussed as follows:

1. Exit burrs (Fig. 9-7A) — Excessive burring from a dull bit will distort terminal pads, encourage foil delamination, cause soldering difficulties, and produce metal particles that can separate and fall into critical electronic circuitry.
2. Entry burrs — The entry dimple is a sure sign that a drill bit is becoming dull. Although not necessarily a problem, dimples detract from PCB appearance and in extreme cases may prevent flush mounting of components.
3. Deburring (Fig. 9-7B) — The easiest method for deburring is to use fine sandpaper, 400 to 600 grit. However, sandpaper may scratch and ruin solder coatings. An alternative, more tedious method is to manually deburr each hole with an oversize drill bit, such as a sharp one-eighth inch bit. This approach tends to enlarge the exit hole by tearing away metal, as seen in the photograph; in extreme cases, pads may tear out to their edges, making soldering of leads difficult.
4. Sharp bits (Fig. 9-7C) — A sharp drill bit gives clean holes with very little burr, minimizing the need for deburring.

In summary, the problems created by minor burring are more than offset by the practicality of using high-speed drill bits instead of carbide bits for printed circuit drilling. Most drilling imperfections will never be

(A) Exit burrs from a dull bit.

(B) Results of deburring.

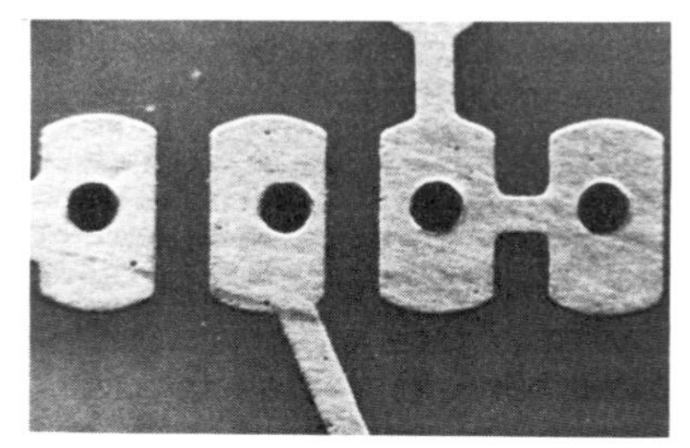

(C) Sharp bits produce little burring.

Fig. 9-7. Burrs after drilling.

noticed once holes are deburred and soldering is complete, because solder fillets will thoroughly cover the pads. In the case of single-sided PCBs, boards should always be drilled from the etched side to minimize burring, which is much less noticeable on the entry side of a hole. The lack of copper traces on the exit side eliminates the problem of exit burrs for single-sided PCBs.

Drill Press • A small, inexpensive ($150) workshop drill press from Rockwell is shown in Fig. 9-8. The useful features of this power tool include the following:

1. Drilling reach — A 5-inch distance from spindle to the back of the stand allows PCBs up to 10 inches wide to be drilled at any point.
2. Adjustable stand — The stand can be moved up and down to accommodate various backup materials and drill bit lengths to give comfortable operation.
3. Adjustable spindle stops — By confining the spring-loaded spindle to a preadjusted range of travel, optimum drilling time is obtained without wasting time by raising and lowering the drill bit excessively.
4. Adjustable speed — Spindle speeds range from 600 to 2800 rpm. Large bits should be run at slower speeds to limit cutting rate, but small PC bits in the range of 0.030 inch (0.762 mm) diameter can be turned as fast as possible. Drill presses that operate at higher

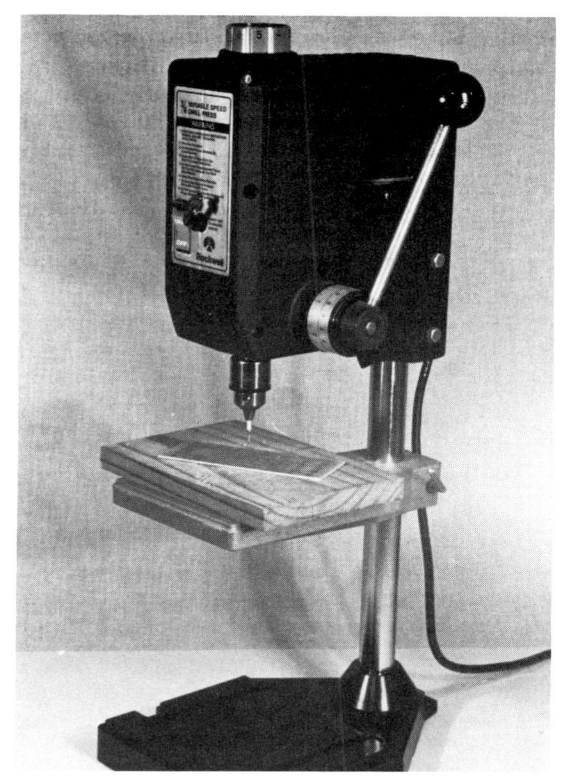

Fig. 9-8. A workshop drill press suitable for printed circuit work with HS bits.

speeds, such as 10,000 rpm, are excellent for use with miniature twist drill bits.

As mentioned in Chapter 1, other small drill presses are available from Dremel, Unimat, and GC Electronics.

Drilling Procedure • Standard drill press chucks will not close down securely around miniature straight-shank bits, so special methods are used to insert them. The simplest approach is to carefully wrap a strip of masking tape around the top half of the bit until its diameter is built up to the point where the chuck jaws can compress around it. A second method is to mount the bit in a miniature *pin-vise* and insert it in the larger chuck. A pin-vise with assorted bits was shown previously in Chapter 1, Fig. 1-32; this device contains a flexible, tapered *collet* which tightens around the bit as a nut is turned to compress the collet. Unless a drill press chuck is very poorly constructed, the masking tape approach will usually center a drill bit as well as a pin-vise.

During drilling, 0.0292 inch No. 69 holes are made in one-sixteenth inch epoxy-glass laminates allowing about 1 second for the bit to complete each hole. The work must be held down securely with one hand, preferably on top of some hard backup material, such as scrap laminate. It will be found that miniature drill bits tend to wander, or "walk" across the smooth copper surface unless a center-punched hole is present to guide the bit when it starts. This is the reason that prototype PCB artwork should always use terminal pads containing a center dot; after etching, a tiny hole is present in the center of the copper pads, which serves as an excellent guide during manual drilling. If center holes are not formed for some reason, each hole should be center-punched before attempting to drill. Miniature, spring-loaded punches are available for this purpose, or a large sewing needle may be used to pierce the copper surface.

Machining of epoxy-glass materials creates a fine dust that is harmful to breathe, since it contains tiny glass particles. There is also danger from flying debris if a drill bit should shatter. Always wear safety glasses and a dust mask when drilling, filing, or sanding circuit boards.

Pattern Registration • The fabrication approach presented in earlier chapters for double-sided PCBs requires that each side be printed and etched separately while protecting the opposite surface. The procedure was described as "Print-Etch-Drill-Print-Etch" because an intermediate drilling step is necessary to allow for exact visual alignment of the two patterns. After etching the first side, registration holes are drilled near the four corners of the board. These holes should be made very carefully in small area round pads, such as those used for dedicated through-holes. Select a small, new, sharp drill bit, such as a No. 70 (0.028-inch) bit, to drill burr-free holes in the exact center of several widespread pads, providing guides for precise alignment of the second photographic mask when the board is turned over for the next image transfer step. Registration holes must be drilled perfectly vertical to the surface of the board or accurate side-to-side alignment cannot be achieved.

After printing the reverse side image with photo resist techniques, pattern registration can be verified by drilling some additional randomly spaced holes across the surface of the board, entering from pads on the component side. If alignment appears to be acceptable, each hole should be touched up with resist paint in preparation for etching. However, if alignment is not satisfactory, the reverse-side image is stripped away and applied again more carefully, using the same holes again for visual alignment. The "Print-Etch-Drill-Print-Etch" approach is practical for prototype PCB construction because it allows fabrication to be carried out in short, incremental steps under careful control. Many operations (such as image transfer) are reversible, permitting problems to be corrected without affecting results from previous steps.

Large Holes • Some circuit boards will require that a few relatively large holes be drilled, such as corner mounting holes and holes for securing heavy devices with machine screws. In most cases these larger holes are not connected electrically to etched circuit paths, but they may still require accurate positioning on the board, as in the case of semiconductor heat sinks. Holes of this nature should be included in the original artwork as isolated, round printed circuit pads, somewhat smaller in diameter than the final desired holes. Positioning small pads on the original grid pattern allows large holes to be accurately located and drilled. During drilling, a small center hole is first made to act as a guide for a larger bit. The final hole is then drilled, located in exactly the right position, but obliterating the original etched pad in the process.

SOLDERING ● Soldering is the electronic assembly technique used to mechanically and electrically join metal surfaces by the use of a molten filler metal, commonly 60/40 tin-lead solder. Hot, molten solder wets the surfaces to be joined, and after solidification a permanent bond is formed. The final bond is strictly metallic in nature, with no chemical bonding reactions occurring except the natural attraction of solder for the basis metal. However, some alloying may occur if the base metal is soluble in solder. Since solder is a very soft metal, the soldering process does not by itself supply much mechanical strength to connections; where strength is needed, wires and leads must be clinched, wrapped, or otherwise secured to provide support before soldering.

Materials to be joined do not themselves become molten during the soldering process. This is an important point, because it means that bonding occurs in a thin interfacial region of two metals, and is therefore strongly dependent upon the characteristics and cleanliness of metal *surfaces.* The single most important factor contributing to a good solder joint is proper preparation of the surfaces to be joined.

Solderability • *Solderability* is the ability of a metal surface to be wetted by solder, and is dependent upon the metal composition, surface coatings, and surface impurities. Materials such as grease, oil, paint, wax, and dirt will interfere with the soldering process and must be completely removed. Heavy tarnish or oxidation will gradually appear on metals, such as copper, causing solderability problems. To preserve solderability, electronic component leads are often coated with protective metals (such as gold) that resist oxidation. Printed circuit boards are commonly coated with 60/40 tin-lead solder for the same reason. Clean copper metal has excellent solderability, but rapid tarnishing produces a coating of copper oxide or sulfide which changes surface properties considerably.

Wetting • *Wetting* is the ability of hot solder to form a smooth film of uniform, strongly attracted molten metal that flows quickly and evenly around metal surfaces to be joined. A highly solderable surface is one that wets easily. If solder sticks to itself and tends to clump up instead of flowing smoothly, that is evidence for poor wetting. Although solder flux is used to promote surface wetting action, readily solderable metal parts should wet *without* the use of a chemically active flux.

Materials for Hand Soldering • The basic elements of hand soldering include the solder alloy, the solder flux, and the soldering iron. These materials will be discussed as they are applicable in printed circuit and other electronic soldering, such as point-to-point wiring.

Solder — The metallic composition of electronic *solder* is almost always Sn-60, or 60/40 tin-lead. Other compositions of tin and lead may work, but they are higher melting and pass through a wide semisolid transition stage when changing from liquid to solid. Sn-63 (63/37 tin-lead) is actually the lowest melting tin-lead alloy at 183 °C (361 °F), the *eutectic point,* but for practical purposes Sn-60 is equivalent, melting from 183 to 191 °C (361 to 376 °F). A sharp transition is desirable because if any movement occurs during solidification, a weak, grainy, or fractured joint will be the result.

Electronic solder for hand assembly is used in the form of a spool of wire containing a core of 3–4% rosin flux imbedded during manufacture. Small-diameter wire solder is preferred for printed circuit application to get into tight spots and to control the amount of solder delivered; typical wire diameters are AWG 18 (0.039 inch) or AWG 20 (0.032 inch).

Flux — Soldering *flux* is a liquid, resin, or paste material that is applied to metal surfaces before and during soldering to promote wetting action. Two types of flux are commonly seen: *acid-* and *rosin*-based compositions. Acid fluxes (such as zinc chloride) should *never* be used for electronic soldering because they are much too active even at room temperature. Metal surfaces will gradually corrode and deteriorate following the use of acid core solder, which also leaves an electrically conductive, ionic film on surrounding surfaces. Rosin flux, on the other hand, is designed to become chemically active only at higher temperatures, just before solder melts. The activity of rosin-based flux is classified at three progressively stronger levels: rosin (R), rosin mildly activated (RMA), and rosin activated (RA). After soldering, rosin flux leaves a yellow film that can be removed with solvents, but which is not generally conductive or harmful to electronic circuitry. Although rosin flux is included in the core of electronic wire solder, many workers brush on additional rosin flux in critical or hard-to-reach spots, using an isopropyl alcohol solution that quickly evaporates.

The use of flux is a tremendous aid in soldering because it improves the speed and reliability of making a connection. To form a good solder joint, all surfaces involved must be free of impurities and the molten solder must flow freely to wet the metal completely. Rosin flux acts in the following ways to help the soldering process:

1. Metal oxides are dissolved and removed from surfaces by hot flux.
2. Flux coats the surfaces to prevent re-oxidation as temperature increases during soldering.
3. Wetting is promoted by the flux acting as a vehicle for smooth solder flow.
4. Flux acts as a heat-transfer agent to bring all parts quickly to the temperature needed to melt solder.
5. Rosin flux is noncorrosive after use and is easily removed with solvents.

***Soldering Iron* —** The *soldering iron* is the device used to heat connections so that molten solder may be applied. The following characteristics are important when choosing a soldering iron:

1. Wattage — Typical soldering irons used for printed circuit work are rated at 25–35 watts. The wattage of an iron does not directly determine its tip temperature, but it is an indication of the amount of heat that can be transferred to the joint without a substantial drop in tip temperature occurring. High-wattage irons are not desirable for printed circuit soldering because the amount of heat transfer is too difficult to control. Copper foil is bonded to PCBs with an adhesive that decomposes under excessive heat, leading to delamination of etched pads and traces. Also, higher wattage irons tend to be larger and clumsier than the lightweight "pencil" irons favored for maneuverability around miniature electronic components. When soldering large metal areas, such as lug and chassis connections, a 60–100 watt iron may be necessary to bring the joint up to proper temperature. Even larger irons and soldering "guns" are available for working sheet metal and braided copper ground straps.
2. Temperature — The typical tip temperature used for soldering 60/40 tin-lead alloy solder is 700–800°F (371–427°C). The equilibrium temperature is a function of tip size, tip geometry, and wattage of the iron. When working with delicate, heat-sensitive components, many workers prefer a high-temperature 900°F (482°C) tip in combination with a low wattage (15 W) iron, because this allows small joints to be soldered very rapidly without spreading heat to surrounding areas. In general, wattage is more important than tip temperature; the user can adjust his soldering technique to complement any temperature in the 600–900°F (316–482°C) range. A *controlled temperature* soldering iron is

illustrated in Fig. 9-9, the popular Weller WTCPN system with grounded tip. In addition to limiting maximum tip temperature, the advantage of this iron is that it gives a fast recovery time after each solder joint is completed; power is applied as soon as a temperature drop is detected.

3. Tip — Soldering iron tips come in many shapes and sizes to suit various applications; most tips are replaceable and different styles can be inserted in the same iron. The user should select a tip design that provides maximum contact area between the surfaces to be joined, and that also allows accessibility to the joint. A thin, chisel-shaped tip is usually preferred for printed circuit work, while a larger, conical-tapered tip is more suited for general purpose wiring. Soldering iron tips are almost always made of copper for good heat transfer, but the best designs have an iron coating that greatly improves tip life by preventing oxidation and pitting that would occur rapidly on bare copper. Bare copper tips must be frequently filed, sanded, and tinned to maintain a smooth, clean, bright soldering surface; iron-clad tips require only a dab of solder and flux followed by a quick wipe with a wet sponge to keep them in top shape. Note that filing a clad tip quickly converts it into the copper variety!

Whatever soldering iron system is chosen, the important thing to control is cleanliness. The tip must be frequently removed from the

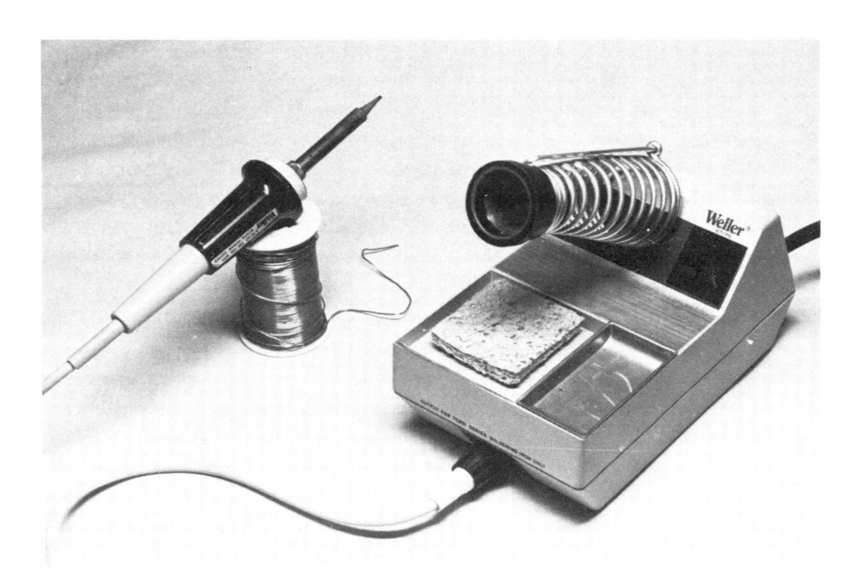

Fig. 9-9. A controlled temperature soldering system.

heating element to loosen oxidation on the screw threads, which prevents good heat transfer from element to tip. A frequent symptom of this problem is a handle that begins to overheat. The tip itself must be kept in perfect shape at all times. Black, oxidized flux encrusted on the tip will prevent good heat transfer to connections, and will introduce grime into solder joints. Tip surface should be bright and clean where it contacts the molten solder. The following procedure is used to "tin" a soldering iron tip with fresh solder:

1. Bare copper tips — If liberal amounts of rosin-core solder will not wet all areas of the tip at operating temperature, unplug the iron and allow it to cool. Then remove all oxidized tip coatings with a file and sandpaper until a bright, copper surface is visible. Plug in the iron and reapply solder/flux before maximum operating temperature is reached; this technique will allow liquid rosin to coat the copper surface before it can become re-oxidized by heat. Remove excess solder from the tip by shaking it towards an expendable surface, where the solder will splatter and solidify. Keep a liberal amount of solder on the tip whenever it is not being used for extended times, since that will protect and preserve the tip surface. Copper metal is slightly soluble in molten solder, so copper tips will gradually become smaller with use.
2. Iron-clad tips — Apply a liberal amount of solder/flux to the hot tip, and then wipe it clean with a wet sponge. The thermal shock should loosen any oxidized coatings easily to give a bright, clean surface. It is interesting to note that the temperature of an iron-clad tip can be estimated from its color. After first wiping with a sponge, the silver color indicates a temperature of 700 °F (371 °C) or less. As the tip continues to heat, it will take on a gold sheen at 800 °F (427 °C). At 900 °F (482 °C), the color becomes blue/purple, and at 1000 °F (538 °C) the tip darkens to gray/black. Iron-clad tips should never be filed or abraded because this will ruin the coating and expose the less stable copper metal.

Soldering Techniques • Cleanliness of metal surfaces is the key to successful soldering. Molten solder will flow quickly and evenly around clean component leads, and the manual techniques required for obtaining good joints will become obvious with a little practice. In general, solder connections should be made with a minimum amount of solder and as quickly as possible. Printed circuit connections should be completed in 3–4 seconds each. A good solder joint is obvious from its bright, shiny appearance and a smooth, even contour of solder around the joined surfaces. If components are heat-sensitive, they must be protected during soldering. The best approach for heat-sinking is to attach alligator clips to leads on the component side of the connection, thereby draining heat away from the sensitive device. The PCB sockets are always desirable for mounting integrated circuits and transistors that are heat-sensitive and difficult to remove from PCBs because of the number of soldered leads.

Printed Circuit Soldering — Before attempting to hand-solder components into a blank circuit board, clean the foil areas thoroughly. Bare copper should be scrubbed with fine steel wool and dipped in dilute (5%) HCl to activate the surfaces. Rinse well with water to remove all traces of acid, and brush on rosin flux/isopropyl alcohol solution across the board for best results. Copper foil that has been coated with protective solder is ideal for preserving solderability; if additional soldering is required later after assembly is complete, the coated surfaces are still easily wetted with solder.

Proper technique for soldering components into PCBs is illustrated in Fig. 9-10. It is assumed that component leads have been formed or bent into proper shape with needle nose pliers, and that the device has been inserted into the board. A circuit board holder (see Chapter 1, Fig. 1-31) is helpful for tilting the board at a convenient angle for comfortable soldering. The proper technique for soldering is as follows:

1. A small dab of solder is first applied to the iron tip, which promotes good heat transfer to the joint.
2. The copper foil and component lead are heated simultaneously with the iron tip (Fig. 9-10A), bringing them up to temperature in about 2 seconds.
3. Leaving the iron tip in contact, small-diameter rosin-core wire solder is then pushed into the hot junction area (Fig. 9-10B) and allowed to melt and flow around the joint. Molten solder will flow in the direction of the iron tip.
4. When the solder puddle is smooth and well formed, both solder and iron are removed and the joint is allowed to cool without disturbing it (Fig. 9-10C). Excess component leads may be cut off with electronic pliers, but the best flush-cutting tool is a pair of fingernail or toenail clippers.
5. Good solder joints are illustrated in Fig. 9-10D for a DIP IC socket.

Most sources recommend that component leads be *clinched* (bent flush) on printed circuit pads to hold the device in place during soldering and to provide mechanical support after soldering. However, that approach requires that leads be preclipped to size. This author does not recommend clinching unless the component is particularly heavy, such as a large electrolytic capacitor. Most PC components are small and the solder joint provides adequate support. Also, a full-length lead serves as a heat sink during soldering, and thus protects the component. Finally, components with clinched leads are difficult to remove

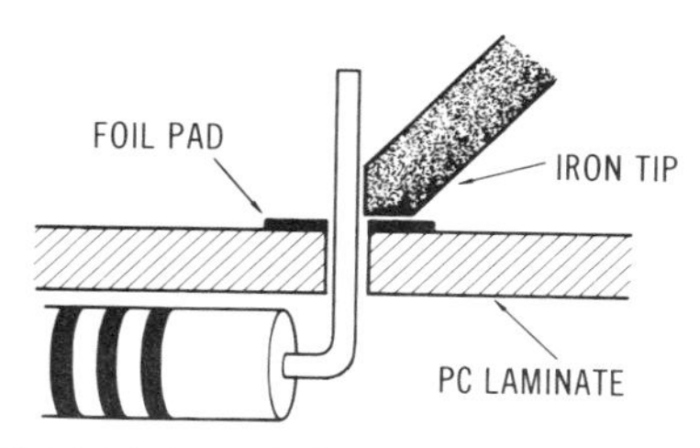

(A) Joint is heated above melting point of solder.

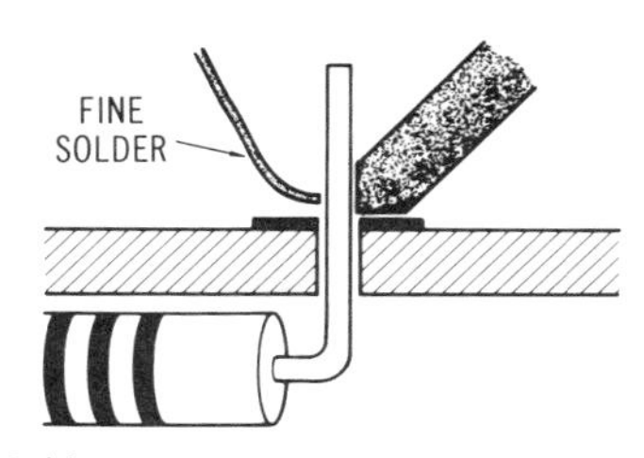

(B) Solder is fed into hot junction where it melts and spreads.

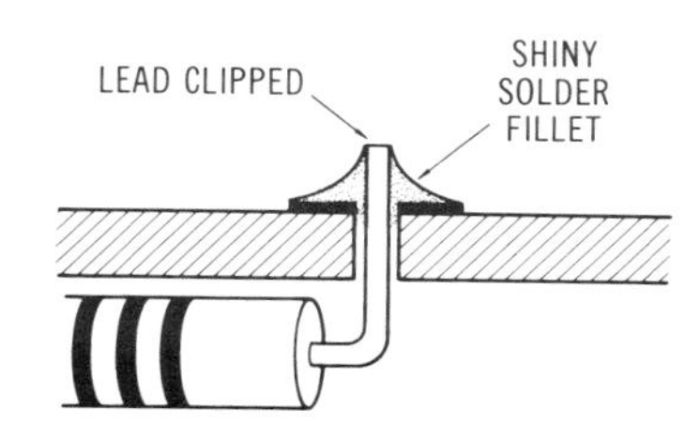

(C) Solder forms a neat puddle around the connection, and solidifies.

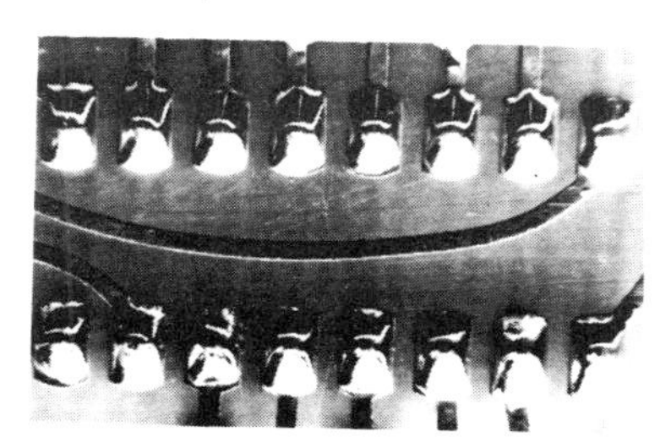

(D) A good solder joint is smooth and shiny.

Fig. 9-10. Techiques for printed circuit soldering.

from PCBs, an important consideration in prototype construction.

When preparing components for PCB assembly, bend the leads with a slightly wider span than necessary, so that the part must be press-fitted into the board, making it self-supporting during soldering. The board can then be turned over without losing the part. Components that operate hot electrically should be raised slightly above the PCB surface to permit good air circulation.

PCB Through-Holes — Double-sided PCBs must have some means of making electrical connections from one side of the board to the other. As discussed in Chapter 2, commercial fabrication processes use plated-through-holes where the walls of each hole are coated with copper, providing continuity between the two sides. Some plated-through-holes are dedicated for just that purpose, while others contain component leads as well. However, the fabrication of plated-through-holes is not practical for low-volume prototype PCBs, so some other means of completing the through-connections must be devised. In most cases, techniques for achieving through-connections should be considered during the design stages of PCB construction, although the actual connections are made during solder assembly. The following solder methods are used to complete through-holes in prototype PCBs:

1. Soldering component leads — When component leads are accessible on both sides of the board, they can be soldered to pads

on both sides to form a through-hole connection. This method is most often used on two lead axial devices, such as resistors, capacitors, and diodes. Care must be exercised when soldering on the component side of the board to avoid damaging sensitive devices; a heat sink is advisable. Integrated circuit leads are rarely soldered on both sides of the board, because there is too much chance for heat damage, and because the device would be very difficult to remove at a later date.

2. Dedicated soldered-through-holes — If dedicated through-holes are designed into the board, they can be filled with short pieces of AWG 22 solid wire and soldered on both sides during assembly. Several dedicated soldered-through-holes are evident in the foreground of Fig. 9-11A. With sufficient use of these connections, the need to solder component leads on both sides of the board can be avoided.
3. Socket pins — Machined socket pins are available that can be soldered on both sides of the board, and then used to mount an IC or other component. A row of soldered socket pins is shown in Fig. 9-11A.
4. Solder cream — *Solder cream* (Fig. 9-11B) is a helpful aid when soldering in tight spots, such as the component side of an IC socket. Before inserting the socket, a dab of solder cream is placed on each etched pad to be soldered. After insertion, socket pins are heated from the reverse side to complete the connections; this was the technique used to form the soldered socket joints illustrated in Fig. 9-11A. Solder cream is typically composed of finely powdered metal solder alloy mixed with rosin flux and polyethylene glycol. Aftr soldering, residues are easily removed with the cleaning methods discussed at the end of this chapter.
5. Eyelets — Small eyelets may be installed in printed circuit holes like rivets, and soldered to pads on both sides of the board. The component lead is inserted through the eyelet center hole. This approach calls for extra large terminal pads and holes in the original design. Eyelets for this purpose are termed "funnel-flanged eyelets."

Wiring — When soldering hookup wire, the same basic principles apply as in printed circuit soldering — work with clean metal surfaces, and heat the junction with an iron to allow rosin-core solder to flow smoothly around the connection. Insulated wire should be stripped just before soldering to ensure a clean surface. Stranded wire should be tinned with solder before securing around a lug or eyelet, because this keeps the strands together neatly and allows solder to penetrate the entire joint. Steps in soldering hookup wire are illustrated in Fig. 9-12:

(A) Socket pins and dedicated through-holes.

(B) Solder cream helps in soldering inaccessible socket pins.
Fig. 9-11. Soldered through-hole connections.

1. Fig. 9-12A — Wire is stripped with a suitable tool. If the metal surface is coated or appears tarnished, the wire should be cleaned with sandpaper or steel wool and dipped in liquid rosin-flux solution.

2. Fig. 9-12B — Stranded wire is tinned. Use enough solder to penetrate the wire, but still keep individual strand outlines visible.

3. Fig. 9-12C — Attach wire to the lug or eyelet, and secure it by clinching with a pair of needle nose pliers. This keeps the wire stationary during soldering, and gives the joint mechanical strength. Do not wrap wire leads excessively around a connection or you may have considerable difficulty removing the wire later. If several leads attach to one point, secure all of them and solder in one step.

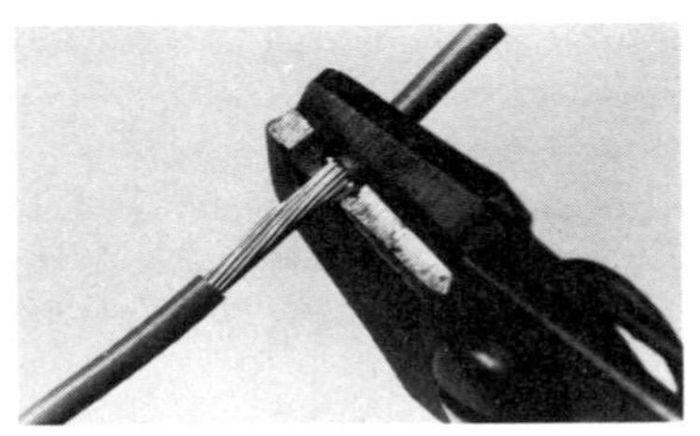

(A) Wire is stripped to expose clean metal surface.

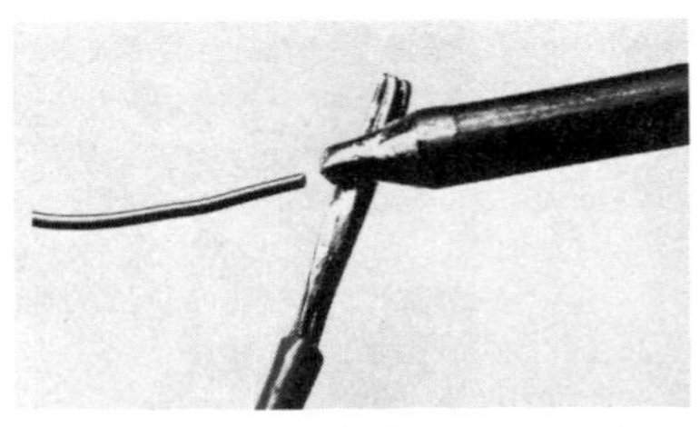

(B) Stranded wire is tinned with solder.

(C) Wire is secured to joint.

(D) Final solder connection.

Fig. 9-12. Steps in soldering hookup wire.

4. Fig. 9-12D — Solder the connection using the same principles illustrated in Fig. 9-10, giving a smooth shiny joint with the minimum amount of solder needed to cover all surfaces. Wiring involves larger masses of metal than printed circuit work, so it is helpful to use higher wattage irons (40–60 watts) with larger tips, and thicker rosin-core wire solder (AWG 14, 0.064 inch). Because there is less problem with heat damage, 5–10 seconds may be allowed per solder connection. Remember that if a joint becomes overheated, flux will burn instead of melting and will not flow freely.

Soldering Problems — The following problems may occur during soldering:

1. Cold solder joint — This type of joint appears rough, dull, grainy, and frosted instead of smooth and shiny. It results from inadequate heat, or from moving the parts while the hot solder solidifies. The joint may be internally fractured and it is always weak. Cold solder joints can be repaired by touch-up with a clean, hot iron.

2. Solder contamination — Impure solder will also take on a dull, grainy appearance, but the joint appearance does not improve with touch-up. This is the result of metallic impurities dissolving from the base metal surfaces into the solder, and is almost always due to too much heat or excessive soldering time. Copper, zinc,

and gold are the worst culprits, and the problem can be avoided by keeping soldering time as short as possible. Once the problem occurs, weakened solder must be removed and clean solder re-applied.

3. Delaminated foil — Etched copper traces that separate from the PC board are difficult to repair, and the condition is a result of excess heat. The problem can be avoided by using a low wattage iron (35 watts or less) and by cleaning foil and parts thoroughly before attempting to solder. Clean surfaces give short solder times. With care, a printed circuit pad can be soldered many times before delamination occurs.
4. Nonwetting of leads — This problem is evident when the molten solder flows evenly around a PC pad area, but does not envelop the component lead, and actually seems to draw away from it. Electrical connection may not occur at all. The cause is a dirty lead, having a coating of grease, oil, or oxide tarnish; this is often a problem with old components and old hookup wire. In many cases wetting can be *forced* by applying enough heat, solder, and flux, but that approach causes other problems, such as delamination. Do not continue heating a junction that does not accept solder readily. The joint should be desoldered, and the component removed for lead cleaning. Isopropyl alcohol is excellent for dissolving dirt, and the component leads can be pre-tinned with clean solder before re-inserting into the PCB. Non-wetting can also occur on PC *pads* if they are dirty or tarnished.
5. Excess solder — When first learning to solder, there is a tendency to force extra solder into the joint, rather than allowing a minimum amount of solder to flow evenly and cover the surfaces naturally. Although excess solder is not in itself harmful, large unsightly blobs of solder may mask underlying problems, such as nonwetted leads and cold joints. A simple, shiny joint is your best assurance that solder connections are trouble free.

DESOLDERING • Techniques for desoldering connections were discussed previously in Chapter 1, but a few additional pointers will be given here. Desoldering requires a great deal of patience, particularly if a number of wires or component leads are involved. It is always best to reduce the problem to removing *one* lead at a time, by snipping wires and crushing components sacrificially. Mount the assembly in a vise to leave both hands free to work.

Removing Point-to-Point Wiring • The joint should first be cleared of as much solder as possible by using a suction device or braided copper wicking material (see Chapter 1, Fig. 3-10). Then, heat the joint with an iron while bending and working individual wire leads so that they are straightened and can be pulled free of the connection. It is usually helpful to cut wires close to the joint, leaving individual ends free to grasp with needle nose pliers; in this way the wire can be extracted from either direction.

Removing Printed Circuit Components • Desoldering components mounted on PCBs is complicated by the need to work on opposite sides of the board simultaneously, and by the danger of delaminating copper foil with excessive heat. If components are physically destroyed so that their leads can be individually removed, this is a sounder approach than risking damage to the board. When constructing your own prototype circuitry, always mount valuable parts in sockets to avoid future desoldering problems.

The first step in PCB desoldering is to remove as much solder as possible from the pad area, using copper braid dipped in liquid rosin flux. (Most suction devices are not recommended because they tend to pull etched foil away from the board.) If component leads are clinched, they should be straightened with pliers, a small screwdriver, or tweezers before proceeding further. The component is then worked loose from the opposite side while heating each lead with a clean soldering iron tip. A third hand is often desirable, and a practical device for this purpose is shown in Fig. 9-13. This *parts puller* consists of a U-shaped metal bracket and a small spring, which is hooked between the bracket and the component lead under tension. As soon as the component lead is free on the reverse side of the PCB, the spring will pull it away from the hole. Commercial parts pullers are available to tug on integrated circuits while their leads are being simultaneously heated on the opposite side.

Fig. 9-13. A simple parts puller for desoldering.

Destroy integrated circuits by carefully crushing them, or by snipping away their individual leads from the IC body. Leads are then removed one at a time with needle nose pliers and a soldering iron. Following parts removal, clean the exposed pads with copper braid and clear the holes with a miniature drill bit.

When a simple two or three lead component must be replaced on a printed circuit board, it may be possible to accomplish this without desoldering and risking damage to the board. A well-accepted technique known as *piggyback mounting* is illustrated in Fig. 9-14. A defective resistor was removed by cutting its leads close to the body on the component side of the board. The protruding leads were then bent to support a new part, whose leads were crimped and soldered to the old leads. During soldering, the old leads were protected with alligator clip heat sinks to prevent solder from melting on the reverse-side PCB connections.

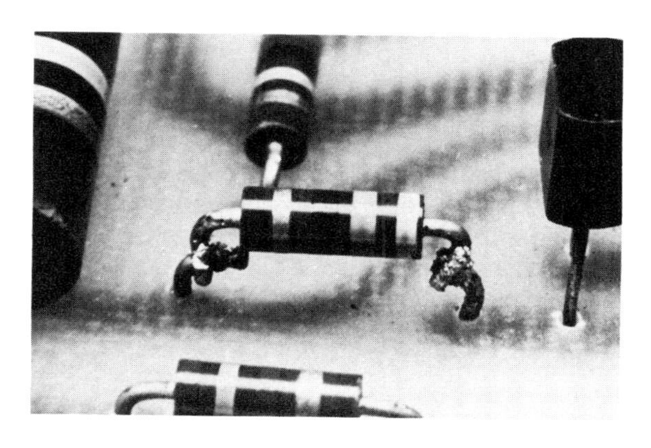

Fig. 9-14. Piggyback mounting of replacement part avoids desoldering problems.

CLEANING PCBs ● After PCB solder assembly, it may be desirable to remove flux residues with a solvent. Although rosin-core flux should not cause corrosion problems, residues detract from the appearance of the finished board. If the flux is burnt during soldering, a dark-brown, carbonized material is formed that can be slightly conductive. Problems such as solder bridging and cracked traces are easier to spot after the board has been cleaned. Various organic solvents are used to dissolve flux residues, either by dipping or by aerosol spraying.

Solvent Dipping ● The best solvents for rosin fluxes are alcohol, aromatic, and chlorinated hydrocarbons. Rosin flux is often used in the form of an isopropyl alcohol solution, but this alcohol is not always completely effective in dissolving flux residues, which become partially polymerized as a result of soldering heat. If isopropyl alcohol is used by itself for cleaning, traces of insoluble white residues are usually noted afterwards on PCB surfaces. Xylene is a better solvent for complete removal of rosin flux residues; note that xylene is the base solvent in Kodak ortho resist (KOR) developer, which can be used for PCB cleaning. Agitate the board in a tray of xylene for 30 seconds, loosening stubborn dark residues with a brush. Then immerse the board in clean isopropyl alcohol to displace the dirty xylene and leave a volatile, fast-drying solvent. The board can also be rinsed clean by spraying it with a squeeze bottle containing alcohol. Final drying in a warm oven is recommended to thoroughly evaporate solvents, which tend to become trapped in nooks and crannies (such as the inside areas of IC sockets).

Most modern electronic components are made from materials that are resistant to solvents, such as xylene and isopropyl alcohol. This is due to the widespread use of solvents to clean printed circuit assemblies commercially. If there is a chance for solvent damage or trapping, devices are usually packaged in hermetically sealed cases. However, some plastics *are* attacked by solvents and if a part is suspect, it should be tested before soldering into the board. Some parts, notably mechanical devices such as potentiometers and switches, should always be assembled after the rest of the board is completed and cleaned. Transformers, coils, and some connectors tend to trap solvents and may contain internal coatings that dissolve in xylene. Delay assembly of these components also. Common parts such as resistors, capacitors, and semiconductors rarely present problems in solvent cleaning.

Spray Cleaning • Small amounts of flux residue are conveniently removed with an aerosol spray cleaner, such as GC Electronics 22-270 Flux Remover and Cleaner. Avoid cleaning agents that evaporate and leave a thin film of lubrication behind, such as electrical contact cleaners. Aerosol flux removers are useful for touching up a PCB after repair, or for cleaning areas that are soldered after the main assembly has been completed. After spraying the board, remove residues by rubbing with a cotton swab or by wiping with a soft cloth.

PCB MODIFICATIONS AND REPAIRS • Despite the inflexibility of printed circuit construction methods, a certain amount of circuit alteration is possible if major component additions are not required. Techniques for repairing damaged PC traces are similar to those required for making alterations, so they will both be discussed in this section. (Removal of defective components was covered in the Desoldering section.) Basic operations are possible for PCB modifications and repairs.

Repairing Traces • *Discontinuous* traces that are damaged during etching may be repaired by bridging with a piece of solid AWG 22 wire, as illustrated in Fig. 9-15A. The wire is cut long enough to overlap each trace end at least one-eighth inch, and the ends are tinned with solder. The ends of each trace should also be tinned. The wire is positioned in-line with the traces and tacked in place with a soldering iron; additional solder can be added once the wire is mounted. If the wire length is much longer than one inch, insulated wire should be used and the wire can be held in place at several points with small dabs of epoxy adhesive.

(A) Bridging a damaged trace with solid wire.

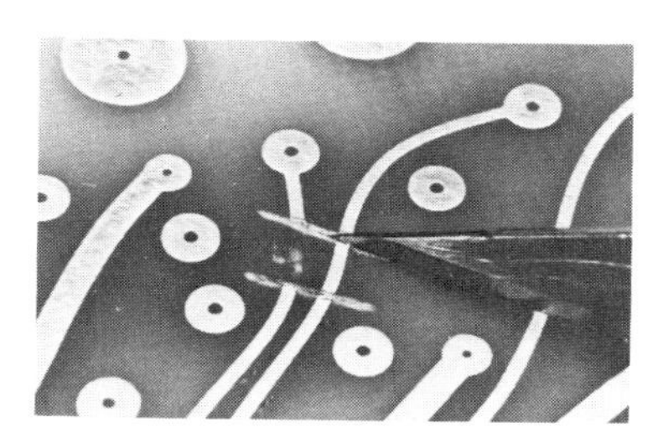

(B) Introducing breaks by removing strip segments.

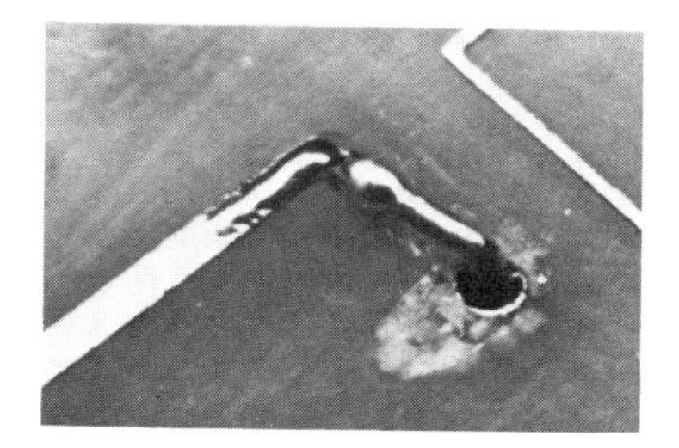

(C) Working around a damaged terminal pad.

(D) Stick-on foil patterns for replacing damaged pads.

Fig. 9-15. Printed circuit repairs and modifications.

Breaking Connections • A *break* can be introduced in an etched trace by removing a short segment of foil (Fig. 9-15B). Using a sharp razor knife, cut through the trace at two points spaced about one-fourth inch apart. Place the tip of a hot soldering iron firmly in the center of the segment to be removed. Due to the knife cuts, heat will be confined and concentrated in the isolated foil strip section, which will quickly delaminate and can be pushed away. The strip can also be peeled away with a thin knife edge.

Eliminating Shorts • A frequent problem seen after etching is incompletely removed copper foil between closely spaced traces. The extra copper shows up as *spurs* that may completely bridge and connect adjoining traces. Using a metal straightedge for a guide, a firm knife cut through the problem area is usually enough to eliminate the short. If the spur is very wide, two parallel cuts can be used to remove a thin strip of foil, which is peeled away with heat or a knife blade.

Lifted Terminal Pads • Terminal pads can become delaminated from the board during soldering, particularly if they have been soldered many times or if excessive heat is applied. A copper foil pattern that is separated but still physically intact can be reattached to the board with epoxy adhesive. First, remove all solder from the pad area with copper braid. Then, clean the area thoroughly with a swab soaked in flux remover. If possible, rub the surfaces to be joined with steel wool. Protect adjoining areas with masking tape to confine epoxy to the damaged area. Epoxy adhesives consist of a resin and an activator that are mixed just prior to use; fast-setting formulations become solid in 5–10 minutes after mixing. Apply adhesive to both surfaces to be joined, and hold the pad firmly in place with weights or a clamp until the epoxy has cured. The pad surface can be cleaned with steel wool or a knife blade after epoxy repair is complete.

Damaged Terminal Pads • Terminal pads that are delaminated and torn from their connecting traces may also be repaired. However, in many cases it is simpler to bend the component lead over against a trace and solder it directly without replacing the original pad. Such a repair is shown in Fig. 9-15C.

If it is decided to replace the damaged terminal area, a similar pad in good condition must be found. Many manufacturers of printed circuit artwork materials supply stick-on copper foil patterns that can be used to repair damaged traces (see Fig. 9-15D). Another source of etched foil patterns is scrap circuit boards, from which pads can be cut and removed for transplant. In either case, the new foil should be glued in place with epoxy adhesive as previously described. At least one-eighth inch of foil overlap is necessary to allow the pad to be electrically reconnected to its original trace by soldering the overlapping foil ends together.

Replacing Connector Tabs • Gold-plated tabs on PCB edge connectors may be damaged after extensive wear or improper seating of the card. A gold strip typically becomes gouged and separated from the board. Tab strips may be repaired by cutting a similar gold-plated foil piece away from a scrap board, and gluing it in place with epoxy adhesive. Extra length is allowed at one end to overlap the original foil pattern, where a solder connection is made. After repair, the connector end of the metal strip should be filed smooth to match the beveled tab edge.

Adding Jumper Wires • Circuit modifications can be made to PCBs by adding jumper wires on either or both sides of the board. Most circuit alterations involve a combination of breaking trace connections and adding jumper wires. Solid AWG 30 Kynar-insulated (wire-wrap) wire is commonly used for this purpose. Four methods used to install PCB jumpers are illustrated in Fig. 9-16:

1. Fig. 9-16A — While heating a component lead on the reverse side of the board to melt solder, a prestripped wire end is pushed through the hole from the component side. The wire is then soldered in place on the reverse side pad. This method depends on the hole being large enough to accommodate both the component lead and the jumper wire, which is usually possible using small-gauge wire-wrap wire. Long jumpers can be secured to the board with adhesive.
2. Fig. 9-16B — Holes are drilled in strategic spots next to etched traces. Stripped wire ends are inserted through the holes from the opposite side, bent over across the traces, and soldered in place.
3. Fig. 9-16C — Insulated wire jumpers can be soldered directly to etched pads and traces on one side of the board, in the same manner that discontinuous lines are repaired.

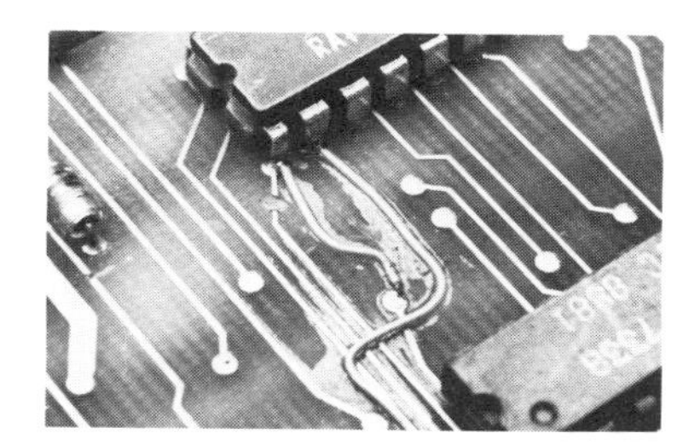

(A) Wire ends inserted in existing holes.

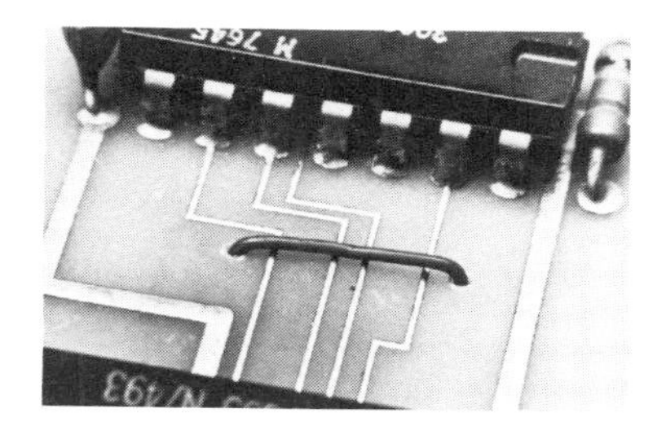

(B) New holes drilled for jumper insertion.

(C) Wire end soldered directly to etched trace.

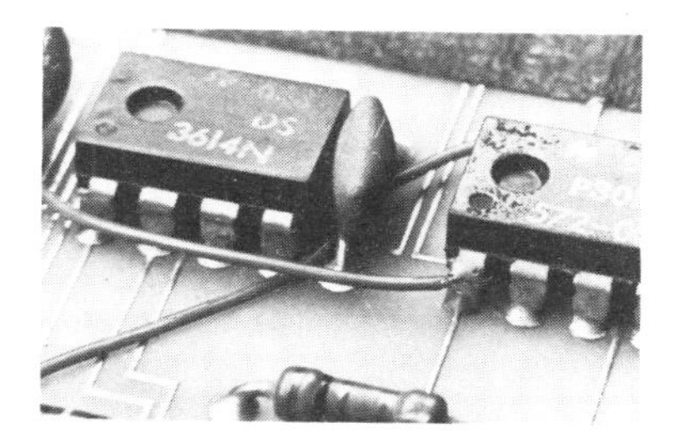

(D) Wire end wrapped and soldered around component lead.

Fig. 9-16. Adding jumper wires for circuit alterations.

4. Fig. 9-16D — Jumpers can be attached to protruding component leads and pins on either side of the board. Leads should first be cleared of solder using copper braid. The stripped jumper end is then wrapped around the lead end, and the joint is soldered.

Adding New Components • As space on the PCB permits, new components may be added by drilling holes. In most cases this involves mounting a socket or part to the board with adhesive. Small devices like ¼-watt resistors may have their leads soldered directly to traces on the reverse side, but larger parts will require wire-wrap or pencil-wired connections to attach on-board jumper wires.

Chapter 10
High Density PCBs

INTRODUCTION • Using the design and fabrication techniques described previously in this book, prototype double-sided PCBs can be routinely constructed with an average component density of about one IC per 2 square inches of board space. This density figure assumes that the components are primarily 14- or 16-pin DIP ICs, with a reasonable number of discrete support devices and dedicated through-holes. The industrial processes outlined in Chapter 2 are able to produce higher density double-sided boards, with typically one IC per square inch. In evaluating the basic differences which limit prototype PCB density, the following factors are seen to be most important:

1. Pad size — Commercial designs make use of very small terminal pads, often 0.040 inch (1.016 mm)-diameter circles with 0.030 inch (0.762 mm) center holes. Small pads allow more routing space for conductors, and permit greater use of dedicated through-holes. Large terminal areas are desirable in prototype work because of the greater tolerances needed for side-to-side registration and for manual drilling. Also, hand-soldering of small pads is difficult due to the hazards of foil delamination.
2. Trace density — Thin 0.015 inch (0.381 mm) lines and closely spaced conductors contribute to the density of commercial PCBs. Wider, liberally spaced traces are recommended for prototype fabrication because of the greater margin of error in image transfer and etching processes.
3. Through-holes — Due to the lack of plated-through-holes, through connections in prototype PCBs must be accomplished by soldering leads on both sides of the board. However, many device leads are inaccessible on the component side of the board; to get around this problem, extra dedicated soldered-through-holes are needed and hence less space is available for conductor routing.

If greater prototype PCB density is to be obtained without tightening process tolerances, it is obvious from the previous discussion that additional conductor routing space is what is needed. The purpose of this chapter is to describe how a *two-layer* circuit board can be constructed to provide an additional plane of etched conductor patterns, allowing an average packaging density of one IC per square inch to be easily achieved by experimenters and prototype circuit designers. The extra layer is obtained by using two one-thirty-second inch thick epoxy-glass laminates in place of a single one-sixteenth inch thick board, creating a new center plane surface for interconnections and thus a 50% increase in routing space. Design and construction techniques for two-layer prototype PCBs will be found to be straightforward extensions of the methods previously described for double-sided boards.

OBJECTIVES ● The objectives of this chapter are as follows:

- Describe a *two-layer* construction approach for achieving greater packaging density of PCB prototypes.
- Note the drawbacks of *multilayer* PCB construction.
- Discuss a design process for two-layer PCBs.
- Show how suitable artwork and photographic masks are generated.
- Outline fabrication steps for two-layer construction.
- Discuss assembly and soldering techniques for two-layer PCBs.
- Illustrate two-layer design and construction with a real example.

THE TWO-LAYER PCB ●

General Concept • The diagram of Fig. 10-1 illustrates the principle of *multilayer* PCB construction, which may be carried out to many levels commercially. Note that the three surfaces shown represent planes of etched copper foil, and not laminate material. Two layers of one-thirty-second inch thick PCB laminate are sandwiched in between three planes of conductor patterns, giving a total board thickness equal to the conventional one-sixteenth inch size. In practice for *two-layer* PCBs, two foil surfaces are etched on the top one-thirty-second inch double-sided board, and one surface is etched on the one-thirty-second inch bottom board. All three surfaces have a similar set of terminal pads drilled in registration.

For simplified circuit layout, most designers assign foil surfaces to specific trace functions as follows:

1. Top plane — This is the component side, and it is assigned to *vertical* signal traces. Vertical traces are defined to be those that run at right angles to DIP IC packages.
2. Middle plane — This area is assigned to *power supply* and *ground* networks. Because of the need to carry substantial current and to minimize electrical noise, power supply traces are usually etched as very wide lines on PCBs. Therefore, a great deal of routing space can be saved by moving power traces to a dedicated plane. Power distribution lines are conveniently routed (horizontally)

down the middle of IC packages and connected to wide borders at the edges of the PCB. Most digital IC circuits require +5 V and ground connections. If additional power levels are needed, such as ±12 V, these patterns can also be worked into the middle plane design; however, extra jumpers on the top plane may be necessary to complete some multiple power supply connections.

3. Bottom plane — This is the reverse side of the final assembly. It is assigned to *horizontal* signal traces, or those lines that run parallel to DIP IC packages.

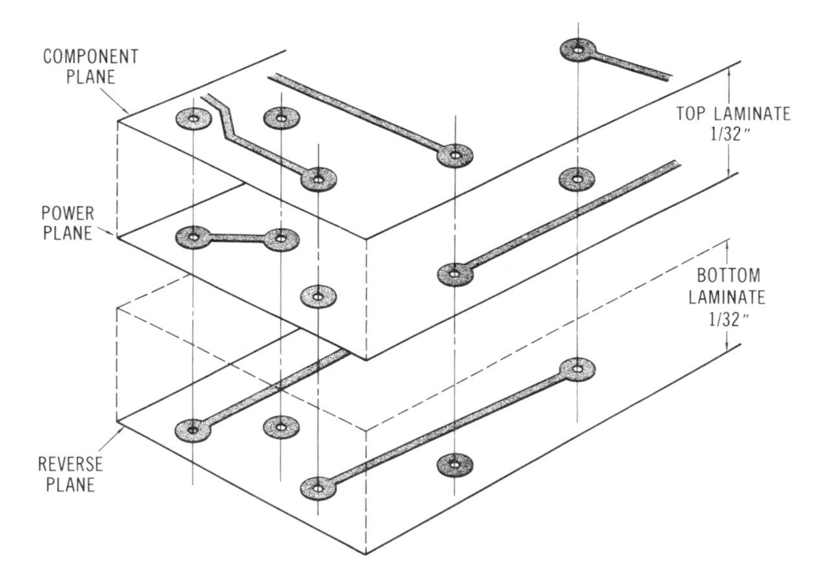

Fig. 10-1. Multilayer PCB construction with two layers of laminate and three planes of foil.

Through-Hole Connections • The multilayer approach is used commercially to achieve high density PCB packaging, often with as many as 12 thin laminated layers. In most cases, plated-through-hole technology is necessary to interconnect various layers, whose etched traces meet at common holes where they are fused to the plated hole walls (see Fig. 10-2). This approach is not practical for low-volume prototype fabrication because of the complexity of plated-through-hole processes.

Clearance Holes — The use of *clearance holes* is another connection technique that is compatible with prototype multilayer PCB construction and hand-solder assembly. Clearance holes are drilled as

Fig. 10-2. Photomicrograph of a plated-through-hole in a multilayer PCB. *(Courtesy Branson International Plasma Corp.)*

necessary to completely expose inner terminal pads for soldering, which is the method used to interconnect pads on the various foil layers. The clearance hole principle is illustrated in Fig. 10-3. If several layers are to be joined, each successive layer requires a larger clearance hole and a larger terminal area surrounding it for connections to be continued on towards the outermost surface of the board. During assembly, clearance holes are filled with solder, which forms through-hole connections by overlapping onto successive pad areas.

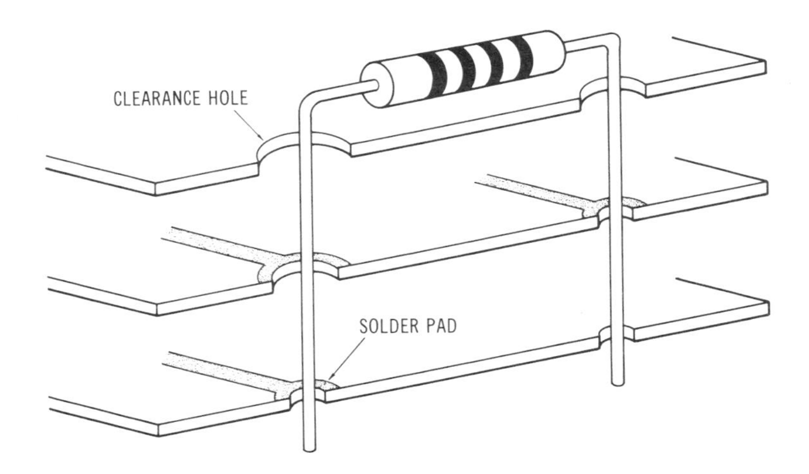

Fig. 10-3. The clearance hole principle for connecting etched patterns in multilayer PCBs.

Two-Layer Conventions — The obvious disadvantages of the clearance hole approach is that terminal pads become progressively larger and larger, moving outwards from the innermost foil layer; that is necessary to expose pads on multiple layers for soldering. The use of large-area pads results in a loss of conductor routing space, which is in direct opposition to the original goals of multilayer construction.

Fortunately, in practice large terminal pads are seldom necessary for two-layer PCBs, particularly if the middle foil layer is reserved for power supply connections. The need seldom arises to join pads on the middle and bottom foil layers, because the bottom layer is assigned to vertical signal traces which do not involve power and ground connections. As a result, clearance holes in the bottom laminate are the same diameter as normal terminal pads, and no extra foil surrounds them on the bottom board surface. Clearance holes are used exclusively to solder component leads to the power plane, and occasionally to form soldered-through-hole connections between the top and middle foil layers.

A great number of through-hole connections are required between the outer two foil surfaces, joining vertical and horizontal signal traces to form a conventional X-Y routing system. Clearance holes are not required to complete these connections, which are made by the standard methods used in double-sided prototype PCB construction — component and socket leads are soldered on both sides of the board, or individual pins are soldered into dedicated through-holes.

Limitations • The main disadvantage of two-layer construction for prototype PCBs is the difficulty in making repairs or alterations to the middle foil layer once assembly is complete. Separation of the two one-thirty-second inch laminates requires that almost every component and pin be desoldered and removed from the board, a tedious task. This is the reason that the middle layer is reserved for power supply patterns, which are uncomplicated as compared to signal traces. Modifications to signal interconnections are readily made because these lines are accessible on the outer surfaces of the two-layer sandwich.

Some designers do switch functions of the middle and bottom conductor layers for a specific purpose. Solder connections to the bottom surface of the board are the ones that hold the assembly together, assuming that the two laminates are not glued together. By placing power supply connections on the bottom layer, these relatively few solder joints can be left until last. Instead of soldering, all power and ground points can be connected daisy chain fashion with pencil wiring, and the board can be tested under temporary wiring conditions. If problems are encountered, the bare pencil wire is easily removed and the layers can be separated. Once troubleshooting is complete, permanent power supply connections are made.

Concept Expansion • The two-layer system is easily expanded to a three-layer, four conductor plane assembly by etching patterns into both sides of the one-thirty-second inch laminates and inserting a dielectric sheet between them. However, one should be careful not to increase the total board thickness much beyond one-sixteenth inch, because many component leads are not long enough to extend the extra distance. Solder connections to the lower middle layer are accomplished by drilling clearance holes in the top laminate.

Designers will find that a fourth foil layer is most useful for incorporating multiple power supply patterns into dense circuit boards, and for adding ground plane designs. A ground plane (see Chapter 3, Fig. 3-28) can be etched into any layer, and is useful for controlling the characteristic impedance of conductor traces, and for shielding critical areas in high-frequency circuits.

Design Considerations • The design of a two-layer PCB proceeds through steps similar to those discussed in Chapter 3 for double-sided boards. By assigning power supply traces to a separate foil layer, more space is available for routing conductors and designs are often simplified. The result is that two-layer boards can often be designed faster than double-sided PCBs, and with an increase in packaging density. Work begins with a well drawn schematic having all power connections clearly indicated. As an example for two-layer construction, the circuit diagram in Fig. 10-4 shows an EPROM programmer for the Apple II microcomputer.

Board Dimensions — The schematic of Fig. 10-4 contains the equivalent of about 12 DIP IC packages, and a total of approximately 260 component leads. Our goal is to allow one square inch per IC, or about one square inch per 20 component leads. The final PCB should therefore have an area of about 12 square inches. The optimum height for an Apple II plug-in PCB is 2.5 inches (63.5 mm), so the final board dimensions will be 2.5 by 5 inches (63.5 by 127 mm). Fitting all of the required circuitry into such a small space represents dense PCB packaging.

Parts Layout — Parts layout follows the general principles of schematic viewpoint, and of grouping components around key areas where many connections converge. Of particular importance in two-layer PCB design is to strictly organize ICs into rectangular geometry. Unless this approach is followed, layout of power traces in the middle foil plane may become quite complex. A simple organization of IC packages is also helpful when signal traces are routed because it allows clear interconnection routes to be established in parallel directions across the board.

The proposed parts layout for our construction example is pre-

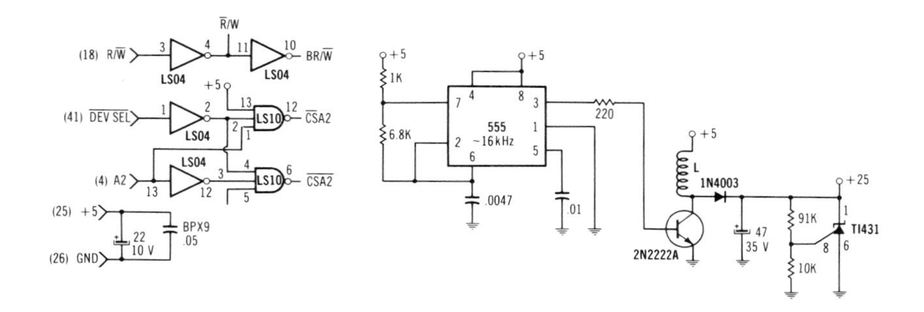

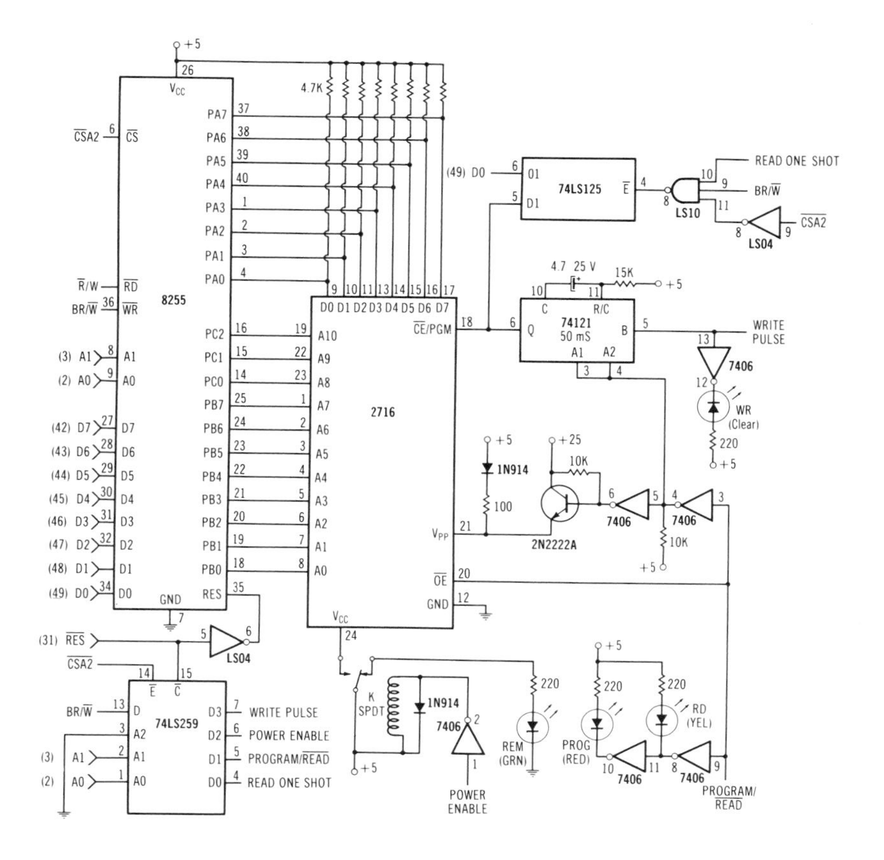

Fig. 10-4. Schematic to be converted into a two-layer PCB design.

sented in Fig. 10-5. It was arrived at through the following steps, using mobile Bishop Graphics Puppets to represent components while testing various parts configurations:

1. The most critical area was around the 50-pin edge connector because of its fixed position and its large number of connecting signals. The 8255 and 74LS259 ICs were positioned close to the edge connector because of their many common signals.
2. The next key area was around the 8255 IC. Most of its signals connect to the 2716 EPROM IC, which was positioned close by.
3. The 2716 IC became the next critical component around which to group most of the remaining parts.
4. The four LEDs are function indicators, so they were arranged in a row at the top of the board for neatness and visibility.
5. The analog circuitry for the dc-dc converter was grouped out of the way in the top left corner of the board, since it has only one output signal (+26 V).
6. All ICs were arranged in neat rows and columns with their No. 1 pins oriented in the same direction. This approach simplifies power trace layout and it makes troubleshooting and testing less confusing after the board is assembled. Orientation was based on the optimum geometry for the 8255 IC, the most critical component.

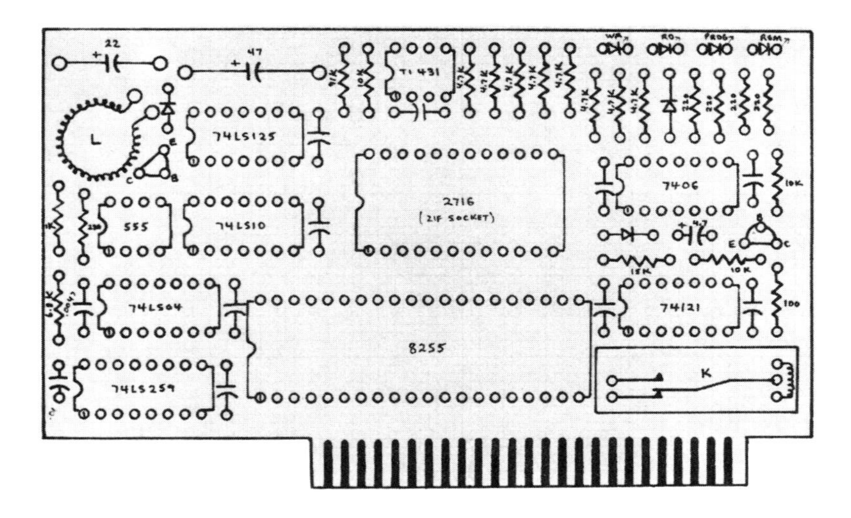

Fig. 10-5. Parts layout for the two-layer PCB, with a packaging density of one IC per square inch.

7. Discrete support devices were arranged around central ICs so as to least obstruct open signal pathways. Bypass capacitors were positioned at the head of each IC package.

After a satisfactory parts arrangement was found with the Puppets, a permanent drawing (shown in Fig. 10-5) was made on a 2× scale, using templates and 0.1 inch (2.54 mm)-gridded drafting vellum. Note that component leads are identified for every device whose lead geometry is ambiguous, such as polarized capacitors and diodes. This information forms a basis for line layout, although parts geometry may be changed later to simplify trace routing.

Line Layout — Line layout proceeded through several revisions, each one based on the previous results and on an updated parts layout. Line layout drawings should be made on a 2× scale using 0.1 inch (2.54 mm)-gridded drafting vellum and erasable colored pencils. In general, it is helpful to adopt a color code to maintain line information in a readable form; a typical color system was proposed in Chapter 3. Because of the new power plane, an additional color is needed for two-layer PCB layouts; power distribution traces are conveniently indicated with yellow pencil and black outlines.

Line layout techniques for two-layer PCBs are essentially the same as for double-sided PCBs, but the following considerations may be helpful in designing two-layer systems:

1. Power plane — A basic power distribution trace pattern for the middle foil layer should be established before any other traces are routed. Wide power traces will take up considerable space and no through-holes can be positioned where they occur. Most designers prefer to route power traces between the legs of IC packages where through-holes are seldom needed; additional connecting lines can be routed later after the rest of the board is completed. The reader should examine various general-purpose wire-wrap boards with etched power supply patterns, because these boards illustrate well how power distribution networks can be conveniently arranged. A simple pattern was chosen for the EPROM construction example, as shown in Fig. 10-6, which is typical of circuits requiring a single +5 V supply voltage and ground. When additional supply lines are needed (such as ±12 V), they may be included in the middle power plane as isolated horizontal strips, which are interconnected on the top plane with vertical strips and through-hole connectors.

2. Top plane — The top plane is used for vertical signal traces that cross at right angles to IC packages. It is important that vertical traces be assigned to the component side of the board, because in high-density designs lines are frequently routed between IC pins spaced only 0.1 inch (2.54 mm) apart. Space between IC pins

is limited on the reverse side of the board because many pads must be drilled out as clearance holes. However, small-diameter pads can be readily used on the component side and this allows extra clearance between IC pins for vertical trace runs.

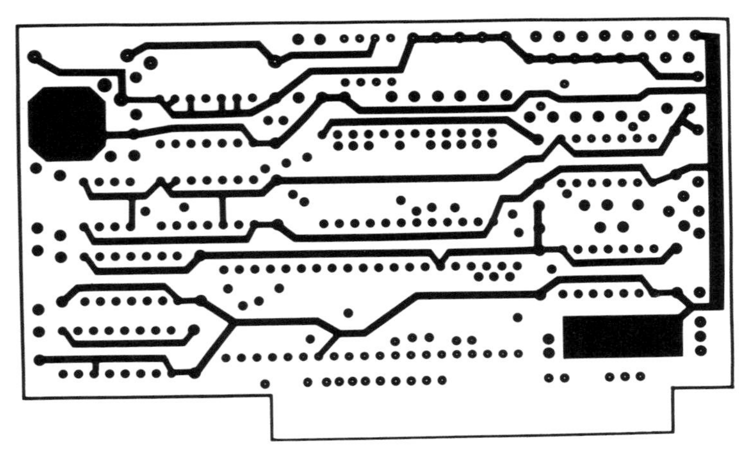

Fig. 10-6. Power distribution pattern for the middle foil plane.

Another good reason for choosing the vertical orientation for component-side traces is that power distribution lines in the middle foil plane are consequently at right angles to the top traces. The two planes together form an X-Y routing system that can be used (with through-holes) to complete certain difficult power connections. On the other hand, it is much more difficult to connect the middle and bottom planes for this same purpose, because bottom pads become obliterated when clearance holes are drilled for soldering.

3. Through-holes — Most through-holes are established between the top and bottom wiring planes as a complete X-Y system for signal interconnections. Through-holes should be used liberally during initial attempts at line layout, and eliminated where possible in the final revision using the optimization techniques discussed in Chapter 3. When all through-holes have been permanently established, miscellaneous power supply connections are completed in the middle plane by routing them through the maze of component and through-hole pads.

A final line layout drawing for the EPROM two-layer PCB is shown in Fig. 10-7. The complexity of this drawing illustrates the importance of using a color scheme to visually separate traces intended for different foil planes.

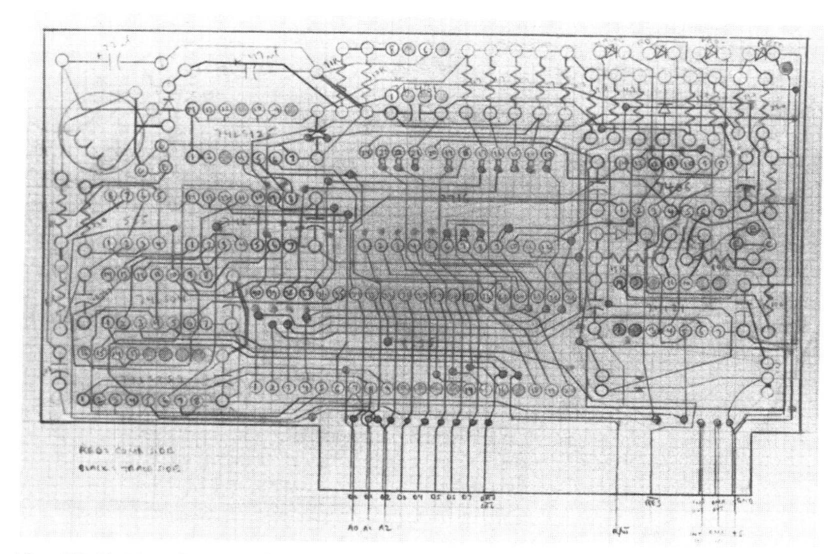

Fig. 10-7. Line layout drawing for the two-layer PCB.

Artwork • Master artwork is prepared on a 2× scale from the final line layout drawing, using the same methods detailed in Chapter 3 for double-sided boards. The only significant difference in two-layer PCB artwork is that an additional overlay will be created to represent the middle power plane. Essential artwork therefore consists of a base pad sheet, a component-side overlay, a power plane overlay, and a reverse-side overlay. Each overlay will be photographed in combination with the base sheet to give three final masks for photofabrication, one for each foil layer. Additional masks may be created for optional images, such as screen-printed labels. Final artwork patterns for the EPROM two-layer PCB are presented in Figs. 10-6 and 10-8.

One additional point concerning two-layer PCB artwork should be made. Certain patterns present in the base pad sheet are not needed in the photofabrication mask for the middle power plane, notably edge connector and plating strip patterns. These patterns can be removed (as in Fig. 10-6) from the final negative using photographic opaque and a brush. They can also be removed *before* photography by covering them with white paper attached to the power plane artwork overlay sheet. However, this latter approach requires that photographic lighting be from the front with no backlighting.

Photography • Standard photographic techniques discussed in Chapter 4 are used to prepare photofabrication masks from artwork; masks may be positives or negatives depending upon the subsequent image transfer process. The component-side mask is right-reading and

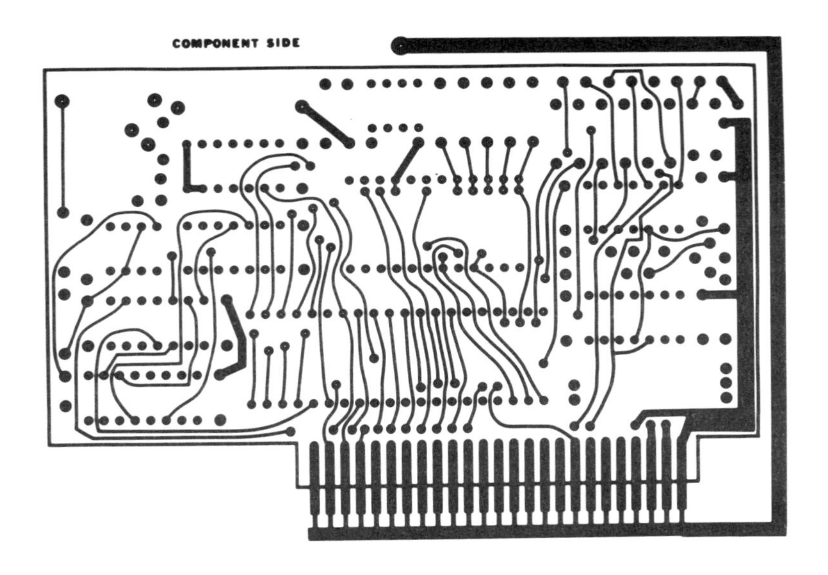

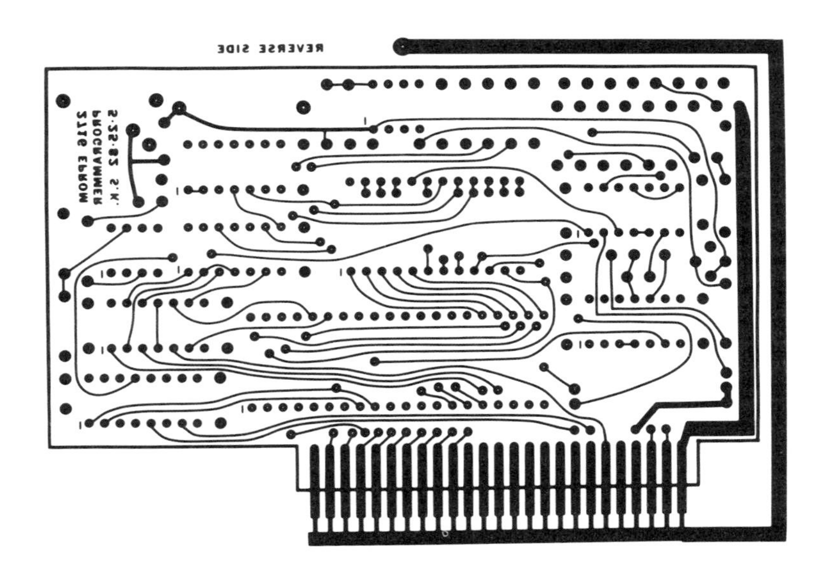

Fig. 10-8. Final artwork patterns for the component- and reverse-side planes.

will be used to form etched patterns on the top surface of the top laminate piece. The middle and reverse-side masks are both wrong-reading because they are used to form etched patterns on the bottom surfaces of the top and bottom laminate pieces. Final film sheets should be checked on a light table for proper pattern registration, which is critical for multilayer PCB construction.

Fabrication • Construction of two-layer PCBs is simplified by the ready availability of one-thirty-second inch thick epoxy-glass laminate material. Thin precut stock is available from Kepro with 1 ounce copper foil on one or both sides, sensitized or unsensitized with photoresist. The one-thirty-second inch material is less expensive than corresponding sizes of standard one-sixteenth inch thick laminates.

Two one-thirty-second inch thick circuit boards are etched from the photofabrication masks, using various techniques described in earlier chapters. It may be convenient to etch both boards from a single piece of blank copper clad laminate. One board is double-sided, the other is single-sided. After etching, the boards are coated with solder. Edge connector fingers are plated separately on each board. The laminates are then temporarily joined by soldering in several through-hole pins randomly located across the board. At this point the board can be cut to final shape and edges filed smooth; it should appear exactly like a normal double-sided one-sixteenth inch thick PCB.

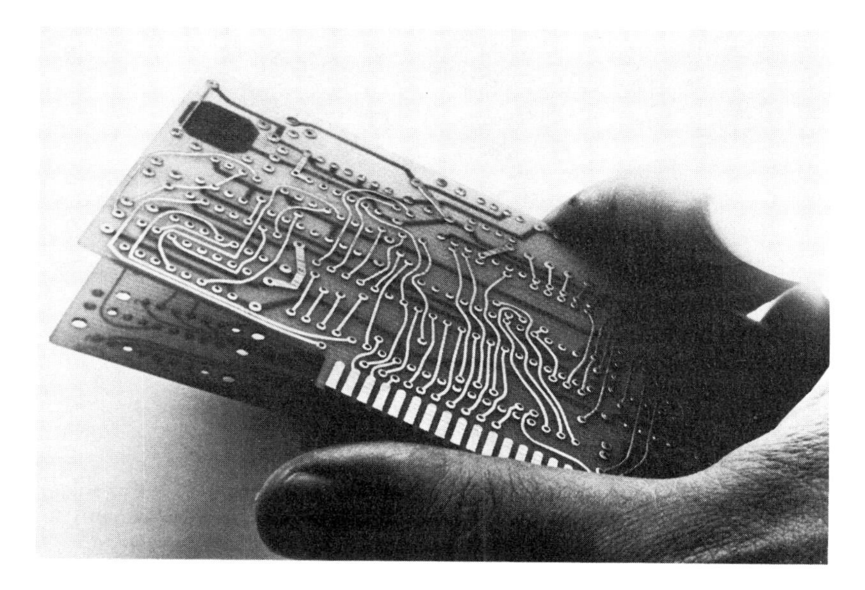

Fig. 10-9. Joining the two laminate pieces comprising a two-layer PCB.

Drilling • Holes are drilled from the top board surface to ensure that small-area pads on the component side are not torn apart due to exit burrs and slight misregistration of patterns. After drilling is complete, the two halves of the board are separated and holes are completely deburred. At this point, large clearance holes are drilled in the bottom laminate to provide solder access to certain terminal pads on the middle foil surface. (Clearance holes are evident in Fig. 10-9.) The boards can then be joined together again by soldering in through-hole pins, or a permanent assembly can be made by gluing together the two halves with epoxy adhesive. If adhesive is used, be careful to keep it away from middle layer terminal pads because smeared epoxy will make soldering difficult. Clamp the boards securely together while epoxy is setting to prevent a warp from occurring. The author prefers not to use adhesive, but to rely on solder joints to hold the two halves of the board together. This approach retains the possibility for disassembly of two-layer circuit boards at a later date, admittedly an unwelcome task.

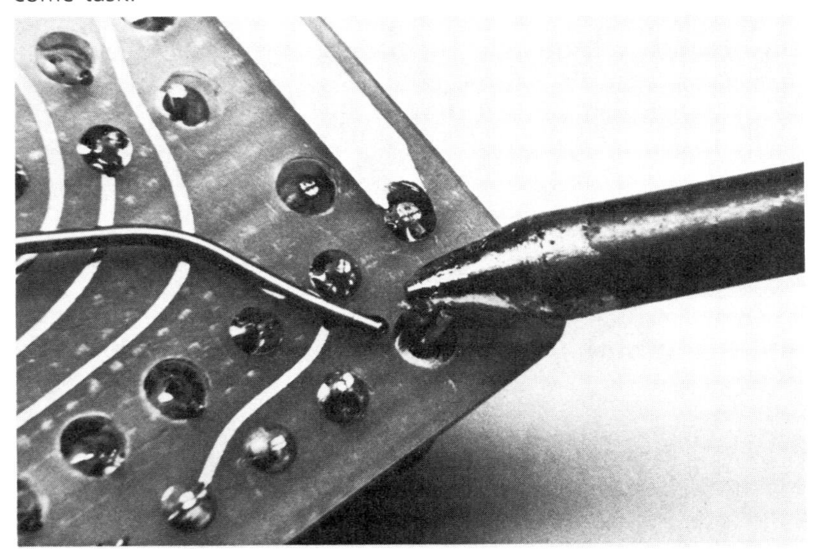

Fig. 10-10. Soldering a connection through a clearance hole.

Assembly • Components are mounted on the board and soldered into place as if it were a normal double-sided PCB. The only difference in assembling a two-layer board is that certain solder joints on the reverse side are actually made to the middle foil plane by poking the tip of the soldering iron into a clearance hole. A common mistake that is often made is to forget to drill all clearance holes, a difficult problem to correct once the board is assembled. If the two one-thirty-second

Fig. 10-11. Final assembled two-layer PCB.

laminates are not glued together, they should be tightly clamped while installing the first few components, which then serve to hold both layers securely together. A closeup view of clearance hole soldering is shown in Fig. 10-10 and the final EPROM two-layer PCB is pictured in Fig. 10-11.

CHAPTER 11
Electronic System Packaging

INTRODUCTION • The modern basis for electronic packaging is the printed circuit board, and that is why the primary focus of this book has been on PCB construction. However, the total electronic system includes some additional important elements: a suitable enclosure, panel-mounted controls and displays, chassis-mounted devices, a wiring system to interconnect internal components and circuit boards, and external finishes. In most cases the actual usefulness of electronic equipment depends on how well it is packaged[27] into a practical system with an appropriate control panel; circuit performance and reliability may also depend heavily on component and hardware placement. Therefore, this chapter is concerned with those aspects of electronic construction dealing with the total system package.

OBJECTIVES • The objectives of this chapter are as follows:

- Discuss the basic planning of an electronic package.
- Illustrate some typical electronic enclosures and their applications.
- Show how controls, switches, displays, and other devices are mounted in an enclosure.
- Review methods for wiring and interconnection.
- Discuss the tools and techniques for working sheet metal.
- Outline various techniques for labeling and finishing cabinets and front panels.

BASIC PLANNING • In order to perform usefully and reliably, electronic equipment must be constructed with the end use in perspective. Equipment that is intended to operate outdoors or in mobile applications must be designed to withstand atmospheric conditions, thermal shock, and vibrations. Some devices are used in highly critical situations, such as in military and aircraft controls, where protection against component failure may be much more important than serviceability. The designer must balance many opposing factors to arrive at a final system package; space, size, weight, reliability, appearance, serviceability, and cost are all important final parameters that are frequently at odds with each other. For purposes of this book, we are interested in *experimental* or *prototype* equipment with fairly well defined characteristics as follows:

1. Indoors operation — Typical environment will be a workshop table or laboratory benchtop.
2. Serviceability — The ability to make changes and circuit adjustments will be important.
3. Portability — The final package should not be too bulky to move around easily.
4. Low cost — Packaging will be designed around common, readily available hardware and components.
5. Front panel design — Considerable emphasis will be placed on mounting control and display devices in front panel fashion to make the equipment practical for experimentation.

Printed Circuit Impact • As discussed in Chapter 2, the miniaturization of electronic components led directly to the development of lightweight printed circuit wiring. Other similar methods, such as perforated board and wire-wrap assembly, are basically circuit board techniques adapted to the needs of experimenters and prototype designers. Even before the solid state era took hold, small, lightweight vacuum tubes were being mounted on PCBs instead of open chassis.

In addition to its impact on assembly automation and miniaturization of equipment, the circuit board has led to basic changes in total system packaging. In effect, the PCB has replaced the conventional chassis by providing independent support and dimensions for the majority of electronic components. The modern electronic package is a combination of circuit boards, control panels, power supply, and an occasional chassis module mounted in a suitable enclosure. Wiring interconnection systems are important as a means of combining the system elements, but most discrete wiring now takes the form of multiwire cables terminating in standardized connectors at each modular unit inside the cabinet. The *motherboard* is another new concept in packaging — small, modular circuit boards are plugged into connectors on a larger board, whose etched traces combine the PCBs into a complete electrical system.

Modularity • The current trend towards modularity in commercial electronic packaging has produced good results because it simplifies all aspects of design, construction, and equipment service. The prototype designer should adopt this same approach because it is equally successful for one-of-a-kind construction projects. As circuitry expands and becomes increasingly complex, it becomes more and more important to break up designs into logical sub-units that can be designed and assembled separately. Final results will be improved through organization, but modular construction is also the most effective way to simplify system packaging. Instead of devising mounting and connection methods for a mass of individual parts inside an enclosure, it is much simpler to choose a cabinet that will house the system as three or four neat subassemblies wired in circuit board or chassis style.

Packaging Priorities • In commercial electronic design, the final size, weight, and appearance of a product may be as important as how well it functions. For example, portable test equipment must be lightweight and compact to compete successfully in the marketplace. On the other hand, audio products for the hi-fi enthusiast are often designed with appearance in mind; large, spacious cabinets with an elaborate, impressive control panel are best sellers. (It is not clear whether such equipment is intended to be conspicuous, or whether it actually satisfies the customer's latent desire to become a jet fighter pilot.) In any case, the commercial designer must often consider the final package before attempting to lay out individual circuit boards. If space is at a premium, it may be necessary to choose component styles to fit a specific enclosure geometry. That is the reason why large items such as transformers and filter capacitors are available in a variety of shapes and mounting configurations. Many packaging limitations may be imposed before circuits are actually assembled and tested.

Prototype designers would prefer that final packaging be the lowest priority thought in their designs. Obviously, that is not always possible; for example, an interface card for the Apple II microcomputer is fairly restricted in its size, shape, and connection scheme because it must fit into an existing cabinet and socket system (see Chapter 1, Fig. 1-37). However, in most cases experimenters can be flexible in choosing enclosures. The following considerations make it desirable to leave packaging as the final step in prototype construction, choosing an enclosure to suit specific parts rather than the reverse approach:

1. Cost — Special-order components to suit a specific packaging system are expensive in low quantity.

2. Availability — Most designers only have a few reliable sources for readily obtaining parts, and component styles are fairly limited.

3. Time — Special parts require a long time for delivery. The hobbyist does not want to wait a month before receiving a notice that the parts have been back-ordered. In other cases, upon arrival it is discovered that the item does not fit your impression from the catalog picture at all. The experimenter ideally wants to look at and hold the parts before selecting them.
4. Design — The complexity of the design process increases significantly when packaging restrictions must be considered before basic circuitry can be assembled and tested.
5. Space — Compactness must be sacrificed for working space. The designer cannot hope to minimize equipment size when assembled circuits are fitted into off-the-shelf enclosures. However, extra size is desirable for prototype construction. Packaging is easier, space is available for your hands and tools to work, parts are more accessible for troubleshooting, and thermal problems are less likely to occur. If additional circuitry must be installed at a later date, this should be possible without designing a complete new enclosure.

STANDARD ENCLOSURES • Many sizes, shapes, and styles of electronic enclosures are available as off-the-shelf items. They are commonly known as boxes, chassis, cabinets, cases, consoles, instrument cases, mini-boxes, utility boxes, cages, and racks. A well-known manufacturer of commercial enclosures is Bud, whose products can be seen in a Newark Electronics catalog. Most electronic distributors carry enclosures, including Digi-Key, Jameco, and Radio Shack. Some typical configurations are illustrated in Fig. 11-1. The following basic styles are popular:

1. Two-piece mini-box (Fig. 11-1A) — Two U-shaped metal pieces snap together to form a simple, lightweight box. Also known as a utility box, this enclosure is used for small, lightweight assemblies that do not require much physical strength or a front panel plate.
2. Plastic box (Fig. 11-1B) — A simple enclosure consists of a molded plastic box with an aluminum cover, which can serve as a front panel. Circuitry installed in the box must be connected with long wires to parts mounted on the panel so that the cover is easily separated for construction and testing.
3. Finished cabinet (Fig. 11-1C) — Complete cases may include a chassis, painted cover, front panel, handle, and rubber feet. The cover is often louvered to provide air circulation. Complete cabinets are the best choice for electronic construction when a sturdy, accessible enclosure with front panel space is needed.
4. Sloped case (Fig. 11-1D) — This variation of the box cabinet is

(A) Utility or mini-box.

(B) Molded plastic box with aluminum cover.

(C) Finished cabinet.

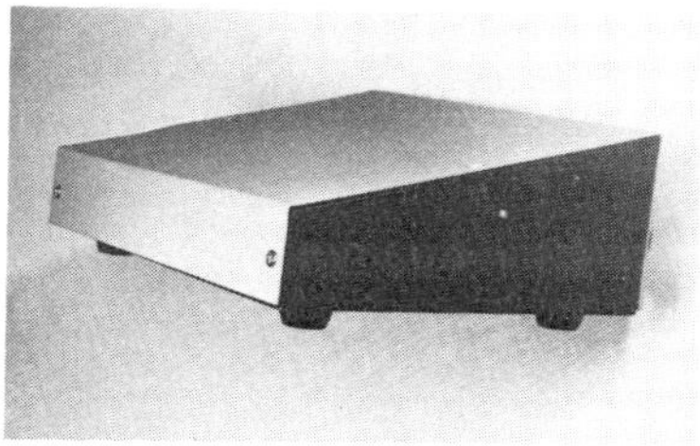

(D) Console.

Fig. 11-1. Typical electronic enclosures.

practical when a great deal of panel space is needed for controls and displays. Cabinets with a sloped front are also known as consoles.

5. Open chassis (Fig. 11-2) — A simple chassis provides a metal frame for mounting components, but does not serve as a complete enclosure. The chassis is usually mounted inside a larger cabinet. (One or more sides of the cabinet are often open to allow part of the chassis to show through, such as a control and display panel.)

6. Industrial cabinets (Fig. 11-3) — Large, industrial electronic equipment is packaged in tall, upright cabinet racks, complete with casters, blowers, and hinged rear doors. Modular cabinets slide into the racks, with standard front panels 19 inches wide. Industrial cabinets can be pulled forward on rails for maintenance, and cable connections exit out of the way to the rear.

7. Card cage (Chapter 1, Fig. 1-28) — A framed cage assembly is often used to package circuit boards inside a larger cabinet. A motherboard or other interconnection system must be attached at the back of the cage to distribute power and signals between the boards.

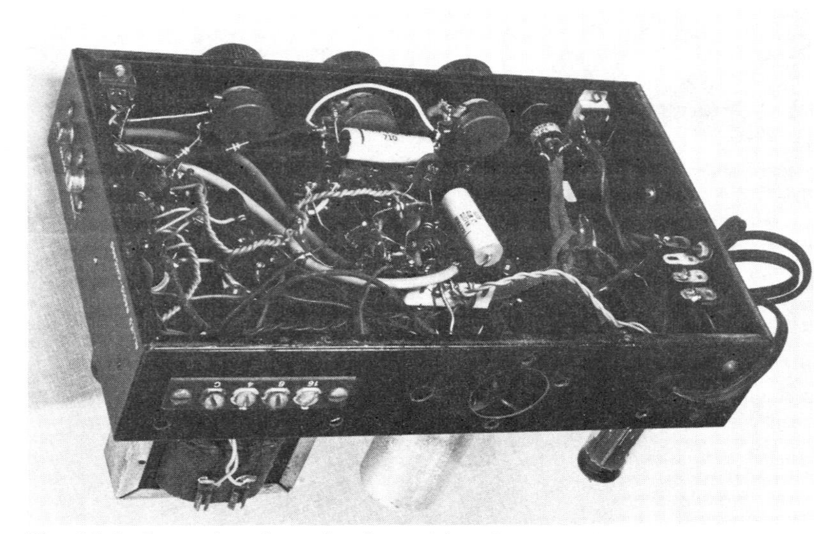

Fig. 11-2. Open chassis packaging with point-to-point wiring.

Electronic enclosures are fabricated from sheet metals, such as aluminum and steel, and from molded plastics. Steel chassis are the most rugged, but they are difficult to work with using hand tools and must be given a protective finish to prevent rusting. Plastic cases are inexpensive and lightweight, but they are subject to cracking if holes are improperly drilled. Aluminum enclosures made from 16-, 18-, or 20-gauge metal are probably the best all-around cabinets and chassis for prototype electronic construction. Aluminum is easily machined and filed, it provides sturdy support, and it has a nice appearance with or without a special finish.

PARTS LAYOUT ● After choosing an enclosure, component and hardware placement become the next phase of packaging design. Both electrical, mechanical, and human factors must be taken into consideration to obtain reliable system performance. As an example of the problems that can occur, the author was recently involved in the construction of a digital thermometer kit from a mail-order electronics company. The circuitry performed well after assembly and calibration, and it was mounted in the small plastic case provided. However, within a few hours the temperature scale drifted considerably away from the calibration point. Two variable resistors on the internal circuit board were adjusted repeatedly for proper range scale and zero, but the calibration would always change once the device was returned to its enclosure. Suspecting thermal problems, the thermometer box was placed directly in front of a wall heater, and the readout began immediately to drift. Three design flaws were responsible for the poor performance of the system:

1. The current reference circuitry was not sufficiently compensated for ambient temperature changes.
2. The voltage regulator produced too much heat for a small, compact case with no provision for cooling.
3. Calibration adjustments could not be made from outside the assembled case. When the case was opened, its steady-state operating temperature dropped considerably.

Any or all of these problems could have been corrected to improve the ability of the thermometer to stay calibrated. In actuality, a simple packaging change was made. A large number of holes were drilled in the plastic case to provide air circulation, and also to make the internal adjustable resistors accessible with a small screwdriver. The thermome-

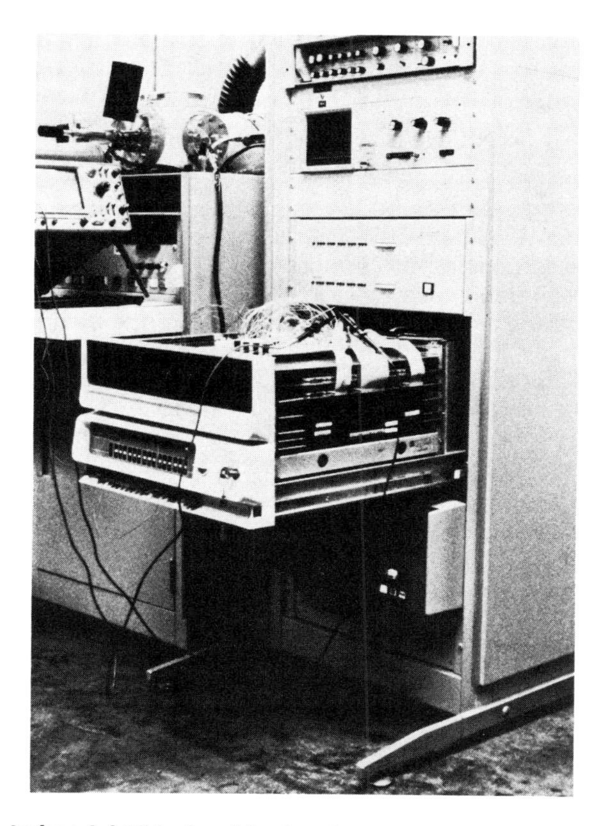

Fig. 11-3. Industrial 19 inch cabinet rack.

ter is now performing reliably, it will have a longer life due to a lower operating temperature, calibration adjustments can be made without disassembling the case, and the owner is pleased. Because no electronic improvements were made to assure current reference stability, it is expected that ambient air temperature changes will still affect calibration, but the device will be used indoors where room temperature is fairly constant year-round.

We will now discuss the various enclosure packaging considerations involved in component and hardware placement.

Human Factors and Safety • *Human factors* relate to design considerations that influence the ease and accuracy with which electronic equipment can be operated by the user. *User* is defined as the person who operates the controls, as well as the service technician who works inside the cabinet during troubleshooting. *Safety* is an important part of human factors because electric shock is a routine hazard associated with line-operated circuits. If a device is to be acceptable, it must be:

- Simple to control and adjust
- Easy to read
- Physically acceptable (size, weight, appearance)
- Simple to connect to associated equipment
- Serviceable
- Safe

Most of these qualities are established during packaging, assuming that the circuit design itself is valid and stable, and that provisions for user-interactive elements have been included in the original plans.

Control Selection — Controls and displays are selected to fit an end application, and to provide the user with the level of adjustments and feedback that he or she needs to operate the equipment efficiently. The following examples will illustrate some typical control design decisions:

1. LED (light-emitting diode) readouts are chosen over LCDs (liquid crystal displays) when the display must be visible in a darkened room. A darkroom timer or bedroom table clock with LCD digits is fairly useless.

2. Some displays must furnish accurate information, while others are only necessary to alert the operator to set conditions. An analog temperature meter provides the driver with really useful information. It lets the driver know the thermostat operating point, when the engine is warm, how long it takes to get warm, when the car is overheated, the extent of overheating, and the general state of the cooling system. If the meter is observed routinely throughout

the life of an automobile, the driver can notice trends in performance and is alerted to potential problems before they occur.

3. Short-circuit and over-current protection are basically provided for with fuses if protection is not built into the electronic circuitry itself, but several hardware possibilities exist. If the line is at high voltage, such as incoming 117-V ac power, then a panel-mounted fuseholder that minimizes shock hazard is chosen. If short-circuits are expected to occur frequently, such as in a battery charger, then a circuit breaker is more desirable than a fuse for convenience sake. If protection is needed but problems are seldom expected, a simple internal fuse block is generally chosen.
4. If a great deal of power must be controlled by a single switch, a large heavy duty switch may be replaced with a miniature switch connected to a relay coil. This decision saves space on the front panel and avoids the need to route heavy wire or a high-voltage line to the panel assembly.
5. If precise control is needed in setting a potentiometer shaft position, a 10-turn potentiometer may be much easier to use than the common single-turn version. Another alternative is to use a large- and a small-resistance potentiometer in series as a *coarse* and *fine* control system.
6. Older minicomputers and microcomputers were often packaged with built-in front panels containing lights and switches for each data or address line. However, these controls are rarely seen in modern computers. The reason is that the average user is seldom operating at the machine language programming level any more, and consequently has no need to single-step the processor. In the event that maintenance is required, or the gutsy user is writing machine code, a logic monitor is usually available with better capabilities than a traditional computer front panel.

This last example is important because it brings up the question of including self-test functions in your equipment. Do not include superfluous controls if they are seldom needed and if test equipment is available to serve the same purpose; instead, design your package so that test points are accessible and so that test equipment can be readily connected, saving both money and packaging space.

Control Placement — The human factors to consider in arranging controls are mainly accessibility, ease of operation, and logical organization. Panels should not be so crowded that individual adjustments are difficult to make without disturbing other controls. Critical alignment or service controls should not be so easily accessible that they could be accidentally moved; a conscious effort should be necessary to alter their settings. Displays must be mounted where they can be easily seen, and common conventions should be observed when arranging switches and potentiometers — up is on, down is off, clockwise rotation is increased output level, and so on. Certain hardware elements, notably cable connectors, are most conveniently mounted at the rear of the enclosure to keep accessory wires and line cord out of the way. Do not mount control hardware at the top of a cabinet if you want to stack other equipment above it.

Safety — Safety precautions should be built into a package design both internally and externally. If the cabinet is metal, it should be connected to earth ground with a 3-wire line cord. Always fuse the 117-V ac hot wire as soon as it enters the enclosure to protect internal electronics and to prevent the possibility of fire; a short-circuited power transformer can overheat to the point of flames. Individual fuses are also advisable for each line originating at power supply outputs, such as the +5-, +12-, and −12-volt supply lines. Choose a fuse that is rated 50% greater than the maximum expected current load, and use a slow-blow type if surges are created when the equipment is turned on. Power supply fuses are extremely important for prototype equipment when it is first assembled and tested because of the likelihood for miswired connections and shorted circuit board traces; the fuses protect both the circuitry and the power supply. By limiting overload conditions to a split second, a fuse can limit damage to the single weakest component or can possibly avoid damage completely. The experienced designer will use fuses liberally and will not have to cover his eyes when the prototype assembly is first powered up (see Fig. 11-4).

It may be assumed that the average bystander will not open up an electronic enclosure and poke around, but internal safety features should allow parts to be accessible for service without the danger of electric shock or skin burns. All high-voltage points should be insulated or covered to minimize accidental contact with a screwdriver or hand. Common hazard points are power transformer input connections, and 117-V ac connections to the line fuse, on/off switch, and pilot light. When an entire circuit contains many high-voltage points, it should be enclosed in a separate grounded chassis attached to the main assembly, and clearly labeled *High Voltage.*

Geometry • The enclosure is initially chosen to give as much room as possible for component placement, construction work, and troubleshooting access. When hardware layout is being devised, many interrelated factors will affect the results; electrical effects, heat, accessibility, safety, and interconnection methods will all influence the final results. Perhaps the most important layout approach is to mount all parts on *one section* of the enclosure; interconnections between different sections become awkward when the cabinet is disassembled for construction or testing. The optimum packaging system is composed of two basic parts, which can be termed the *cover* and the *chassis*. No parts are mounted or connected to the cover, and when it is removed, all of the internal parts should become accessible. The chassis may be composed of a number of sub-units, such as the bottom half of the cabinet, a power supply frame, a PCB card cage, and a front control/display panel. However, the sub-units should never require disassembly unless major repairs or basic construction changes are needed.

Fig. 11-4. Powering up a prototype assembly.

Internal layout geometry for a typical prototype assembly is illustrated in Fig. 11-5, a digital pulse counter. All parts are mounted on the bottom half of a two-piece utility box. The basic sections include the following:

1. Circuit board — The majority of electronics is packaged on a PCB, which is mounted on the base of the box using standoffs.
2. Bulky components — Large, heavy parts are attached to the base and back of the box with machine screws, and are positioned around the PCB as space permits. These parts are related to power supply functions. Four rubber feet are attached to the bottom of the chassis, giving clearance for machine screws extending

Fig. 11-5. Typical prototype parts layout.

through the base. Bulky components should always be mounted so as to produce an even layout with balanced weight distribution.

3. Front panel — The front side of the chassis serves as a control/display panel. Items mounted on this panel are parts that must be routinely accessible to the equipment operator, including an LED display, control switches, and input/output signal connectors. Note that the display components are all mounted on a small, separate PCB, following the principle of modularity.

4. Back panel — Certain parts are routinely mounted on the back side of a chassis, and they are usually input/output connectors which must be occasionally but not routinely used by the equipment operator. Typical back panel items include the 117-V ac power cord, a line fuse, and external signal connectors.

5. Interconnection system — Internal wiring consists largely of hand-soldered hookup wire connecting the main circuit board to other sections — back panel, front panel, and chassis-mounted power supply components. The display PCB is permanently connected to the main PCB with a short, 60-wire flat cable.

An example of undesirable hardware layout is illustrated in Fig. 11-6. Circuit board and transformers were mounted on the bottom half of the enclosure, and the top half was used to mount front and back panel components. That approach is awkward because the box cannot be opened without disgorging a tangle of interconnecting wires. Packaging could be improved somewhat by neatly connecting the enclosure halves with a single, multiconductor cable having a plug and socket at one end, but wire lengths become long and signals may become cross-coupled unless discrete, shielded cable is used to separate critical wiring. In general, poor planning in hardware placement can only be salvaged by a corresponding increase in wiring complexity.

The specific geometry chosen for mounting electronic hardware inside an enclosure will depend on many factors besides space, size, and weight, but it is usually best to position the largest parts first, and smaller ones are fit in around them. Physical strength of the chassis should be considered when locating heavy parts and items that involve torque or pulling forces. For example, some jack connectors may subject a front panel to considerable pulling and pushing pressures when plugs are inserted and removed. Front panel design will be discussed more in a later section, but be sure to allow clearance for front panel hardware to protrude inside the enclosure.

Fig. 11-6. Undesirable layout and wiring.

Heat • *Heat* is a major physical problem in electronic systems. If its effects are not properly accounted for in packaging, the performance, reliability, and life of equipment will be drastically affected. Manufacturers rate the reliability of individual components in terms of *mean hours before failure,* and this characteristic is directly related to operating temperature. Dynamic circuit performance can also be affected; for example, the oscillation frequency of crystals is dependent upon temperature. Heat effects can be controlled with three common methods: convective cooling, forced-air cooling, and proper component placement.

Convective Cooling — Convective cooling is the natural tendency for air to circulate and remove heat from hot surfaces, and in many cases it is sufficient to ensure cool, trouble-free operation of electronic equipment. A typical guideline for convection is to allow 6 cubic inches of open space for each watt of heat power to be dissipated inside an enclosure. However, this rule assumes that components are fairly evenly distributed in the space, and that vents or louvers are provided in the enclosure to allow air to circulate. Extra volume must be allowed if the cabinet is completely sealed, and localized heating will occur if components are too tightly spaced. Hot air rises, so air holes should ideally be located at the top of an enclosure; however, due to the tendency for dust to accumulate, side holes are a frequent compromise position.

Miniaturization of circuitry and printed circuit interconnection techniques have led to some new problems in heat removal. Localized heating often occurs when the high planar density of PCB packaging is combined with the vertical density of mounting many parallel PCBs in a card cage assembly. This problem has been one of the motivations for developing new types of integrated circuit elements that consume less power, generate less heat, and require smaller, lightweight power supplies.

Some individual semiconductor components generate a great deal of heat, such as voltage regulators, power transistors, and SCRs (silicon controlled rectifiers). A metal *heat sink* is used to dissipate heat from these devices through the convection process, preventing localized heating and component destruction. Most power transistors have their collector element internally bonded to the metal case for this purpose. A heat sink may consist of a specially machined part, or the metal chassis itself can serve as a heat sink. Several types of heat sinks are illustrated in Fig. 11-7. Fig. 11-7A shows small heat sinks directly attached to components mounted on PCBs. Fig. 11-7B shows a large chassis-mounted heat sink used to dissipate considerable power. Power devices are commonly mounted on the external side of an enclosure to limit the need for internal cooling.

Heat sink theory is complicated because it involves three processes: transfer of heat from the device to its case, from the case to the sink, and from the sink to the air. Heat sinks are usually made from anodized aluminum with fins to increase surface area. Parts are attached tightly to the sink with machine screws, and silicon grease is used to increase thermal conductivity. If an electrical problem would be created by at-

(A) For low-power devices.

(B) For power transistor.
Fig. 11-7. Heat sinks.

taching a device directly to metal at chassis ground potential, then an insulated mica washer and plastic spacers are used to insulate the leads from surrounding metal parts.

Forced-Air Cooling — *Fans* and *blowers* are installed inside electronic enclosures when simple convection does not provide enough cooling capability. The popular 4 inch 117-V ac box fan is illustrated in Fig. 11-8. The use of fans may create new packaging problems because *filters* are often required to prevent dust accumulation inside cabinets. The fan is mounted so that air is sucked in through holes in the cabinet, and then through the filter material. Other cabinet holes allow hot air to exit under positive pressure. Fans are effective cooling devices if air filters are kept clean.

Component Placement — Components that produce heat must be located away from devices that are temperature sensitive. This was

Fig. 11-8. A 4 inch box fan.

previously noted in Chapter 3 concerning PCB layout, but it applies equally in positioning chassis-mounted parts. Power transformers, rectifiers, and power transistors should all be located at a distance from circuitry whose dynamic electrical performance depends on operating temperature. In such cases, temperature *fluctuations* may be just as undesirable as *high* temperatures. Precision crystal oscillators in communication gear are often packaged in a temperature-controlled metal can with an internal heater to hold the circuitry at a steady operating temperature above ambient.

The user should keep in mind that hot air rises. If electronic equipment is stacked in industrial racks, the hottest units should be placed at the top. Similarly, inside a small enclosure it is poor practice to mount a power supply at the base and circuit boards above.

ACCESSIBILITY • Accessibility is important during three phases of the life of an electronic product: construction, operation, and service.

Construction • Accessibility during construction means simply that there should exist enough working space to assemble the equipment without extraordinary efforts or uncommon tools. Extremely compact assemblies make compromises in this area by requiring parts to be assembled in a specific order. A common difficulty during assembly is being unable to reach or hold a nut while a machine screw is being tightened. For appearance sake, screws are inserted from the outside of the chassis, which also ensures room to hold a screwdriver. However, the nuts may be well-concealed inside a tightly packed enclosure. One solution to this problem is the use of *sheet metal screws* that do not require nuts, but these screws do not form strong connections and the sheet metal hole is easily stripped. A better approach is to consider mechanical attachments as an important part of packaging design, and keep nuts accessible. When problems are unavoidable, nuts can be permanently attached to metal surfaces by spot-welding, soldering, or gluing with epoxy adhesive.

Operation • Certain devices must be accessible during the routine operation of an instrument. Typical examples include switches, potentiometers, fuses, and connectors. Once it has been decided that a particular device needs to be available to the operator, it is mounted on the front or back panel of the enclosure, depending upon expected frequency of need. Input/output connectors are usually positioned on the back because this avoids a tangle of wires exiting from the front of the cabinet. However, if a connector is used regularly, such as the input signal jack for a piece of test gear, then accessibility dictates that this item should be front-mounted.

Control panels can become overly cluttered with unnecessary hardware that hinders construction and wiring. The designer must, therefore, exercise judgment when deciding which items will be operator-accessible. A good illustration is the use of the following potentiometers:

1. Primary controls — When a potentiometer is used frequently in the routine operation of equipment, such as with a volume control, it should be accessible on the front control panel as a shaft and knob.
2. Calibration controls — Some potentiometers are used to make calibration and operating point adjustments infrequently, but they must not be disturbed once they are set. These controls are accessible on the back panel (see Fig. 11-9), and a screwdriver is specifically needed to turn the shaft. An example would be the coarse gain control for an instrumentation amplifier. When a variable resistor is only used for maintenance purposes or infrequent calibration, it should not be readily accessible to the operator. Such controls are often mounted inside the cabinet, but with an access hole that permits the technician to insert a small tool from outside the cabinet. A typical example would be the zero adjustment for a meter display.
3. Alignment controls — Some potentiometers are never adjusted after initial circuit alignment, unless an active component in the circuit is replaced. These variable resistors should be confined to

the circuit board, completely inaccessible unless the cabinet is opened. The trim resistors illustrated in Chapter 3, Fig. 3-7 are used for offset adjustment of an IC op-amp circuit.

The reader can probably think of other devices where hardware style permits a range of accessibility. Fuses may be available as a panel-mounted fuse holder, a chassis-mounted fuse block, or a PCB-mounted clip. Switches are accessible all the way from front panel toggles to miniature PCB-mounted DIP switches.

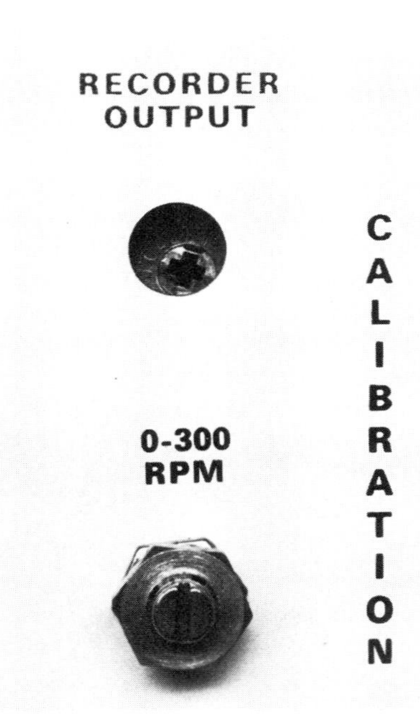

Fig. 11-9. Limited accessibility of calibration potentiometers.

Service • Electronic equipment will always require testing, adjustments, calibration, troubleshooting, and repairs. These service operations are particularly important for experimental and prototype designs that are rarely expected to function perfectly when first turned on. Be realistic when packaging new designs and assume that modifications will be needed. One of the first requirements is that the circuit be *electrically operational* with the cover removed and all parts accessible for testing. Unfortunately, the compactness of some equipment requires that electrical connections be broken in order to bring certain circuitry into the open; in such a case, special wire harnesses and circuit board extenders may be needed to permit *in-circuit* testing. Try to avoid the need for such equipment.

A well-designed enclosure should allow service adjustments to be made without complicated hardware disassembly. Balanced, roomy layouts are preferred. All adjustable components, fuses, and test points should be accessible to tools, probes, and visual inspection. If certain devices are unavoidably hard to reach, cut an opening or access hole if adjustments are likely to be needed. Most technicians have stories to tell about spending hours taking apart a chassis in order to change a simple pilot light bulb or a rubber belt.

When problems are discovered and repairs are needed, some disassembly is often necessary. Modular construction and simple wiring will help during repairs. The optimum packaging design will allow any individual module to be removed by detaching a few screws and unplugging a cable connector. Plug-in circuit boards with edge connector contacts are popular because they combine a modular construction approach with simple serviceability. In general, try to design prototype equipment so that desoldering is seldom required to remove basic chassis elements from the enclosure.

PANELS • The areas of an enclosure used to mount externally visible components are termed *panels.* Panels may or may not be separate pieces from the chassis or cabinet, but a flat detachable plate is the ideal design with respect to fabrication and assembly. Commercial cabinets are often constructed with separate panel plates to simplify machining, finishing, screen-printing, and final wiring. That is also consistent with a modular approach in packaging. Simple enclosures that do not have separate flat panels can be modified to incorporate this feature; a large rectangular hole is cut out in the metal enclosure and covered with a flat aluminum plate attached with machine screws.

Aluminum sheet metal is the best material for panel construction because it is easy to machine. A great number of openings, cutouts, and holes will be required to mount electronic hardware. Aluminum also provides a fine appearance without special finishing techniques. For example, the popular *brushed aluminum* finish can be obtained by simply rubbing the surface in parallel strokes with fine sandpaper or steel wool. The actual tools and techniques used to form holes in aluminum panels will be discussed in a later section.

Panel Devices • Panel-mounted devices fall into four categories: switches, potentiometers, displays, and connectors. These parts should be chosen for practicality and for a unified, pleasing panel appearance. Miniature hardware consumes less space, but at a higher cost. One exception is with LED displays, whose small size, simplicity, and low power requirements have taken over the market from traditional incandescent or neon bulbs. In general, choose parts with simple mounting requirements — a round hole, and a machined nut/washer to secure the device. Small rectangular holes are much more difficult to cut neatly than round ones, so choose toggle switches instead of slide switches. Avoid items that may be difficult to remove from the panel after installation, such as press-fitted pilot lamps.

If panel space is limited, try to combine controls and eliminate extras. An on/off switch can be combined with a potentiometer, and a pilot lamp may be unnecessary if other displays become lighted when power is applied. Take advantage of modern miniaturization. Unique LEDs are now available that can display three separate colors, and a small LED bar graph display may serve the same purpose as a large, conventional mechanical meter. If standard panel meters are used, remember that custom scales can be inserted inside the faceplace. Using artwork and photographic techniques described in Chapter 4, special designs can be transferred to photographic paper that is mounted inside the meter movement.

An interesting modern approach for panel displays is illustrated in Fig. 11-10. Instead of attaching individual devices directly to the front panel, they are assembled on a circuit board, which simplifies wiring and extends the modularity principle. This approach is particularly practical for LED devices. The display PCB is mounted flush against the front panel, which contains openings for lights to shine through. Sometimes a single rectangular opening is cut in the panel and is covered with transparent, dark plastic to reveal an intricate display PCB mounted from inside the cabinet. A popular hardware item for mounting digital LED displays is known as a *bezel assembly*. The bezel resembles a miniature metal frame that mounts on the outside of a rectangular panel cutout, concealing imperfections around the hole. The bezel includes an adapter to mount a PCB on the opposite side of the panel.

Fig. 11-10. Panel-mounted PCB display.

Panel-mounted connectors are necessary to bring signals in and out of electronic enclosures. Connector hardware has become fairly standardized and the designer would do well to follow conventional practices; that will ensure compatibility with other electronic equipment. For example, audio signals use *phone* plug/jacks for high-level signals, and *phono* plug/jacks for low-level signals requiring shielded wire. *Feedthrough barrier strips* are one of the simplest systems for connecting external wires to screws, and are usually mounted on the back panel (see Fig. 11-11). Barrier strips are commonly used for speaker wires and multisignal control cables. Radio frequency signals require special shielded hardware, such as the standard *BNC connectors* used on oscilloscope panels and other test equipment.

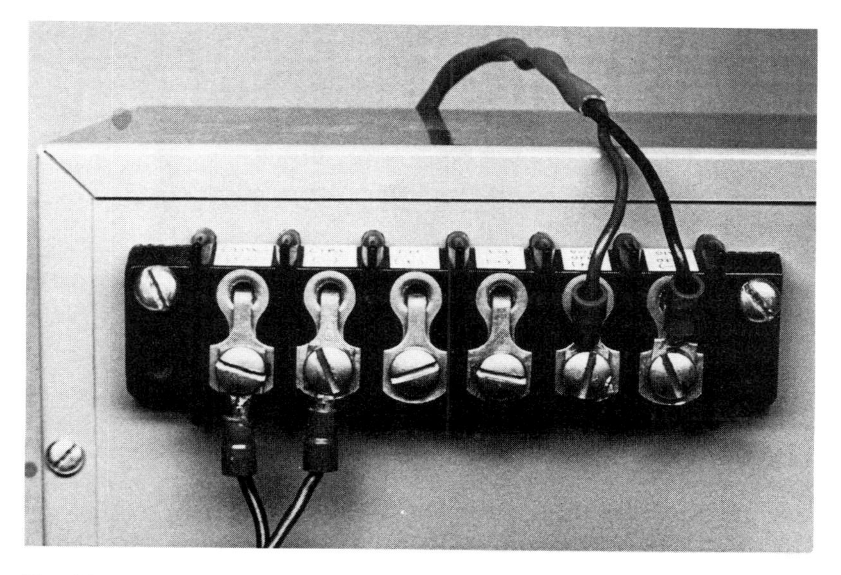

Fig. 11-11. Back panel feedthrough barrier strip.

Computer-related electronics have produced a variety of new connector hardware due to the large number of discrete signals that must be distributed between devices. The majority of this hardware mounts directly on circuit boards instead of panel plates, and was discussed in Chapter 1. However, standard interface connectors for serial communications systems (such as RS-232) are available in panel-mounted hardware form, as well as connectors for flat ribbon cable up to 40 wires wide.

Panel Layout • A typical electronic control panel is illustrated in Fig. 11-12A. Hardware is positioned to give a pleasing, balanced appearance with logically related devices grouped together. Control panels should emphasize ease of operation, and accessibility without crowding. Remember in prototype designs that additional controls may be added to the panel later, so leave extra space. If an adjustable component is seldom used, position it on the back panel or mount it inside the enclosure. The use of miniaturized hardware makes more efficient use of panel space, but it may be more difficult to assemble and wire. If possible, mount display elements in circuit board fashion, using standard PCB layout techniques.

(A) Hardware is positioned to give balanced appearance.

(B) Control panel wired point-to-point style.
Fig. 11-12. Typical electronic control panel.

Panel Wiring • An internal view of a control panel wired in point-to-point style is shown in Fig. 11-12B. All of the wires were individually soldered to panel component terminals, and brought together to form bundled cables that were routed to various circuit boards. This approach is ideal because it allows plug/socket connectors to be installed in-line wth cables, permitting the front panel to be detached as a modular unit. In most cases wiring systems are complicated by the need to separate certain critical signals that could create interference or noise problems. For example, the 117-V ac power line connected to an on/off switch cannot be run next to an unshielded, low-level input signal. Typical front panel wiring, therefore, uses several types of wire cable exiting the area to other parts of the enclosure, where they are soldered or terminated at connectors.

ELECTRICAL CONSIDERATIONS IN PACKAGING • The electrical factors discussed in Chapter 3 for printed circuit layout are equally important in chassis layout, although greater distances between components are involved. The most important electrical design task is to separate components and wiring such that high-level ac signals do not become coupled with low-level signals through capacitive or inductive effects. For example, a power transformer should be positioned away from input leads or an antenna. A common problem in audio circuits is *ac hum* induced into the amplifier because of indiscrete wiring runs. Special types of wiring can be used to reduce cross-coupled signals, such as *shielded cable* for low-level signals, and *twisted pair* wires for canceling electromagnetic radiation from ac signals. Many digital signals are effectively high-frequency ac lines that require special consideration in long wiring runs.

Packaging of radio frequency (rf) circuitry is a difficult problem because of the high degree of signal radiation and the crucial importance of lead lengths. Some rf circuits require a special metal shield for complete isolation, such as the metal boxes used to house radio and tv tuning units.

Grounding is another electrical problem in packaging. Most circuits require frequent connections to a ground reference, usually established from some point in the power supply section. The metal chassis is commonly connected to this same point. If ground connections are made randomly from various circuits to chassis metal, current patterns will be created in the chassis and some areas will be at different electrical potentials than others due to the small, natural resistance of aluminum or steel. The result is known as a *ground loop,* and it causes unpredictable coupling between the various signals involved. Chassis ground loops can be prevented by bringing all circuit ground connections back to essentially a single chassis connection point, and tying that point to power supply ground.

CHASSIS INTERCONNECTIONS • The basic method used for internal enclosure connections is *point-to-point wiring,* as opposed to *printed circuit* techniques at the circuit level. The most effective approach in chassis wiring is to isolate electrical hardware into physical or functional groups, such as PCBs, front panel, rear panel, and power supply. Each group is assembled separately and the final interconnection system consists of cables and wires between the modular groups. The use of suitable connectors at the origin of each module is recommended because it allows whole units to be separately installed or removed for service. An overall goal in wiring should be to design connections such that any discrete component in the entire assembly can be removed from the system, with a minimum need to desolder and resolder wired connections.

Point-to-Point Wiring • Point-to-point wiring is an electronic assembly approach where each connection is completed manually by soldering hookup wire or wire cable between individual component terminals and solder lugs. In many cases the components themselves are soldered between convenient lugs and tie points. This interconnection method was the standard 30 years ago for even the most basic circuitry, before solid state and miniaturized components made it impractical for the bulk of modern wiring. As illustrated in Fig. 11-2, an open chassis was the common packaging medium in the 1950s for securing heavy components, such as transformers, capacitors, and tubes; point-to-point wires and smaller components were attached underneath the chassis using socket lugs and terminal strips as tie points. Because of the physical size of the bulky top-mounted parts, plenty of space was available for this sort of component interconnection. Point-to-point wiring is still used today in heavy power supplies, on control panels, and as a means to connect individual bulky components that cannot be directly attached to circuit boards. However, large, heavy devices are relatively few in number and do not often require a separate chassis for support; power transformers, for example, may be directly mounted to the base of a metal cabinet.

Hookup Wire • Hookup wire for electronic assembly is typically AWG 10–26 in American Wire Gauge units, and is made of solid or stranded copper with a tin coating for improved solderability. Stranded wire is preferred because it is more flexible and less likely to break. Various stranded and solid wire sizes are compared in Table 11-1 along with their recommended current-carrying capacities. A good standard choice for most point-to-point wiring applications is AWG 22 stranded hookup wire with polyethylene or polypropylene plastic vinyl insulation. Other types of insulation are available for special high-voltage or high-temperature applications. Insulation comes in many colors, solid and striped, creating a practical means of keeping track of wire signals; use many different colors in your assemblies to eliminate wiring errors and to simplify troubleshooting.

Wire Cable • Multiwire cable is used for parallel signals having similar origin and termination points. Cable is often a neater wiring

Table 11-1 Current-Carrying Capacities of Solid and Stranded Wire

AWG Number	Solid Wire Diameter (inch)	Stranded Equivalent*	Maximum Current (amps)
10	0.102	37/26	20.5
11	0.091		16.5
12	0.081	19/25	12.0
13	0.072		10.0
14	0.064	19/27, 37/29	8.0
15	0.057		6.5
16	0.051	19/29, 27/30	5.0
17	0.045		4.0
18	0.040	7/26, 19/30, 27/32	3.0
19	0.036		2.5
20	0.032	7/28, 19/32	2.0
21	0.028		1.5
22	0.025	7/30, 19/34	1.0
23	0.023		1.0
24	0.020	7/32, 19/36	0.8
25	0.018		0.6
26	0.016	7/34, 19/38	0.5
27	0.014		0.4
28	0.013		0.3
29	0.011		0.25
30	0.010		0.2

*First number is number of strands, second number is AWG of each strand.

material than many individual wires run separately and then bundled together; cables may also impart certain electrical properties to the internal wires. The following types of wire cable are common:

1. Multiple round wires — This type of cable is simply a number of insulated, specified-AWG hookup wires held together by an outside insulated jacket. Each wire has a different color, and may be solid or stranded copper.

2. Twisted pairs — Multiple twisted pairs enclosed in an outer jacket are commonly used to eliminate noise in parallel, high-frequency digital signals. Each pair is used for a separate signal and ground connection.

3. Shielded cable — Shielded cable is enclosed in a braided wire jacket or wrapped aluminum foil for noise protection of internal signals. The shield may also have an outer plastic insulation, and is

generally grounded in the circuit. *Coaxial cable* is a shielded cable where the characteristic impedance between signal and shield is carefully controlled by the type and thickness of insulation material, a useful feature for critical rf connections. Many combinations of shielding, twisted pairs, and standard round wires may be used to form special-purpose cables.

4. Flat ribbon cable — Multiple round insulated wires may be bonded together to form a flat cable instead of a large, round cable. Ribbon cable is seen in Chapter 1, Fig. 1-26. It was developed for digital applications where cross-coupling between parallel signals must be minimized. Individual wires are very small gauge because current-carrying capacity is not important. Ribbon cable is ideal for circuit board termination, since the wires are spread out in a flat, planar geometry.

Wiring Tools • Common hand tools used in electronic wiring are shown in Fig. 11-13 and described as follows:

1. Wire strippers — The job of wire strippers is to remove insulation without nicking the wire, which would weaken it. When the handles are squeezed together, stripper blades cut part way through the insulation but stop short of reaching the copper wire. It may be helpful to rotate stripper blades around the wire if insulation is particularly tough, and then pull the wire away to strip the plastic. Small wire strippers are adjustable for various wire gauges, typically AWG 12 to 30.

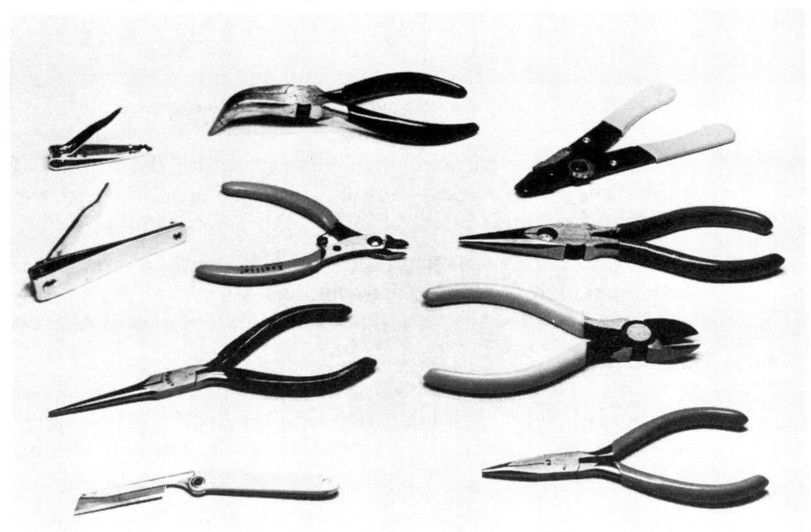

Fig. 11-13. Electronic wiring tools.

2. Needle nose pliers — Designed for bending, shaping, holding, grabbing, and inserting wire and parts in cramped quarters, electronic pliers are available in many sizes, weights, and styles. The tip of these pliers is most important; ends should be precisely machined to identical thickness, and should grip small objects tightly.
3. Cutting tools — Diagonal cutting pliers are the standard wire-cutting tool. Other useful cutting tools for trimming insulation and cutting small wires are the razor knife and fingernail clippers.

Wiring Techniques • Wire soldering methods were described in Chapter 10. In general, the best wiring approach is neat in final appearance and makes use of connectors at appropriate points to allow simple installation and removal of modular circuit assemblies. Point-to-point connection made with hookup wire is the simplest approach for wiring front panels or chassis containing many components with eyelet terminals. A neater variation is known as *square-corner wiring,* where all wires are made to run in parallel or perpendicular directions, as illustrated in Fig. 11-14A. A *dressed* layout of this type is time consuming, but it lends itself well to forming neat bundles with central paths and organized branches.

Cable ties and *clamps* are valuable for keeping wires in neat bundles and for routing them close to chassis walls, secure and out of the way. Waxed linen strips were formerly used to bundle wires into neat harness assemblies with hundreds of tiny hand-made knots. A quicker, more recent approach is to use plastic cable straps, such as the ones shown in Fig. 11-14A.

When a complicated chassis with many parts and many connections must be wired, the neatest assemblies are produced by forming wires and cables into a *harness assembly* before even beginning to solder individual wires. Commercial harnesses are produced on a custom plywood jig with nails representing bends and component terminals. Wiring is laid out on the nails and bundled together with cable ties or lacing; after stripping each wire end, the harness is installed in the enclosure and connections are soldered. A similar approach can be adopted in prototype wiring by using the assembled chassis itself as a wiring jig. Extra long wires are laid out and routed in square-corner fashion to each component terminal, where they are loosely connected without stripping the ends. Wires are then adjusted into neat bundles and tied together until a complete harness is formed. At this point the wire ends can be cut to final length, stripped, and connections completed by soldering.

Heat-shrinkable tubing is a valuable accessory to wiring techniques. This special plastic insulating material is slipped over wires, joints, and connectors, and when heated, contracts to form a tight seal.

Portable electric heat guns are used to apply heat, but in many cases a disposable butane lighter is handier. Shrinkable tubing is useful for bundling wires, covering bare wire, insulating soldered connections, adding mechanical strength to joints, shutting out moisture, and insulating closely spaced terminals on connector plugs and sockets. This last application is illustrated in Fig. 11-14B.

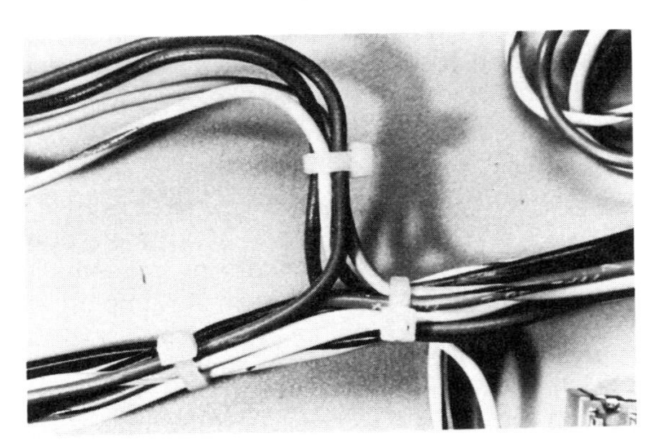

(A) Dressed square-corner wiring with plastic cable straps.

(B) Heat-shrink tubing for protecting connections.

Fig. 11-14. Techniques for neat wiring.

Connector hardware is used whenever wiring connections must be broken without requiring desoldering. Most connection systems can be classified roughly as plug/socket or screw terminal types, and were discussed in detail in Chapter 1. Another useful connection device that is particularly common in chassis wiring is the Molex plug/socket for round wires shown in Fig. 11-15. Molex connectors are available for almost any size and number of wires, and provide a convenient in-line quick-disconnect for wire bundles connecting various modules in a chassis assembly.

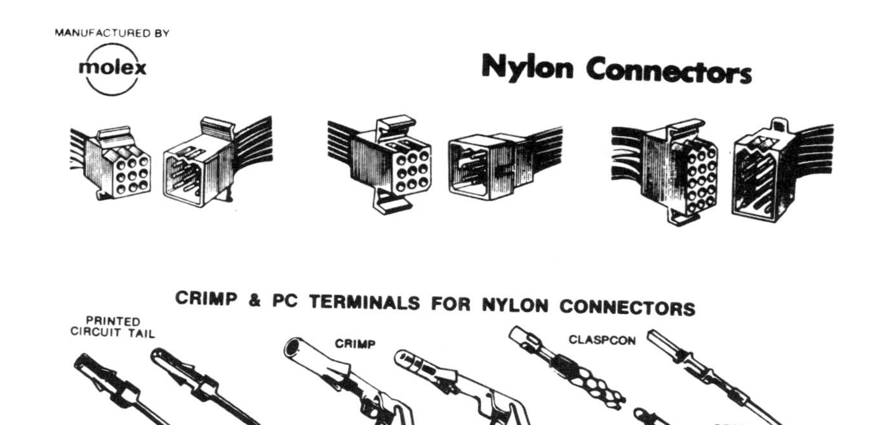

Fig. 11-15. Molex plug/socket in-line connectors.

OPENINGS AND HOLES • Construction of prototype electronic equipment with a professional final appearance depends on the ability to machine neat, accurate openings and holes in enclosures, chassis, and panels. Before discussing the appropriate tools and methods for this task, the following basic principles should be noted:

1. Choose enclosures such that holes are made in aluminum sheet metal, not steel. Working steel with hand tools is an extremely difficult undertaking.
2. Machining a flat panel plate is easier than cutting holes in a wobbly, box-shaped enclosure.
3. Avoid the need to cut small rectangular openings or large round holes. Small round holes and large rectangular openings are easier to form.
4. Choose panel-mounted hardware that overlaps and conceals the edges of mounting holes. Mistakes, gouges, and rough edges can be well hidden with such hardware. Exposed cutout edges do not look good unless they are perfectly cut.
5. Prepare a complete parts layout before attempting to start cutting.
6. Use the proper tools for the job.

Panel Preparation • After designing a panel layout, make a full-size drawing on graph paper or gridded drafting vellum, showing the location and size of each hole. This drawing can be temporarily pasted on the sheet metal with rubber cement, which is easily removed later.

In addition to protecting the metal finish, graph paper will allow holes to be perfectly positioned without scribing lines on the metal surface. The center of each hole to be drilled should then be *center-punched.* The center punch is a tool that produces a small dimple in the metal to serve as a drill bit guide. Flat panel plates are much easier to work with than cabinets because they can be laid out on a bench top without special supports or clamping, and they are easier to secure in a vise.

If front panels are to be screen printed, consider the possibility of screening the design *before* drilling holes, eliminating the problem of aligning printed labels with previously machined openings. Guides in the screened design can then be used to accurately locate and punch holes. If the printed labels are ruined during machining, they can always be stripped and reapplied after holes are cut.

Measuring Tools • Three basic measuring tools for sheet metal work are illustrated in Fig. 11-16. The *metal rule* is useful for scribing lines and making rough measurements. *Calipers* are used to make fairly wide, accurate measurements, or to measure inside hole diameters. The *micrometer* is able to make extremely accurate measurements of electronic hardware dimensions, and is very helpful in determining the proper size for machining mounting holes.

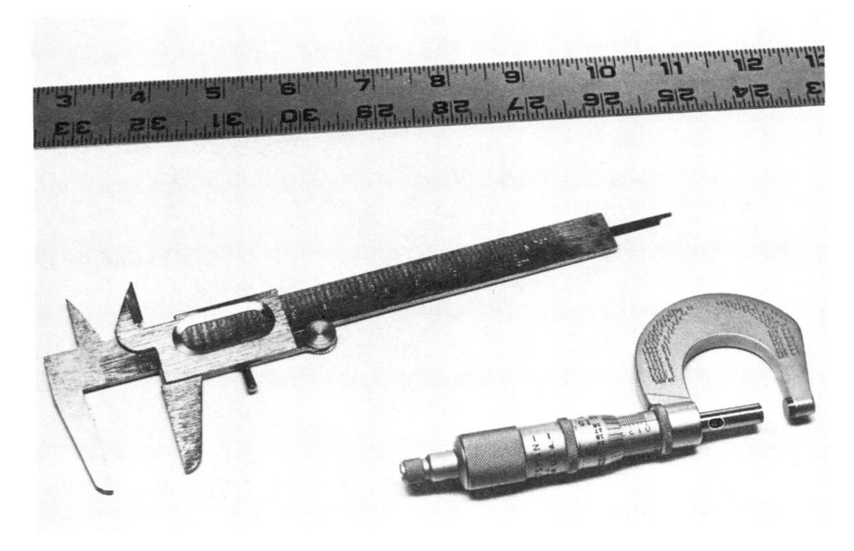

Fig. 11-16. Measuring tools for sheet metal work.

Drilling Holes • The electric hand drill is your basic tool for making holes in sheet metal. A large, expensive drill press is nice to have, but the portable drill may be easier to use in tight spots. In any case, large drill bits are not the ideal cutting tools for making neat chassis holes. The result of drilling sheet metal with a large twist drill bit is shown in Fig. 11-17; the bit tends to vibrate in the hole and an enlarged, squarish opening with excessive burrs is formed. Special sheet metal bits are available to improve the hole quality, but two more common techniques are useful to avoid this problem. First, drill holes by proceeding in *incremental* steps with larger and larger drill bits. This method tends to keep the holes round and centered at the original punch mark. Second, drill holes purposely undersized and enlarge them to final diameter with a *hand reamer,* which is the tool pictured in Fig. 11-17. The reamer gives a nice round final hole with a minimum amount of burring, and it can be used to produce holes from one-eighth to thirteen-sixteenth inch in diameter. As sheet metal becomes thicker, the reamer is less desirable for forming holes because it requires a great deal of manual strength to turn, and the final hole is slightly tapered. Taper can be minimized by reaming alternately from both sides of the aluminum metal.

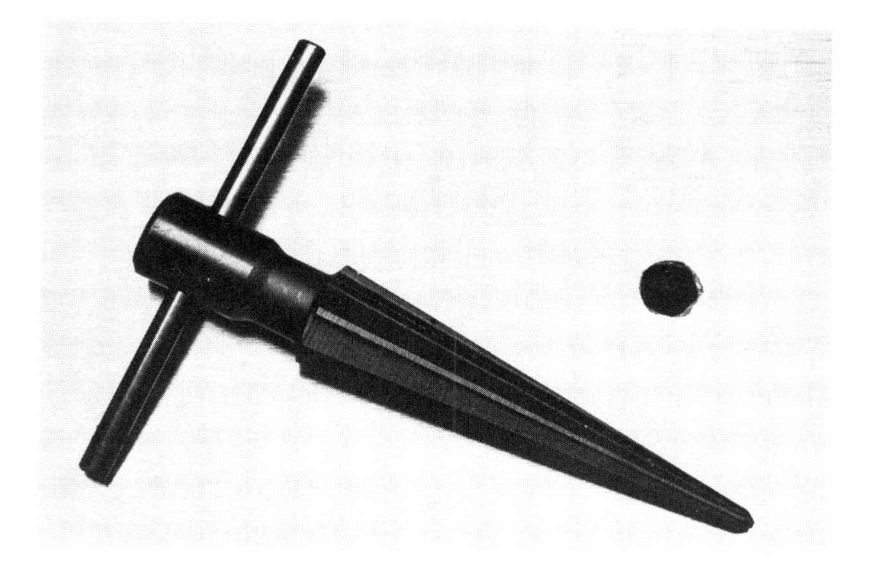

Fig. 11-17. Some holes need to be reamed.

Punching Holes • *Chassis punches* are the best tools for forming small to medium-sized openings in sheet metal. Two types are shown in Fig. 11-18: the Greenlee Tool Co. style, and a Roper Whitney No. XX hand punch. Both types can be used to produce clean, burr-free holes in aluminum or steel sheet metal.

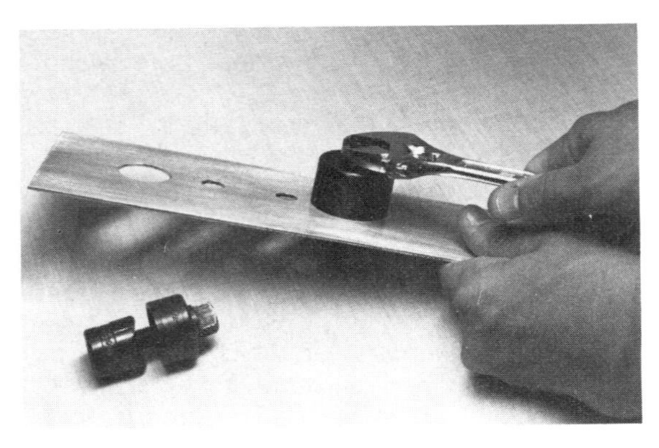

(A) Greenlee type.

(B) Roper Whitney No. XX hand punch.
Fig. 11-18. Chassis punches.

Cutting Holes • Large sheet metal openings can be cut with the *nibbling tool* shown in Fig. 11-19, which is particularly handy for square, rectangular, or odd-sized cutouts. A seven-sixteenth inch hole must first be made to insert the head of the nibbling tool. By repeatedly squeezing the handle, a smooth opening with flat edges can be cut in any direction, with maneuverability similar to that of scissors on cardboard. Guidelines should be drawn on the metal surface for the tool to follow; by cutting on the inside edge, the final opening edges can be shaped and finished with a flat file.

Another approach for cutting large rectangular openings is to drill holes in the four corners, insert a hacksaw blade, and try to cut the proper shape by following each side down to the next hole. Sawing is a cruder method than using a nibbling tool and will require a great deal of filing to produce acceptable results; soft aluminum is one of the few metals that can be worked in this manner. It is often difficult to hold a metal chassis securely in a vise during sawing, which produces a great deal of vibration and bending action.

Unless appropriate punches are used, small odd-shaped chassis holes are difficult to form neatly. An example is the rectangular hole needed for a small slide switch. This hole is too small for a nibbling tool or saw blade. The best approach (although tedious) is to drill away metal inside the hole outline as completely as possible with various-sized drill bits, and form the final opening manually with small files.

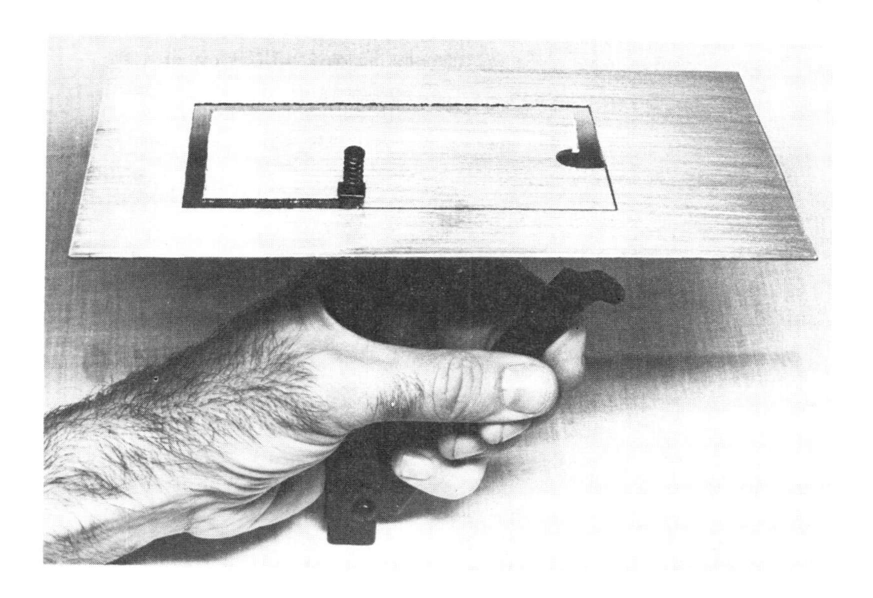

Fig. 11-19. Sheet metal nibbling tool.

Large, round openings are probably the most difficult designs to cut manually in sheet metal. A nibbling tool may be used to get the approximate final shape, but after filing, hole edges rarely appear perfectly round. Luckily, most meters that mount in round openings cover the edges considerably and will mask your mistakes. Some workers recommend *fly-cutters* for making large circular cutouts; this compass-shaped tool must be used in a large drill press, with the work perfectly aligned and securely gripped to the drilling platform, this tool may prove to be a disaster.

CABINET FINISHING • After machining holes and preparing an electronic enclosure for final assembly, coatings are applied for both rust protection and for a professional finish. Coatings may be applied before cutting holes, but finished surfaces could be marred while working with hand tools, and extra care is required when working with prefinished enclosures, such as off-the-shelf instrument cases.

Enamel paint is a standard finish applied to sheet metal cabinets because it is easy to use by spray methods. Many colors and styles of finish are possible depending upon the binder, carrier, dye, or pigment used; common styles are flat, glossy, metallic, and transparent. If the paint contracts or expands during drying, special finishes are possible such as crackle, spatter, hammertone, wrinkle, and flake. For small paint jobs, packaged aerosol spray cans are economical and simple to use. Epoxy paints form a more durable coating than enamels, but are more difficult to apply by spraying. Steel should always be painted to prevent rust, and aluminum may or may not be painted depending on the final appearance desired. In most cases the main electronic cabinet is painted a solid color, and front panels are left as plain or anodized aluminum. The basic steps in painting are surface cleaning, priming, and the use of proper spray techniques.

Cleaning • Cleaning is the most important step in preparing a metal surface to accept paint. Solvents are used to remove grease and dirt, mechanical scrubbing will loosen tough deposits, and acid/alkaline chemical cleaners are most effective for complete dirt removal. Dilute hydrochloric acid and 15–30% phosphoric acid are the standard cleaning agents for steel, and 10–20% sodium hydroxide is the standard cleaning chemical for aluminum. The alkaline aluminum cleaner actually etches the metal surface and thereby improves surface bonding by generating tiny scratches. After etching, aluminum will be covered with a black smut that can be removed by dipping in 50% chromic acid, or a mixture of 1/4/12 sulfuric acid/nitric acid/water with a small amount of hydrochloric acid added. Anodized aluminum should never be cleaned with acids or alkali before painting, since the anodized coating already forms a good base surface. Acid and alkaline cleaning chemicals are hazardous and must be used with caution and proper self protection (see Appendix D).

Primer • A primer coat is required for steel to prevent corrosion, and is necessary for aluminum to ensure adhesion of enamel paint. The paint manufacturer will recommend a primer suitable for his specific products. Aluminum primers usually contain zinc chromate to improve bonding, and steel primers are heavy in pigments to seal and block out moisture and the atmosphere. Primer paints can be applied by brush, and the dry coating is easily smoothed with sandpaper or steel wool. A further advantage of priming sheet metal cabinets is that the coating tends to hide scratches and other surface imperfections.

Spraying • Paint must be diluted to proper consistency for spray application. If the paint is too thick, it will splatter and fail to atomize properly; if it is too thin, the coating will tend to run. Aerosol can paints are already adjusted to the proper dilution and the only precaution to be taken is to shake the can extremely well before use, since it may have been sitting on a store shelf for some time. Good spraying technique involves moving in parallel sweeps across the surface about 12–15 inches (30–40 cm) away, not attempting to get a thick coating too quickly. Start spraying before the spray reaches the metal edge, and continue spraying on past the end to get a uniform finish. If a spray can is held too close to the metal surface, paint will be too thick and will run; if the can is held too far away, the finish will be sandy instead of smooth. Several thin coats will produce much better results than attempting to apply the complete coating in a single step. Each coat should be allowed to completely dry before proceeding to the next one.

ANODIZED ALUMINUM • *Anodizing* is an electrochemical process used to form coatings of aluminum oxide on aluminum metal. The coating is an excellent base for enamel paints and can be dyed to produce brilliantly colored finishes for front panel plates. The anodizing process is very similar to electroplating, except that the workpiece is connected as the *anode* instead of the cathode in an electrolyte, promoting *oxidation* instead of reduction at the metal surface. Basic anodizing steps include surface cleaning, electrolytic treatment, dyeing, and surface sealing. The process is basically very simple, although baths and equipment can grow fairly large if large panels are to be treated.

Cleaning • The best surface finish is obtained if aluminum is pretreated with a light alkaline etch in 20% sodium hydroxide, the same cleaning solution recommended before painting. The aluminum surface may also be buffed, polished, or scratched to obtain specific finish characteristics. Pure aluminum metal gives the brightest oxide coatings; aluminum alloys containing copper and magnesium tend to give duller results.

Electrolysis • Electrolysis can be carried out in a plastic, glass, or stainless steel tank. Since the cathode will be lead, commercial stainless steel anodizing tanks are usually completely lined with lead, which serves as the cathode and provides a large electrode surface area for even coating. The greater the cathode surface, the better. The workpiece is hung in the electrolyte and attached to the positive side of a dc power supply, serving as an anode element. Note that this is the opposite configuration as compared to electroplating. Anodizing creates a great deal of heat and off-gases, so the tank must be lined with cooling coils to control temperature, and it must be operated in a well-ventilated area. Anodizing conditions are as follows:

Electrolyte	15% sulfuric acid (H_2SO_4)
Cathode	lead
Temperature	70 °F (21 °C)
Temperature Control	stainless steel heater and water-cooled coils
Time	15–30 minutes
Agitation	air bubbles
Current Density	10–25 amp/ft^2 (about 0.1 amp/square inch)
Voltage	14–18-V dc
Racks and Clamps	aluminum

Power supplies for anodizing are similar to the ones used in electroplating, such as the circuit in Chapter 7, Fig. 7-3. However, a higher operating voltage is necessary. A good anodizing supply can be made by altering a 12-volt battery charger based on a high-current, 25.2 volt center-tapped power transformer. The center tap is removed, and diodes are added to form a full-wave bridge rectifier circuit, furnishing the extra voltage needed. A large electrolytic smoothing capacitor should be added to the output. Be sure not to exceed the current rating of the transformer, which will be half of the original center-tapped configuration. Approximately 10 amps of current are needed to anodize panels up to 8 by 10 inches (203 by 254 mm) in size, assuming one panel side is masked to limit power requirements. Current is adjusted with an autotransformer ahead of the power transformer, and both current and voltage meters should be used at the output to monitor operating conditions.

Air bubble agitation in the tank is important to obtain uniform current flow, and to promote oxidative conditions at the anode surface. If temperature is not controlled with cooling coils, excessive heat will prevent a thick oxide film from forming, and the coating will be porous. A useful oxide film thickness of 0.001–0.002 inch (0.025–0.051 mm) can be formed in 15–30 minutes if anodizing conditions are acceptable. Although thicker coatings are possible, a maximum point is reached in 1–2 hours.

Neutralization • To prevent carryover of sulfuric acid into the dye bath, the anodized panel should be neutralized in dilute (1–2%) sodium carbonate solution and then rinsed with water. The pH of the dye bath is important for proper dyeing.

Dyeing • The anodized coating should be dyed as soon as possible after electrolysis to obtain good dye penetration. Typical dyes that work well are the aniline-based textile dyes used as a hot aqueous dispersion, 1–2 grams per liter at 150 °F (65.5 °C) for 10–15 minutes. Fabric dyes are available in many colors and tend to be light-fast; some common, inexpensive ones are Rit Gold 23 and Rit Fuchsia. Eastman Kodak is a manufacturer and distributor of many types of dyes, and dyes made specifically for anodizing may be purchased from suppliers listed in the Metal Finishing Guidebook[21]. Assuming that dye does properly penetrate the oxide film, the depth of color produced will depend on time, temperature, dye concentration, and oxide layer thickness.

Sealing • To seal the dyed coating, it is first placed in a hot (200 °F or 93 °C) solution of 0.5% nickel acetate and 0.5% boric acid for 5 minutes to prevent dye from leaching out in the next step. The panel is then placed in boiling water for about 30 minutes, which causes the coating to expand and effectively seal the pores. Expansion occurs due to the hydration of aluminum oxide to $Al_2O_3 \cdot H_2O$, increasing the coating volume. After sealing, anodized coatings are tough, resistant to abrasion, staining, and corrosion; the oxide layer is also an electrical insulator. (It should be noted that electrolytic aluminum capacitors are formed through a process similar to anodizing, where a thin insulating oxide dielectric layer is created between two metal plates.) Anodizing is often carried out to produce a tough, clear, scratch-proof coating on aluminum panels without attempting to dye the film.

Final Surface Treatment • After sealing, the anodized surface may show a small amount of smut coating. The boric acid used in the nickel acetate solution is intended to minimize this problem. If smut is evident, it can be removed by polishing the aluminum with a metal cleaner, or by rubbing it with fine steel wool.

PANEL LABELS AND DECORATIONS • Many methods exist for labeling electronic enclosures, chassis, and front panels, ranging from the use of decals to elaborate engraving machines. We will discuss three approaches, all based on various techniques previously discussed in this book: dry transfers, screen printing, and photo resist methods. All three methods are capable of producing professional results.

Dry Transfers • The use of dry transfer lettering, numbers, and symbols was illustrated in Chapter 3, Fig. 3-23, for the preparation of high-contrast, camera-ready artwork. Dry transfers can also be effectively applied to clean metal surfaces with a pencil or wooden burnishing tool. The method is relatively fast and produces neat results if sufficient attention is devoted to careful placement of transfer figures. An unsatisfactorily positioned figure is easily removed by placing some transparent tape over the design and peeling it away. The principal drawback of the dry transfer approach is that it is limited by the designs that are available. Letters and numbers are distributed in many sizes and styles at art and office supply houses, and specific electronic symbols, words, and markings are available at electronic distributors. If a front control panel requires primarily letters, numbers, and dial markings in black color, dry transfer symbols are ideal. Once labeling is complete, two steps are required to finish the surface. First, wax residues from the dry transfer process are removed with a soft cloth soaked in toluene. These residues are not noticeable on white paper, but they show up on bright metal surfaces. Be careful not to damage the transfer characters themselves by rubbing too rigorously. Second, the entire surface is sealed with a clear acrylic or vinyl spray lacquer to protect the transfers from abrasion.

Screen Printing • Screen printing is probably the most versatile technique for applying paint, creating custom labels, and obtaining multicolored designs on flat enclosure surfaces. The method is not limited by enclosure dimensions, but as in all control panel work, the process is more convenient if the workpiece is a flat plate. Screen printing was described in detail in Chapter 8. High resolution patterns and precise alignment are possible if photographic screen printing methods are used (see Chapter 8, Fig. 8-1). Important considerations in panel screening are the type of paint used and surface cleanliness, since a primer paint cannot usually be applied underneath the design. In general, enamel paints are the easiest to use and epoxy paints form the most durable coatings. Since aluminum is one of the more difficult metals as far as paint adherence, a brief caustic etch is helpful to prepare plain aluminum panels. If the aluminum can be anodized (either clear or dyed), this forms an ideal coating for adherence of screen-printed images. A standard industrial approach is to paint the metal with a primer and a light-colored enamel, followed by screen-printed labels and designs over the paint. The main disadvantage of screen printing in prototype finishing is the long setup time required, due to the need for artwork preparation, photography, screen preparation, and actual printing. The process is not very time effective for low volume applications, but it is the best approach available to the prototyper who needs sharp detail and multiple colors in a front panel design.

Photo Resist Methods • The electronic experimenter who uses photo resist for PCB fabrication has a remarkable tool at his or her disposal for applying labels and designs to panel surfaces. Results can be as attractive and precise as screen printing, but the need to prepare and use a special screen stencil is eliminated. Photo resist may be used by itself, with etching, in combination with paints, or in combination with anodized aluminum. Basic techniques for artwork preparation, photography, and image transfer are completely discussed in Chapters 3, 4, and 5.

Simple Photo Resist — Photo resist by itself can be used as a final coating for labeling panels, if it is dyed black after application. Any smooth surface that is resistant to the photo resist developing solvent can be used as a basis material for resist coatings. Negative-acting KPR 3 photo resist is tough and strongly adherent to most metal and plastic surfaces if it is baked well after developing and dyeing the image. The final coating is similar to a black, matte-finish enamel in durability and appearance. Flat panel plates are labeled with exactly the same photo resist procedure used for PCB image transfer: the metal is coated with monomer, exposed through a negative photographic mask, image is developed, dyed black, and the final positive pattern is baked to give a durable, attractive, extremely sharp design.

Chassis, boxes, and odd-shaped metal enclosures can also be labeled with photo resist if the surface is clean and smooth, despite the fact that the workpiece cannot be completely immersed in a developing bath. Photo resist is applied to these objects *selectively* with a brush, confining it to areas where a pattern will appear. (It is only necessary to get a coating thick enough so that the final dyed film layer will appear opaque.) The surface is exposed to UV radiation through a mask, and the image is developed by simply spraying the proper areas with developing solvent from a squeeze bottle. The enclosure is held so that solvent will run off the metal without streaking across areas where it might dry and leave traces of resist. After drying the image for a few minutes, black photo resist dye is applied with a medicine dropper to bring out the pattern in appropriate areas, and excess dye is washed off with water. A disadvantage of using photo resist as a labeling material is that some solvents will dissolve and leach out the black dye if they accidentally splash on the finished panel.

Etch-and-Fill — The next logical extension for using photo resist to label panels is to *etch* the pattern into the metal surface, and fill the depressions with a paint or other filler material. This process gives a remarkable engraved finish to the panel, and is the common technique used to decorate nameplates, belt buckles, and other objects by *photofabrication.* In this case, the photo resist coating must be developed as a *negative* image on the panel surface to expose pattern areas for etching. Flat metal plates are easiest to work with because their open areas are easily protected with paint, and plates are easily immersed in an etching bath.

After baking and touching up the resist coating, the panel is etched as deeply as necessary to obtain the desired results. Aluminum metal is commonly etched with 20% sodium hydroxide solution at room ternperature; higher temperatures will give a faster etch rate, and lower concentrations will slow it down. Steel or stainless steel surfaces are commonly etched with ferric chloride, of the type available for low-volume PCB etching. However, the ferric chloride solution should not contain any copper residues when etching steel because unsatisfactory results will be obtained with a low etch factor. After the pattern has been etched to the desired depth, the plate is rinsed in water and the resist coating can be removed with the use of acetone and light scrubbing with steel wool.

In some cases the etched patterns will be left unfilled, but colored enamel paints will create attractive images if they are used to fill the depressions. Excess filler material is removed by two methods, *wiping* and *abrading*. If the patterns involve fairly narrow lines, excess paint can be wiped from the surface with a soft cloth while it is still wet. Wiping works best if a flat object, such as a block of wood, is used with the cloth attached. For wide patterns, or patterns that are not etched very deeply, the abrasive approach is more successful. In this case the paint is allowed to dry, and a sanding block with fine sandpaper is used to remove excess filler.

When wiping or abrading are not appropriate methods for removing excess coloring material, a variation of the etch-and-fill technique involves the use of epoxy-based paint for the filling agent. This technique is applicable to very wide etched areas, or patterns that are only very slightly etched into the metal surface. After etching, the photo resist coating is left intact and epoxy paints are carefully applied to the open pattern areas, allowing as little overlap as possible with the surrounding photo resist. The epoxy is allowed to dry and cure thoroughly, at which point it becomes completely solvent resistant. The metal plate is then immersed in a photo resist stripping solvent that causes the resist to swell and lift with a little help from some light rubbing with a soft cloth. Excess epoxy covering the resist will also be removed by the swelling action, leaving behind the desired painted patterns intact. Two precautions must be followed to obtain good results with the epoxy paint filling technique. First, photo resist cannot be baked excessively before etching, because this hinders solvent-stripping at the end. Without a maximum-tough resist coating, deep etching will probably not be possible. Second, epoxy paint must not be brushed on too thick, or it will hinder the ability of stripping solvent to attack and swell the resist.

Photo Resist and Anodized Aluminum — Aluminum plates that combine photo resist imaging, etching, and anodized finishes are widely used to make nameplates and electronic panel plates. An example is the Fotofoil product line from Miller Dial Corporation. Fotofoil products are principally aluminum sheets available in thicknesses from 0.003 to 0.020 inch (0.076 to 0.508 mm) and in many sizes. The sheets are anodized, dyed with brilliant colors, and precoated with photo resist. In most cases, users transfer a resist image to the sheet, etch through the anodized layer with alkali, and produce a final

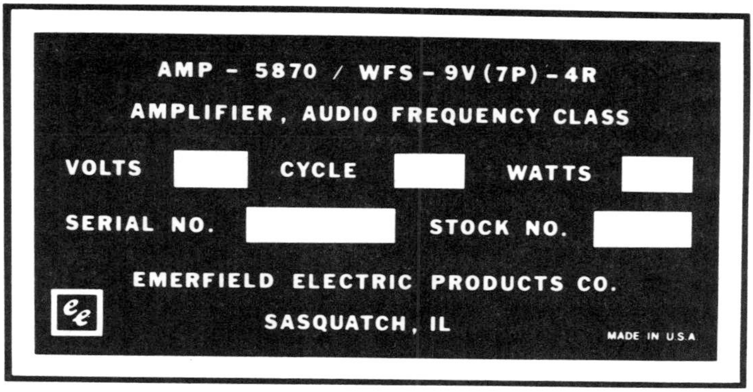

Fig. 11-20. Anodized serial number plate constructed with Fotofoil materials.

design composed of negative and positive images formed by colored and natural finish aluminum. The foil is cut to final shape and transferred to other surfaces with standard or pressure-sensitive adhesives. An equipment identification plate etched from black anodized Fotofoil sheet is illustrated in Fig. 11-20.

Another possibility for forming intricate, multicolored aluminum anodized plates is to combine photo resist masking with anodized dyeings. Each color is applied by anodizing the entire plate, dyeing, and sealing; resist is applied to selected areas, and the remaining unprotected anodic coating is stripped with alkali. This complete cycle is then repeated for other desired colors.

APPENDIX A
Notes and References

1. Kynar material is polyvinylidine flouride, an insulating polymer with good flexibility and excellent abrasion and cut-through resistance.
2. Solderless breadboards will not be discussed in this chapter, because they are used as circuit design tools rather than for final construction. For more information on solderless breadboards, the reader should consult catalogs from Jameco and Digi-Key, which illustrate products from AP Products Inc., and Global Specialties Corporation.
3. C. F. Coombs, Jr., Editor, *Printed Circuits Handbook,* 2nd Ed., McGraw-Hill Book Co., New York, 1979.
4. W. Sikonowiz, *Designing and Creating Printed Circuits,* Hayden Book Co., Rochelle Park, NJ, 1982.
5. D. Lindsey, *The Design and Drafting of Printed Circuits,* 2nd Ed., Bishop Graphics, Inc., Westlake Village, CA, 1982.
6. R. S. Villanucci, A. W. Avtgis, W. F. Megow, *Electronic Techniques: Shop Practices and Construction,* 2nd Ed., Prentice-Hall Inc., Englewood Cliffs, NJ, 1981.
7. A. Kosloff, *Screen Printing Electronic Circuits,* The Signs of the Times Publishing Co., Cincinnati, OH, 1980.
8. *Circuits Manufacturing,* Benwill Publishing Corp., 1050 Commonwealth Ave., Boston, MA 02215.
9. *Electronic Packaging and Production,* Cahners/Kiver Publishing Co., 222 West Adams St., Chicago, IL 60606.
10. *Metal Finishing,* Metals and Plastics Publications Inc., One University Plaza, Hackensack, NJ 07601.
11. Forrest M. Mims III, *Engineer's Notebook II,* 1982, Radio Shack, Cat. No. 276-5002.
12. Arthur Tyrell, *Basics of Reprography,* Focal Press, 1972.
13. *Basic Photography for the Graphic Arts,* KODAK Publication No. Q-1, 1981, Eastman Kodak Company, Department 454, 343 State St., Rochester, NY 14650.
14. Basic Developing, Printing, Enlarging in Black-and-White, KODAK Publication No. AJ-2, 1979, available from photographic dealers and from Eastman Kodak Company, Dept. 454, 343 State St., Rochester, NY 14650.

16. *KODAK Photosensitive Resist Products for Photofabrication,* KODAK Publication No. G-38, 1981.
17. *Characteristics of KODAK Photosensitive Resists,* KODAK Publication No. P-173, 1978.
18. *Photofabrication Methods with KODAK Photo Resists,* KODAK Publication No. P-246, 1979.

19. Kodak pamphlets are available from Graphic Data Markets, Eastman Kodak Company, Department 454, Rochester, New York 14650.
20. Persulfate etchants are manufactured by FMC Corporation.
21. *Metal Finishing, 48th Guidebook-Directory Issue,* 1980, p. 272, Metals and Plastics Publications, Inc., Hackensack, NJ 07601.
22. F. A. Lowenheim, *Electroplating Fundamentals of Surface Finishing,* McGraw-Hill Book Co., New York, 1978.
23. *Metal Finishing, 48th Guidebook-Directory Issue,* 1980, pp. 230–236, Metals and Plastics Publications, Inc., Hackensack, NJ 07601.
24. A. Kosloff, *Screen Printing Techniques,* Signs of the Times Publishing Co., Cincinnati, OH, 1975.
25. Coined by Albert Kosloff in 1942, taken from the Greek *mitos* and *graphein.*
26. R. Fossett, *Screen Printing Photographic Techniques,* Signs of the Times Publishing Co., Cincinnati, OH, 1973.
27. G. J. Wheeler, *Electronic Assembly and Fabrication,* Reston Publishing Co., Inc., Reston, VA, 1976.
28. *Safety in the Chemical Laboratory,* N. V. Steere, Editor, 1968, reprinted from *The Journal of Chemical Education,* published by the Division of Chemical Education of the American Chemical Society, Easton, Pennsylvania 18042.
29. *Prudent Practices for Handling Hazardous Chemicals in Laboratories,* National Research Council, National Academy Press, Washington, D.C., 1981.
30. *Guide for Safety in the Chemical Laboratory,* 2nd Ed., Manufacturing Chemists Association, Van Nostrand Reinhold Co., New York, 1972.
31. *The Merck Index,* 9th Ed., M. Windholz, Editor, Merck & Co., Inc., Rahway, NJ, 1976.
32. N. Irving Sax, *Dangerous Properties of Industrial Materials,* 5th Ed., Van Nostrand Reinhold Co., New York, 1979.

APPENDIX B
Distributors, Manufacturers, and Sources

Ace Glass, Inc.
P.O. Box 688
1430 Northwest Blvd.
Vineland, NJ 08360

American Scientific Products
1210 Waukegan Rd.
McGaw Park, IL 60085

Augat
Interconnection Systems Div.
40 Perry Ave.
P.O. Box 779
Attleboro, MA 02703

Bishop Graphics, Inc.
5399 Sterling Center Dr.
P.O. Box 5007
Westlake Village, CA 91359

Branson International Plasma Corp.
31172 Huntwood Ave.
P.O. Box 4136
Hayward, CA 94544

Chartpak Graphic Products
16 River Rd.
Leeds, MA 01053

Cincinnati Screen Process Supplies Inc.
1111 Meta Dr.
Cincinnati, OH 45237

Circuit-Stik, Inc.
24015 Garnier St.
Torrance, CA 90510

Cole Parmer Instrument Co.
7425 North Oak Park Ave.
Chicago, IL 60648

Coval Industries Inc.
2706 W. Kirby
Champaign, IL 61820

Datak Corp.
65 71st St.
Guttenberg, NJ 07093

D. F. Goldsmith Chemical & Metal Corp.
909 Pitner Ave.
Evanston, IL 60202

Digi-Key Corporation
P.O. Box 677
Hiway 32 South
Thief River Falls, MN 56701

Douglas Electronics
718 Marina Blvd.
San Leandro, CA 94577

Dyna/Pert Division
Emhart Corporation
Elliott St.
Beverly, MA 01915

Eastman Kodak Company
343 State St.
Rochester, NY 14650

Eastman Kodak Company
Laboratory and Specialty Chemicals
Rochester, NY 14650

Electrovert USA Corp.
Suite 104
11126 Shady Trail
Dallas, TX 75229

Engelhard Corp.
Engelhard Industries Div.
70 Wood Ave. South
Iselin, NJ 08830

Excellon Automation
23915 Garnier St.
Torrance, CA 90509

Feedback, Inc.
620 Springfield Ave.
Berkeley Heights, NJ 07922

FMC Corporation
Box 8
US Highway 1
Princeton, NJ 08540

GC Electronics
400 South Wyman St.
Rockford, IL 61101

Gerber Scientific Instrument Co.
P.O. Box 305
Hartford, CT 06101

Global Specialties Corp.
Box 1942
New Haven, CT

Hollis Engineering, Inc.
15 Charron Ave.
Nashua, NH 03063

Injectorall Electronics Corp.
98 Glen St.
Glen Cove, NY 11542

International Printing Machines Corp.
5 Campus Dr.
Somerset, NJ 08873

Jack Spears, Inc.
3169 Maple Dr., N.E.
Atlanta, GA 30305

Jameco Electronics
1021 Howard St.
San Carlos, CA 94070

Kepro Circuit Systems, Inc.
630 Axminister Dr.
Fenton, MO 63026

Kewaunee Scientific Equipment Corp.
5166 S. Center St.
Adrian, MI 49221

Lab Glass, Inc.
P.O. Box 5067
Fort Henry Drive
Kingsport, TN 37663

London Chemical Co., Inc.
P.O. Box 806
Bensenville, IL 60106

MCB Reagents
480 Democrat Rd.
Gibbstown, NJ 08027

Miller Dial Corp.
4400 N. Temple City Blvd.
El Monte, CA 91734

The Naz-Dar Co.
1087 N. North Branch St.
Chicago, IL 60622

Newark Electronics
500 N. Pulaski Rd.
Chicago, IL 60624

OK Machine & Tool Corp.
3455 Conner St.
Bronx, NY 10475

PEC Industries
5780 Carrier Dr.
Orlando, FL 32805

Philip A. Hunt Chemical Corp.
DEA Equipment Div.
1027 West 23rd St.
P.O. Box U
Tempe, Arizona 85281

Radio Shack (Tandy Corp.)
500 One Tandy Center
Fort Worth, TX 76102

Roper Whitney
2833 Huffman Blvd.
Rockford, IL 61101

R. W. Borrowdale Co.
230-270 West 83rd St.
Chicago, IL 60620

Shipley Co., Inc.
2300 Washington St.
Newton, MA 02162

20th Century Plastics, Inc.
3628 Crenshaw Blvd.
Los Angeles, CA 90016

Vector Electronic Co., Inc.
12460 Gladstone Ave.
Sylmar, CA 91342

Wabash Metal Products, Inc.
1569 Morris St.
P.O. Box 298
Wabash, IN 46992

APPENDIX C
Trademarks

Accufilm — Bishop Graphics, Inc.
Accuscale — Bishop Graphics, Inc.
Apple II — Apple Computer, Inc.
Autopositive — Eastman Kodak Company
AZ — Azoplate, Inc.
Blue Poly 2 — Ulano Products Co., Inc.
Chromaline — Chromaline, Inc.
Dacron — DuPont Co.
DCR — Dynachem Corp.
Diamond Chase — Diamond Chase Co.
Encosol — The Naz-Dar Co.
Estar — Eastman Kodak Company
Fotofoil — Miller Dial Corp.
Just Wrap — OK Machine & Tool Corp.
KC Enamel Plus — K.C. Coatings, Inc.
Kodagraph — Eastman Kodak Company
Kodalith — Eastman Kodak Company
KOR — Eastman Kodak Company
KPR — Eastman Kodak Company
Krylon — Borden, Inc.
Kynar — Pennsalt Chemical Corp.
Levelair — Electrovert Ltd.
LSI-11 — Digital Equipment Corp.
Microposit — Shipley Co., Inc.
Molex — Molex, Inc.
Multibus — Intel Corp.
Mylar — DuPont Co.
Naval Jelly — Woodhill Chemical Sales Corporation
PanaVise — Panavise Products, Inc.
Pentel — Pentel of America Ltd.
Permatex — Woodhill Permatex, Div. Loctite Corp.
Photo-Flo — Eastman Kodak Company
Powerstat — Superior Electric Co.
Precision Line — Eastman Kodak Company
Puppets — Bishop Graphics, Inc.
Re-Solv — Coval Industries, Inc.
Rubylith — Ulano Products Co., Inc.
SBC-80 — Intel Corp.
Sealbrite — London Chemical Co., Inc.
Slit-N-Wrap — Vector Electronics Co., Inc.
Soldapullt — Edsyn, Inc.
Super Prep — Ulano Products Co., Inc.
Tech-Wick — Tech-Tool Ind., Inc.
Tefzel — DuPont Co.
TRS-80 — Tandy Corp.
Tygon — U.S. Stoneware
Wire-Wrap — Gardner Denver Co.
WTCPN — Weller-Excelite
X-Acto — X-Acto

APPENDIX D
Chemical Hazards

Because of the many chemicals used in procedures throughout this book, a general discussion of chemical hazards and safe handling practices is included here. All chemicals (including water) can be hazardous if used under certain circumstances; to avoid problems, the laboratory worker must adopt general safe practices and know the specific properties of substances that he or she is dealing with, including their potential interactions with other chemicals. Most manufacturers of chemical materials supply information on the hazards of using their own products. For additional information, the reader is directed to the following references in Appendix A:

General Chemical Safety — An inexpensive guide[28] for laboratory safety is published by the American Chemical Society. A text[29] from the National Research Council on the handling of hazardous chemicals and a laboratory safety book[30] from the Manufacturing Chemists Associations are recommended as further references.

Chemical Properties — The Merck Index[31] is recommended as a comprehensive source of chemical properties for most specific compounds. An even more extensive encyclopedia[32] of hazard information for common industrial laboratory materials is also referenced.

Chemical hazards can be generally classified into two groups: health hazards (such as toxicity) and physical hazards (such as flammability). Ten important types of chemical problems will be discussed, and related to various substances or situations mentioned in this book.

1. Toxicity and Exposure — A person can be exposed to the toxic effects of a chemical by breathing fumes, absorption through the skin, and direct ingestion. Proper ventilation and protective clothing are the keys to avoiding these problems. Some substances are toxic at very low levels, and their use should be avoided where possible. This is the reason that KOR developer is recommended in place of trichloroethylene (a suspected carcinogen) for KPR 3 photo resist image transfer in Chapter 5.
2. Ventilation — A fume hood, such as the one pictured in Chapter 5, Fig. 5-3, is recommended when working with volatile chemicals, such as organic solvents. Good ventilation will reduce both toxic exposure and flammability hazards. The fume hood must be constructed of chemically resistant materials, and the blower motor must be explosion-proof to avoid ignition of flammable vapors.
3. Flammable Liquids — Most solvents mentioned throughout the book are highly flammable, even at room temperature. The liquid temperature at which vapor concentration exceeds the lower flammable limit in air is defined as the *flash point*. The temperature to which a flammable material must be raised to start a fire is known as the *ignition temperature*. In general, fire and explosion hazards can be reduced by proper storage of solvents, good ventilation during use, and elimination of ignition sources. Common sources of ignition are open flames, heater elements, and sparks occurring in electrical equipment, such as motors.

When transferring flammable liquids from one metal container to another, a ground wire strap should be attached to both containers and ground to eliminate static buildup that can produce a spark. In Chapter 5, the reader is cautioned concerning the drying of flammable photo resist products inside an electrically heated oven. Always provide a dry chemical fire extinguisher in your laboratory or workshop.

4. Eye Protection — Safety glasses should always be worn when handling chemicals or using shop machinery. A sink with running water should always be handy in case of an emergency. The eyes are the part of the human body most vulnerable to chemical damage, and the value of your eyesight cannot be underestimated.
5. Protective Clothing — Gloves should be worn when handling toxic, corrosive, or irritating chemicals. The best all-around composition for laboratory gloves is neoprene rubber. A heavy fabric glove may be more suitable when handling hot materials, such as solder reflow oil. A protective labcoat or apron and rubber-soled shoes are recommended when working with strong acids and bases, such as hydrochloric acid and sodium hydroxide (caustic, lye). These substances cause severe skin burns, and in the case of sodium hydroxide, may not be immediately detected. Upon contact with the skin, wash the affected area with large amounts of water. Many fairly harmless chemicals (such as photographic developers) produce skin irritation or dermatitis after prolonged contact.
6. Corrosive Chemicals — Certain chemicals are particularly corrosive, such as hydrochloric acid fumes which will attack and corrode most metal surfaces. Fume hoods and blowers that exhaust acid vapors must be coated with epoxy paint for adequate corrosion protection. In Chapter 6, an old chest freezer is recommended as a containment area for acid etching solutions, and the reader is cautioned against disposing of acid wastes down the drain.
7. Reactive Chemicals — The laboratory worker must exercise particular caution when mixing or combining various chemicals, which may produce an unexpected or uncontrollable reaction. The results could be splattering, fire, or even an explosion. In Chapter 6, the reader is cautioned against pouring water and other substances into strong acid, which produces localized heating and splattering. Acids should always be added to water rather than the reverse operation. In general, oxidizing agents (such as strong acids and peroxides) can present fire and explosion hazards upon contact with organic compounds and reducing agents. For example, KPR 3 photo resist solutions may react violently with strong acids; and persulfate etchants must be segregated from all other materials because of their strong oxidative nature. Common reducing agents found in PCB shops are solvents, resist stripping residues, electroless plating baths, carbon filter material, and metals such as aluminum, iron, and copper. When using commercial substances of proprietary or unknown composition, always consult the manufacturer's literature for safe handling practices.
8. Chemical Storage — Chemicals must be stored in safe containers. In the case of flammable solvents, quantities over one gallon should be stored in a metal safety can with a flame-arresting spout. Many reactive chemicals must be stored in glass or inert plastic bottles. Always ascertain the compatibility of etching and plating solutions with their containers before setting up a large bath.
9. Spills — Three types of hazardous chemical spills occur frequently in the laboratory: acids, bases, and flammable solvents. The acids and bases should be neutralized with appropriate mild agents, such as sodium carbonate or sodium bisulfate, respectively. The spill can then be diluted with water and mopped up or washed down the drain. Flammable solvents should be absorbed with cleanup aids, such as sand or vermiculite for containment and disposal. Commercial spill materials are available from laboratory supply houses that also serve to reduce the vapor pressure of absorbed solvents, reducing the fire hazard.

10. Waste Disposal — Three common methods are available for disposal of chemical wastes: incineration, sewer flushing, and landfill. High temperature incineration is the most environmentally acceptable method of waste disposal, since it converts most materials into harmless by-products of combustion. Chemical waste may be collected and incinerated by contract, since most areas now have suppliers of such services. Simple open burning of wastes may disperse toxic, corrosive, and irritating substances into the air instead of destroying them. Many water-soluble chemicals can be safely flushed down the drain, but certain materials create problems in the wastewater treatment system and their disposal may be regulated by law. Examples of pollutants that may *not* be introduced into sewers are flammable solvents, highly acidic substances (such as etchants), materials that could stop up the sewer flow, and chemicals that interfere with biological systems, such as metal salts. Landfill is an adequate disposal procedure only if chemicals are first converted into an innocuous form such that groundwaters will not leach out toxic materials at a later date. In Chapter 6, it is recommended that ferric chloride etchant be disposed of by neutralization and immobilization with mortar mix prior to landfill.

Index

C

F

G

N

O

P

Q

R